Geochemistry at the Earth's Surface

Andreas Bauer • Bruce D. Velde

Geochemistry at the Earth's Surface

Movement of Chemical Elements

Springer

Andreas Bauer
Karlsruher Institut für Technologie (KIT)
Institut für Nukleare Entsorgung (INE)
Eggenstein-Leopoldshafen
Germany

Bruce D. Velde
Ecole Normale Supérieure
Laboratoire de Géologie
Paris CEDEX 5
France

ISBN 978-3-642-31358-5 ISBN 978-3-642-31359-2 (eBook)
DOI 10.1007/978-3-642-31359-2
Springer Heidelberg New York Dordrecht London

Library of Congress Control Number: 2014939512

Printed on acid-free paper

Springer is part of Springer Science+Business Media (www.springer.com)

Introduction

Geochemistry is a frequently used term in geological studies that defines a discipline which has developed in importance over the past 40 years. The meaning of this term and the substance of the discipline is the relationship between the chemical properties of elements found in the geological sphere and their expression as minerals or other phases found in rocks. Essentially the elements that are the most common and abundant in surface materials are limited in number, such as Na, K, Mg, Ca, Al, Si, and Fe (called the major elements) with concentrations of elements such as Mn, P, S, C, Cl present in smaller amounts and considered as minor elements. The elements of industrial importance, such as the transition metals Cu, Cd, Co, Ni, Ti, and Au, among others, are of yet lesser abundance in rocks and most often occur as inclusions in phases dominated by major elements or as isolated concentrations due to special geological conditions. Other elements of low overall abundance in rocks such as As, Sb, and Sn are more difficult to classify in that their occurrence is limited to small and sporadic concentrations in rocks. In the classical geological chemical framework the elements found as gases in the atmosphere such as Ar, N, Ne, He, and others are little present. Some atmospheric elements such as C and O are present in both the realm of geological occurrence and the sphere of the surface atmosphere. Geochemistry is the study of the occurrence and reasons for these occurrences of different elements.

A further direction of investigation, used with great success to solve some specific problems, is the determination of the relative abundance of the isotopes of the elements present. Isotope geochemistry has proved to be a powerful tool in the determination of many geological processes of rock formation.

Our proposition in the present study is to use the available published information concerning elemental abundances at the surface of the earth, in the zone of atmosphere and rock interaction, to follow the selection and segregation of elements during the surface chemical processes which are dominated by the interaction of atmospheric and biologically determined chemical processes with rocks that have formed and attained chemical equilibrium under other conditions of temperature and pressure.

One fundamental point to remember is that rocks are composed roughly of half oxygen atoms and the atmosphere is still more oxygen rich. Water, the agent of change, is composed of oxygen and hydrogen atoms. Oxygen is the major element of the geological realm and especially that of surface geochemistry. Interactions of cations with oxygen are the dominant modes of chemical interaction. The geochemical actions initiated at the surface are essentially those of hydrogen exchange for cations in rock minerals which creates "hydrated" minerals such as the silicate clay minerals or hydroxy oxides of Fe or Al. Geochemistry at the earth's surface is initiated by phase change through the interaction of water with rocks which sets the stage for geochemical exchange throughout the erosion, transportation, and deposition processes. The activity of oxygen and ensuing oxidation of metal elements are another fundamental constraint on surface geochemical reactions. Since most multi-oxidation state elements are in a more or less reduced state in rocks, oxidation disrupts the former chemical interactions in the minerals and this produces the destabilization of these phases and eventual formation of new minerals. However, biological activity at the interface of surface materials and the atmosphere creates variations in oxidation state of materials in many cases which reorganizes some of the material present and changes the chemical relations among ions and solids. Thus hydration and oxido-reduction are the key concepts to keep in mind when considering earth surface geochemistry. These are the chemical variables or active forces present.

This is the chemical framework. Then the first part of the term "geochemistry" must be considered. In many texts, treatises, and scientific papers, the concept of geological forces at play is not taken into account. Chemistry is essentially static, where the electronic configuration of the elements present is arranged as quickly as possible to create an assemblage of phases that are in chemical equilibrium. Geology by contrast is the realm of movement dynamics driven by the forces of tectonics (mountain building) and the forces of gravity which drive water movement, the major transport vector. The earth is a restless object, especially at its surface. Movement of matter in solid or dissolved form is the general rule.

Then the "geo" part of geochemistry is the study of the chemical accommodation to the dynamic forces of geological energy. However, the inherent instability generated by the earth processes is such that chemical equilibrium is often not attained and materials of contrasting chemistry are found together in the same environment. The work of the geochemist is then that of sorting out the nature of materials in a given sample and their origins in order to understand where they came from and what will happen to them as they approach a chemical equilibrium in a given setting. As a simple example, if one considers the sediments carried in a river and its load of dissolved elements, one must look at the relief and erosion rate of the drainage basin, its climatic variation, and geomorphologic characteristics in order to interpret the chemical variations of materials found in the river load and river water. These will be reflected in the sediment that is eventually deposited by the river along an inland lake or ocean shelf. Surface geochemistry is the interrelationship between chemistry and geology.

In the text that follows we have purposefully left out any reference to isotopic variations of the different chemical species in that there is not a great abundance of such information available and the changes and instability of surface materials render the interpretation of isotopic variation very difficult in many instances. If one studies the isotopic variation of an element in a given stable context, things can often be worked out to a satisfying degree, but such studies are not generally available and using them to interpret the overall variations of chemical isotopic abundance throughout a geological sequence is very difficult indeed. Hence we attempt to find the underlying relations between elements, of major and minor abundance depending upon the local chemical context and the dynamics of the geological setting.

It is clear in reading the following studies that there is much to be done yet in order to understand the many aspects of surface geochemistry; however, we hope to lay out some general principles and relationships that can be used as a basis for further work.

Contents

Chapter 1
Geology and Chemistry at the Surface

1.1 The Geological Framework of Surface Geochemistry

In the mid-sixteenth century Bernard Palissy (1563) mused upon the characteristics of rocks and water and the interaction of the two. He was interested in the movement of water within the surface of the earth and the relations of water and rocks. In his publication *Discours Admirables*, presented in published and verbal form for the royal court in France, he was interested in the durability of rocks and the eventual transfer of material at the surface of the earth. Since at this time the earth was said, dogmatically, to have been created and unchanged since biblical times, he wondered if it was in fact true, being strongly influenced by the thinkers of the Renaissance. He contemplated caves and caverns in rock. Here he saw cavities produced by water–rock interaction and precipitation of matter from aqueous solution. In his account concerning rocks (*pierres*) he noted that water can enter a rock, move within extremely small cavities, and when it comes out, deposit crystalline material. Here rock–water interaction and dissolution is clearly evoked, which is the basis of surface geochemistry. It took a long time, but eventually modern science caught up with his ideas, and those of others, concerning surface geochemical interactions. However no name was given to the phenomenon at the time. Nevertheless the phenomenon of water–rock interaction was realized to be a significant event in the structure of the earth and its history. In Fig. 1.1 one sees the representation of such an event, somewhat romanticized, in the grotto of the Pitti Palace in Florence. Here it is evident that there is dissolution and precipitation of matter, which are the fundamentals of surface geochemistry.

The use of the name Geochemistry applied to the study of chemical relations in different geological situations in or on the earth is somewhat recent, coming into full vogue in the 1950s. It has its roots in the application of chemical principles (what was often termed mineral chemistry, dating to the origins of chemical investigation in the eighteenth to nineteenth centuries). Most of the elements of major abundance were discovered and defined in the period 1750–1850 (Correns 1969). Although the idea of geochemistry is rather old, mid-nineteenth century

A. Bauer and B.D. Velde, *Geochemistry at the Earth's Surface*,
DOI 10.1007/978-3-642-31359-2_1, © Springer-Verlag Berlin Heidelberg 2014

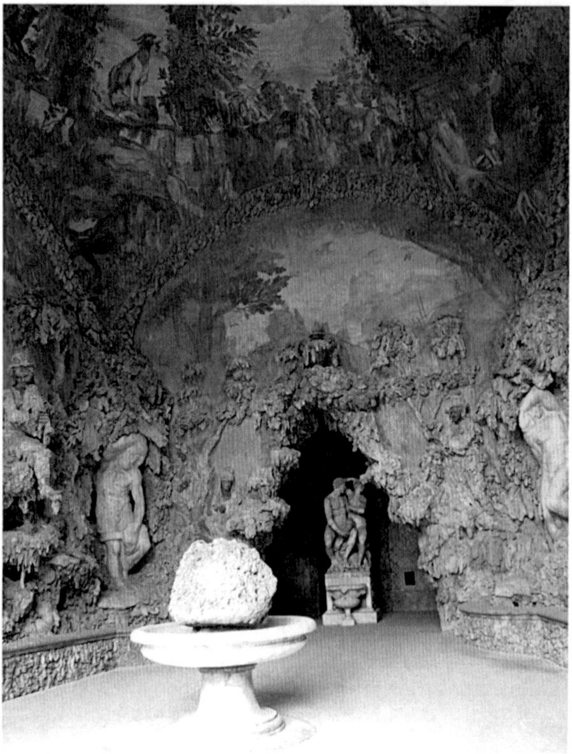

Fig. 1.1 Grotto of the Pitti Palace, Florence showing dissolution and re-crystallisation phenomena, which are the basis of surface geochemistry

(Correns 1969), the concept was not made popular until the fundamental text of Goldschmidt was published (Goldschmidt 1954) followed by several other fundamental texts which opened the realm of chemistry in the geological sphere (Mason 1958; Krauskopf 1967, among others). This set the stage for a new era of geological investigation using well-defined and modern scientific principles borrowed from another field, chemistry. With time geologists became more and more interested in the distribution and affinities of element isotopes as they reflected different chemical and physical properties that formed rocks and minerals. In the twenty-first century one often thinks of this part of the chemical world of geology more than the distribution of chemical elements and the chemical causes of these distributions. Surface geochemistry, as we wish to treat it here, will be uniquely considered from the standpoint of chemical occurrence and the causes of the presence of the different elements found in different surface environments and materials. Isotope geochemistry certainly has an important role to play in understanding the processes that affect the surface of the earth, but it is not our preoccupation here.

We will treat the basic affinities of elements as they experience different chemical environments where earth surface materials are transformed and transported. The geochemistry of the surface reflects the changing chemical environment which leads to an instability of rock forming minerals at the surface and it reflects the forces of nature: plants, rain, gravity, which affect and move surface materials from one site to another.

The most important concept and reality of surface geochemistry is that elements can enter or leave a given mass of matter, forming what thermodynamicists call an open system (see Korshinski 1959). Further the chemical constraints on the material in the system can come from outside. The passage of water into and through a rock will change its chemistry, but in a large part of the reactions the chemical constraints, which change mineral phases, come at least partially from outside of the rock. The interaction of biologic agents, bacteria and others, can influence the chemistry of the rock or geo-materials so that they have different properties and mineral phases. Surface geochemistry is the study of multiple influences, coming from outside of the geologic system (air and water) or from within but where the products of reactions can often be released outside of the initial material considered. This is the case for chemical constraints, but one must consider the physical movement of masses of matter also.

1.1.1 Movement of Materials

As all geologists know, Geology is a four-dimensional problem. The earth has three dimensions, all of which play a major role in the dynamics of the earth. Using the word dynamics implies the dimension of time. Hence the four dimensions. If a mountain is built by rock material being moved upward, the problem is three dimensional since the rock has to come from somewhere and is usually moved by forces described in x, y, z coordinates. The mountain is not build in a day, and hence the dimension of time is extremely important. The movement of materials at the surface of the earth implies an understanding of the four dimensions of geology (see Birkland 1999).

Surface geological materials are composed, to simplify, of rocks and the products of their alteration through the interaction of water and rocks. The thickness of the alteration zone can vary from several centimeters to tens of meters depending upon the length of time of exposure of the rocks to the surface chemical forces and the intensity of alteration, i.e., the amount of water coming into contact with the rocks. Rocks can be moved by gravitational forces, as well as can the alterite material. However the distance of rock transport is relatively small, hundreds of meters, whereas the distance of transportation of altered rock can be hundreds of kilometers. Movement is the key to understanding the surface of the earth.

The earth is in constant movement at the surface with mountain building and the erosion of these mountains. The leveling of geographic highs is done in different manners and the eventual transport of altered materials from one site to another is extremely important. The relative amounts of matter moved by different physical processes have been estimated as follows (Milliman and Meade 1983; Gorsline 1984) in billions of tons/year:

Particles in rivers	13.5
Dissolved in rivers	4.2
Ice transport	3
Dust transport	0.6
Volcanic ejecta	0.15

In looking at these numbers it is clear that the largest part of alteration products due to rock–water interactions moved from mountains to lower areas is displaced as particulate material in rivers. This is the nature of longer distance movement of surface alteration materials, which eventually end up as sediment, which becomes rock as it is buried, hardened, and recrystallized by pressure and temperature.

The movement of mountains to the final "resting place," sediments, is complex and depends upon many factors. The rate at which rocks alter to new materials, essentially oxides and clays, depends upon such factors as climate, rate of mountain building or elevation, and plant interactions, among others. In order to better understand these processes, we will go from large scale to small scale which allows us to consider the mechanisms and impact of geological forces on the geochemistry of the earth's surface. The questions are: how do surface materials form, what are they, and where do they go as they interact with the physical forces of the surface (gravitational forces, aqueous, or wind transport)? These questions define the basic setting for surface geochemistry, which has been and is geological.

Geology is a nonlinear sequence of events. Mountains are built by infrequent vertical movements, signaled by earthquakes. Landslides occur during earthquakes and periods of exceptionally high and episodic rainfall, bringing the materials moved upwards by tectonic action to a lower geographic level of deposition. The movement of most of the surface material occurs during events that are widely spaced in time. However the daily input of fine material into the oceans by major rivers is very important in that it is the system of major transit to the eventual final residence of material moved by surface erosion. This transport is also episodic but less so than major massive short-distance transportation (landslides), which occurs on a smaller geographic scale.

1.1.2 Physical Constraints

It is initially important to understand the physical setting of surface geochemistry. The surface the earth is in a state of continual change, on a geological scale, thousands or hundreds of thousands of years, or periods much shorter, tens of years, when biological activity is involved. In fact the dominant trend of surface geochemistry is change of place.

1.1.2.1 Slope Failure and Transport

The basic paradigm of surface geochemistry is one where rocks formed at higher than surface temperatures are brought in contact with the atmosphere of the earth's surface and its chemical constraints. The reasons for this contact are the upward movement of rock masses by tectonic forces to form mountains. Surface geomorphologic relief produces instability. Here rocks are placed in two unstable situations: physical and chemical. The physical constraint is that the rock mass is higher than the lowest residence point, the ocean, and it is therefore subjected to the forces of erosion directed by gravitational movement of solids. What goes up must come down. A brief illustration of massive transport of surface materials is given in Fig. 1.2. In the case of massive erosion due to structural failure on highly sloped surfaces, the major force is gravitational which, under conditions of exceptional stress such as high rainfall or earthquake tremors, accelerates the inherent unstable materials into downward movement. This produces landslide, rockslide, and slumps of surface material, which is rocks and soils. This type of movement is by far the most important in surface movement as far as the mass of materials is concerned when relief is significant. However the effect is local to a large extent, essentially stopping at the foot of the slope that caused the instability of the rock–soil masses.

Once the slide material is deposited it in turn is subjected to rainwater transport where the small particles, which are newly exposed at the surface, are subject to displacement and eventually find their way into a stream or river. Massive displacement is followed by more small-scale erosion, which continues the movement of materials to lower levels.

1.1.2.2 Resistance to Erosion

One might think that the hardest rocks resist erosion more than others and that those most resistant to chemical dissolution will be also more resistant to transformation and erosion. This is true and not true. In looking at mountains, one sees very clearly that sandstones and carbonates tend to make salient features while granites, basalts, and other eruptive rocks tend to be eroded as well as are shales, and different aluminous schistose metamorphic rocks. Normally carbonate minerals are strongly affected by the acidity of atmospheric aqueous solutions, tending to dissolve with a much greater coefficient of dissolution than silicate minerals. Quartz is the most resistant mineral, even though thermodynamically it should dissolve readily, it resists water attack with great efficiency. In the observations noted above, the sandstones should resist chemical attack and remain chemically stable. They are brittle under differential tectonic stresses, tending to fracture, but not to a great extent. Granites and basalts are more soluble than sandstones, but the soluble elements are less soluble than calcium carbonate. Thus in the sequence of chemical stability one should put sandstone > granite, basalt, shales, and metamorphic rocks

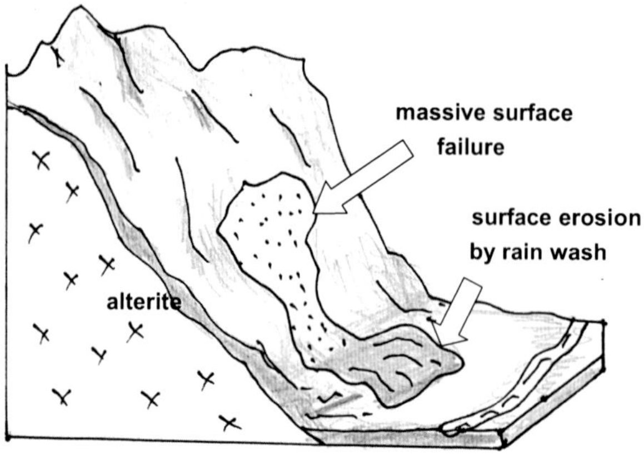

Fig. 1.2 Massive slope failure showing displacement of unstable material on slopes which is deposited at the foot of the slope to be eventually moved by the action of rainwater toward streams by surface erosion

> carbonates. Yet one sees that the most chemically stable and least chemically stable rock types both form resistant ridges and massifs in mountain ranges. The anomaly is the carbonate rock.

If we consider the mechanism of chemical alteration, it is clear that the agent of dissolution is rainwater, which is slightly acidic (pH around 5.2). For rainwater to have a chemical effect it must stay in contact with the rock and mineral matrix it should alter. Mineral dissolution takes a certain time. In mountain ranges, the rocks have been recently subjected to anisotropic forces and confining constraints, which lead to local failure and the development of cracks. These cracks can be large enough to be called faults, hundreds of meters long, or of smaller dimensions down to the size seen in thin section under the optical microscope, several millimeters in length. These failure planes are open to water flow when the confining pressure is released as the mountain finds it final position, at least for hundreds or thousands of years. If all rocks responded to tectonic forces in a similar manner one would expect to find similar patterns of faulting and fracturing in them. However this is not the case. The resistant rocks, forming ridges and peaks in mountain edifices, sandstones, and carbonates, show fewer micro-fractures than do granites, basalts, and to a certain extent metamorphic and sedimentary rocks. In fact carbonates are more resistant to fracturing because they deform in a ductile way, being semi-plastic under great pressures and pressure differentials. Hence they show fewer brittle fractures in their macro- and microstructures. As a result there are fewer small cracks in a carbonate rock that in a granite for example. The micro-fractures are passageways for water entry and hence for chemical interactions, i.e., dissolution and minerals transformation.

There is thus a paradox of the normal laws of chemical reaction and rock stability, which is due to the mechanical properties of the rocks involved. Carbonates are more reactive chemically but less accessible to rainwater chemical interaction, while granites and other rigid macro-crystalline rocks present more passageways to water infiltration increasing the residence time of aqueous solutions which promotes chemical reaction that destroys the mechanical coherence of the rock. Here we see that the weakest can be the strongest. The most chemically fragile and least mechanically resistant rocks are least transformed by aqueous solution attack and fracture failure due to mechanical stress. Despite their inherent weaknesses, carbonate rocks resist the effects of alteration and tectonic weakening and remain among the most stable rock types forming mountains.

1.1.2.3 Water Transport of Dissolved and Suspended Material

The chemical transformation of minerals, where minerals formed at high temperature become out of chemical equilibrium with the chemical conditions of the surface, is that dominated by an abundance of water. The transformation of rock minerals to more stable phases is accomplished by the interaction of water with rocks in an oxidizing atmosphere which promotes the recrystallization of the minerals and at the same time removes some of the elements present from the solids. Generally this process is accompanied by the incorporation of hydrogen or water molecules into the new mineral phases. Surface chemical interaction is essentially one of hydration. Materials are dissolved either integrally or incongruently, forming "residual" alteration minerals. More importantly, the new minerals formed are of smaller dimensions than those of high temperature origin. This change in size allows the new minerals to be transported more easily by moving water down to lower geographic levels, and eventually into the sea. Of course larger chunks of rock are moved as well as particles of mineral grains (sand and silt). The larger the particle, the faster it will be deposited in the trajectory of the water carrying it toward the sea. The principle of faster deposition of larger particles (particle size sorting) can be illustrated by taking a handful of soil material and dispersing into water in a beaker. The finer materials stay in suspension while the larger ones settle out rapidly. The larger they are the more rapidly they settle. This is illustrated in Fig. 1.3.

The further the materials are transported, the lower the transport gradient, and the less energy available for moving particles. Thus the larger grains tend to be deposited along the slopes of mountains, or deposited at their bases. The lower gradient areas, along rivers and streams, are where sand-sized materials and clays are deposited along slow moving and less energetic water. Clay materials are gradually deposited until they reach the ocean where the finest material held in suspension is flocculated in the saltwater and moved along the seacoasts. A significant portion of the fine material reaches the greater depth of the ocean continental platforms where they become buried by materials of subsequent sedimentation.

As the slopes are less important, fluctuations of rain inputs can vary the energy of transport and at times local deposition occurs as the streams overflow or become less active. The deposited material, along streams for instance, before final deposition in seawater can be chemically altered further if exposed again to the rain and plant interactions of the earth's surface along the river banks or on flood plains. With each cycle of alteration some matter is lost from the solids to the solution and the material that remains becomes finer grained.

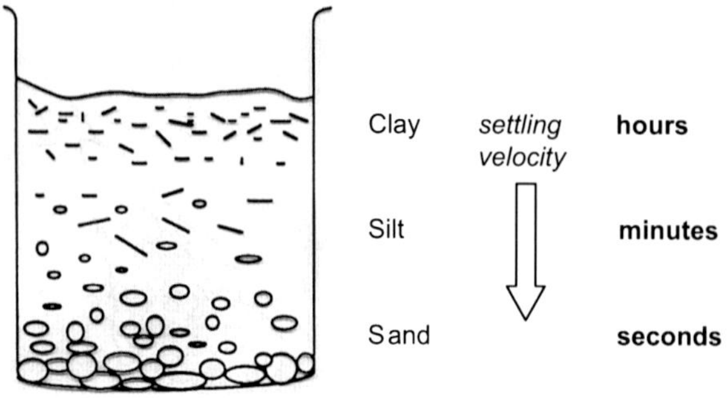

Fig. 1.3 Beaker experiment where heterogeneous alterite material is dispersed in water and allowed to settle. The largest grains are deposited first and the smallest last. Eventually the small grains can be held is suspension for very long periods of time if the water is moving and turbulent

In Fig. 1.4 the relations of slope and grain size for some major rivers are indicated. In general the larger the river, the larger the drainage basin the more fine particles (clay size or <2 μm in diameter) are transported. These are the fundament variables of surface geomorphology, which determines the types of material transported. In general, the further from a mountain chain one is, the more fine particles the rivers will carry.

Figure 1.5 indicates situations where material can be transported and deposited locally, from the slope of a stream bank for example, and deposited, along the flood plain of the stream. Streams are at the same time vehicles of transportation and deposition. When a stream has a high sediment charge under conditions of high intensity flow, material is transported that would not otherwise be carried due to the high energy of the moving water. The importance of floods to the transport of clay-sized material is emphasized by the fact that many rivers transport more than half of their annual sediment load in only 5–10 days of the year (Meade and Parker 1985). When the period of high flow is less important, or when the stream overflows its normal channel, the transport energy decreases and material is deposited along the edges of the stream channel. This deposition is very important in that it is a temporary holding area for fine-grained material, which will eventually reach the sea. However, during its period of temporary deposition, the newly deposited material forms a substrate for plant life and biological activity, which modifies the chemistry of the sediments. This secondary soil forming process is or has been very important to mankind in that many ancient civilizations were founded on stream or river deposits along the major channels of water movement. A classic example is the Egyptian civilization built upon soil renewal through deposition of the fine-grained sediments of the Nile river. The formidable Near Eastern civilizations of the same period were based upon sediments in the Tigris–Euphrates river basin. These holding areas of deposition essentially of soil materials have been a key in the development of agricultural mankind.

Not only is the height of the mountain important as a factor of transport, but also the climate, measured by temperature and rainfall which are the agents of alteration,

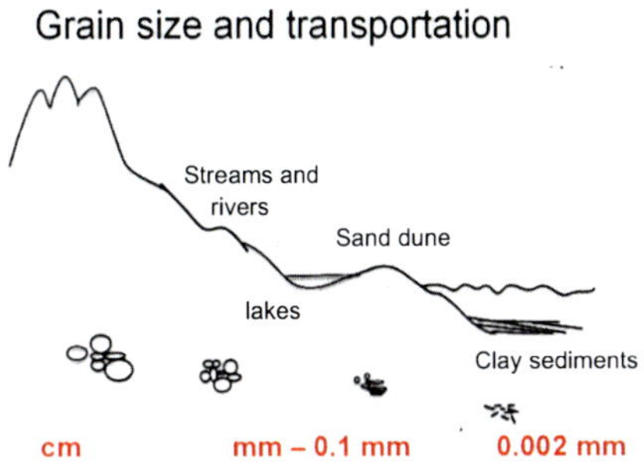

Fig. 1.4 Illustration of transport distance as a function of grains size in a landscape situation

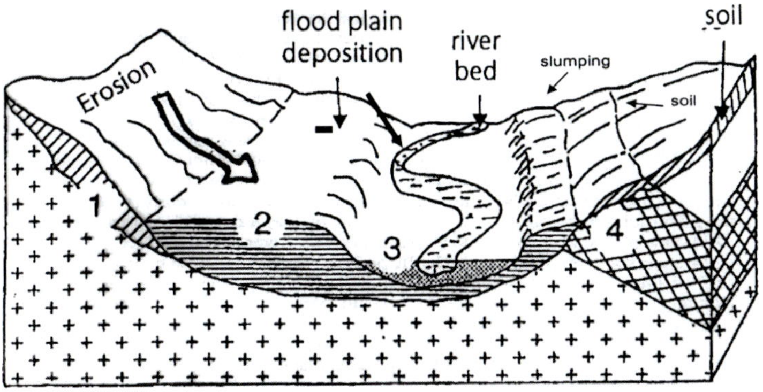

Fig. 1.5 Illustration of massive transport erosion and re-deposition along a river. *1* Surface erosion and slump movement of soil surface material, *2* temporary deposition of slump material which is eroded in its turn by rainfall erosion, *3* deposition of the overflow material from the stream when it reaches low energy water movement, *4* slump movement of poorly consolidated rock material

is an important factor. The more rain and the higher the temperature, the more small particles formed from unstable minerals in rocks and the more fine material there is in the rivers. Sediment yields on a basin scale vary greatly, and there is no correlation between basin size and yield (Fig. 1.6). Nor is there any statistical relationship between variables such as runoff and yield (Selby 1984). The two factors that are important are geomorphologic relief and climate. Gibbs studied the variation of sediment yield in the Amazon Basin which showed that 80 % of the sediment is derived from only 12 % of the basin area comprising the Andean mountains. Gorsline (1984) pointed out that the land use by man since late Pleistocene times also greatly affected sediment yields.

To a large extent the fine materials (clays) are moved by rivers into the oceans and deposited offshore, i.e., several hundreds of meters or kilometers from the shoreline. This action depends upon the configuration of the ocean bottom, depth and rugosity, and the intensity of the river flow which can move material to greater or smaller distances depending on its flow rate (energy). However not all of the

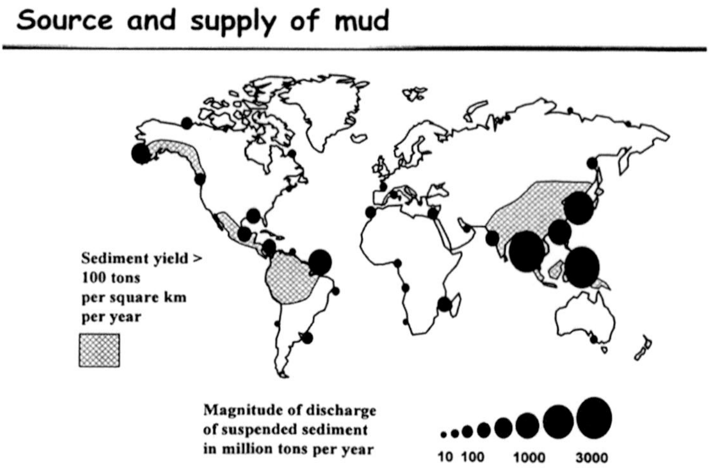

Fig. 1.6 Sediment yield map for major geographic areas [adapted from Hillier (1995)]. The Western Pacific Islands are combined. Highest yield occurs in areas of high relief and/or high rainfall

material finds its way to deep repositories off the coasts. Some is moved along the shore, laterally from the mouth of the river, which brings the sediment to the sea. The clay (or mud) deposits and movement are generally termed wetlands. Slightly more coarse material is often moved by coastal currents on or near the shore forming deposits of sand-sized material, usually composed of quartz (silica) and refractory minerals little affected by the alteration processes of the surface forming sand dunes. In areas of high relief, movement of these un-transformed grains will include minerals less stable or refractory to alteration, such as olivines and pyroxenes in areas of volcanic activity or garnets and amphiboles coming from metamorphic rocks. Sand reflects at the same time chemical resistance to weathering and the importance of slope and gravitational movement of alterite material.

This transport of the finer materials and their deposition occurs along coastlines of large lakes or seas. Initially, suspended or deposited material can be moved by shore currents parallel to the coastline and deposited material through strong turbulent wave action can re-suspend material deposited at some distance from the coast, depending upon the depth of the continental shelf. These zones are of importance to humans in that they concern a large portion of the aquatic life used as a food source. Most often sedentary shore or shallow marine life depends upon the sediments (clays) and adhering organic matter for sustenance. The chemistry of this material is thus involved in the nearshore food chain. These "wetlands" are considered as being very important to ecological balances (Fig. 1.7).

As we all know, one finds sand deposits, as beaches or dunes, along the edges of large water masses as well as deposits of clay rock sediments. The reason why one or the other type of deposit forms, usually well-defined grain size categories, is undoubtedly due to the energy of the movement of water masses along the edges of land. However it is not easy to see why there is such a bi-polar selection of grain sizes as to select grains from rocks that have not reacted with the surface

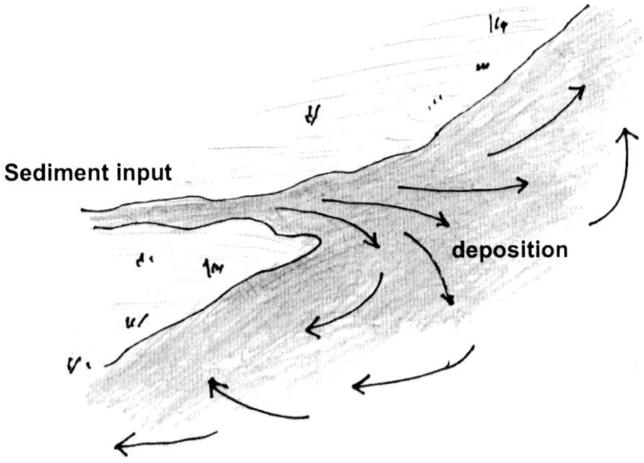

Fig. 1.7 Illustration of deposition of fine-grained stream-borne material, which becomes littoral deposits

environment (essentially quartz) and the newly formed materials that have been the result of water–rock interaction (clays and mud). In any event most clays are detrital, probably more than 90 %. They are supplied to the sedimentary systems from two main sources. One source is rocks, and the other is the soils that develop on them by weathering. Sandstone and shales are usually rather well defined, with some intermediate mixture material, but for the most part there is a strong difference between these sedimentary rock facies.

In any event concerning the clay-rich material deposited and re-worked along a coastline, it forms two types of structures, the slikke and the shorre, i.e., the clay-rich newly deposited and mobile material moved by tidal action (slikke) and this same material which can be fixed in place by plants where the newly arrived clays are stabilized to become wetland deposits or remain mobile along the coast tidal zones (shorre, see Fig. 1.8).

The roots and stems of the peri-maritime plants (grasses and small shrubs) provide a catchment for sediments, clays, moved by tidal action. As they are fixed they form an advancing front of sediment, stabilized by plants, which extends the dimensions of the wetland and hence the continent into the sea. Such structures are often very useful to farmers forming fertile zones where soil clays filled with nutrients and organic matter can be exploited by mankind. When these zones are protected from further intrusion by tidal movements (diking) they become prime land for farming as demonstrated by the Dutch and others in coastal areas. Plant growth fixes clays and as they are fixed plants roots advance seaward forming deposits of stable clays. This is an advance of the land into the sea by plant action.

Thus in general the farther from the mountains and the lower the slopes, the smaller the particles will be which are transported by rivers. Deposition of more coarse materials occurs along the trajectory as a function of slope and river flow intensity. Exceptions are sand-sized material transported under periods of high stream flow intensity, which can be moved along the coastline through coastal currents, forming sand dunes and beaches. When this material is deposited in a deep-water environment, it is buried successively to form a sandstone.

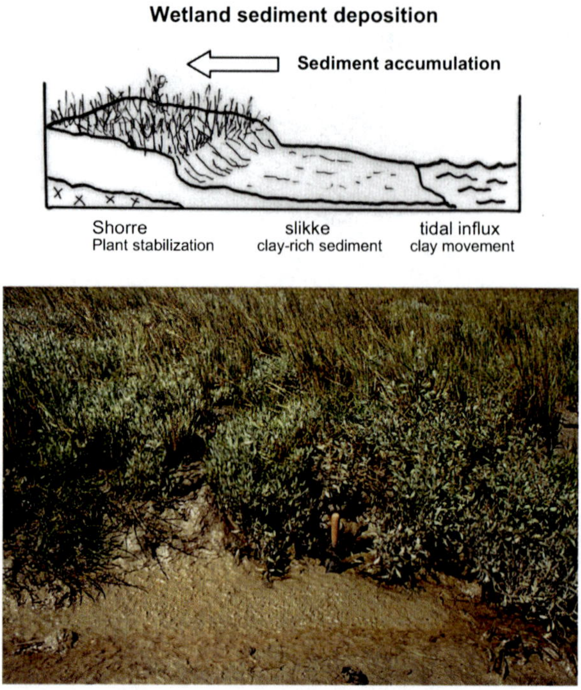

Fig. 1.8 Illustration of the stabilization of fine-grained, clay-sized material along coastlines. The shorre is the area of plant fixation of the deposits stabilized by plant roots and the slikke is the area of recent and continued movement of clay-sized material by tidal and wave action. The stabilization by plants allows more sediment accumulation and an advance of the land area into the tidal deposition area (photo)

1.1.2.4 Wind Transport

Another means of moving matter at the surface is by wind action. In certain areas of the world wind is an important agent of erosion, transport, and deposition of sediments. Dust storms, which are some of most frequent natural hazards in desert regions, can be a quite important source of deep-sea sediments. Impressive pictures of dust storm from the space can be found, for example, at the NASA earth observatory (http://earthobservatory.nasa.gov). The grain size effect is quite limited, mostly in the clay fraction (<2 μm diameter for long-distance transport, hundreds of kilometers). The source of dust material or loess is usually in desert areas where previous or infrequent current movement by water has left the fine material without plant cover and thus vulnerable to the action of wind erosion. Sand seas occur throughout the interior. The large deserts are on either side of the tropical, high rainfall zones near the equator except where the Himalaya Mountains have deviated the normal weather circulation belts moving them to the north and the case of the Americas where the mountains have changed the weather circulation directions especially rainfall patterns. For the most part the transport of loess today is toward the oceans from the deserts except for the case of China where the western deserts at the foot of the Himalayas move dust to the east over the Chinese mainland. In Xian the rate of accumulation is near 2 mm/year, enough to renew the agricultural soil zone in 100 years. Nevertheless smaller amounts of dust move

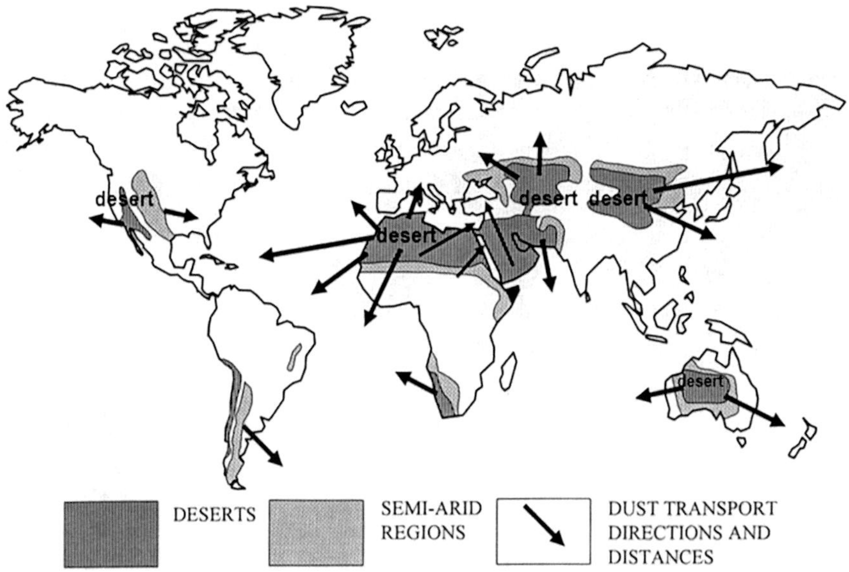

Fig. 1.9 Illustration of major movement patterns and transport distances of dust material and loess. Loess (dust borne by wind currents) originates in desertic areas on either side of the tropical zones of high rainfall. Long-distance transport involves particles with grain sizes of <20 μm [adapted from Hillier (1995) after data in Pye (1987)]

to the north from the Sahara into Europe and eastward across the central United States. Figure 1.9 indicates the major movements of dust over the globe.

Wind transport is often not considered in studies of soils, for example, but the input to the surface environment can be very important especially concerning the minor element content of wind-borne material. It is a very important aspect of modern surface geochemistry in that wind-borne material from industrial sites can become an important factor in surface geochemical interactions.

1.1.3 Chemical Effects

Along the coastlines and in deep oceanic waters, one finds effects of biological activity which can extract materials from the dissolved part of transported materials, especially calcium and magnesium which can be combined with CO_2 which were dissolved in the water to produce carbonate material. These deposits are most often dominated by the carbonates, which become sedimentary rocks concentrating the dissolved part of the sediment load (Ca and Mg notably).

In some instances stream water is concentrated in closed basin lakes (i.e., without an outlet to the sea) where it is evaporated depositing dissolved material as sulphates, chlorites, and borates. Evaporate deposits are very important for human activity in that they often contain life-sustaining sodium chloride, as well as potassium chloride which is essential to aid in plant growth. Other useful elements such as boron are found in such deposits also.

The coastal concentration of sand, which is essentially un-reacted rock material refractory to surface chemical processes such as quartz, the deposition of clays in the nearshore area, and the formation of carbonates produce the strong chemical segregation of materials found in sedimentary rocks; shales which are potassium, iron, and alumina-rich, sandstones silica-rich, and calcium, magnesium-rich carbonate materials.

Magmatic materials →	Al, Fe, K	Shale (essentially altered rock material)
	Si	Sandstone (refractory non-altered minerals)
	Ca, Mg	Carbonate (mostly biological deposits)
	S, Cl, Br, S	Evaporite deposits
	Na	Seawater
	P	Shallow sea deposits

Other elements found in rocks will follow these major mineralogical, sedimentary groups as a function of their chemical affinities with the minerals present in these sediments. This is the fundamental geochemical effect of surface interaction of rocks with the chemistry of weathering and alteration of rocks which dissolves the rock materials, and forms new minerals of different compositions.

1.1.4 Alteration: Rock to Soil Transformation

1.1.4.1 Alteration and the Development of Alteration Profiles: Water Rock Interaction

The initial framework of surface geochemistry is that of a chemically unstable assemblage of minerals in a compact material (rock), which will re-adjust to the chemistry of the surface producing small, new minerals, which are the product of mineral–water reactions under conditions of oxidation. The chemical alteration process occurs at two distinct levels in nature: the water–rock interaction zone and the plant–soil interaction zone. The second level is in fact the surface of the contact between mineral and atmosphere where plant and other biological forces control and modify the chemistry and physical presence of alteration materials. The initiation of alteration can be of either mineral or biological interaction. Bare rocks, in areas of some moisture, usually show the presence of some form of living material, such as moss or lichens, where the interaction begins through, either integral dissolution or partial dissolution and formation as new phases. The initial interactions are initiated by the fixation of lichens on the rock, and then the development of mosses. A gradual accumulation of wind-borne dust (clay and silt) material allows the formation of grasses and the development of a very thin soil zone (Fig. 1.10).

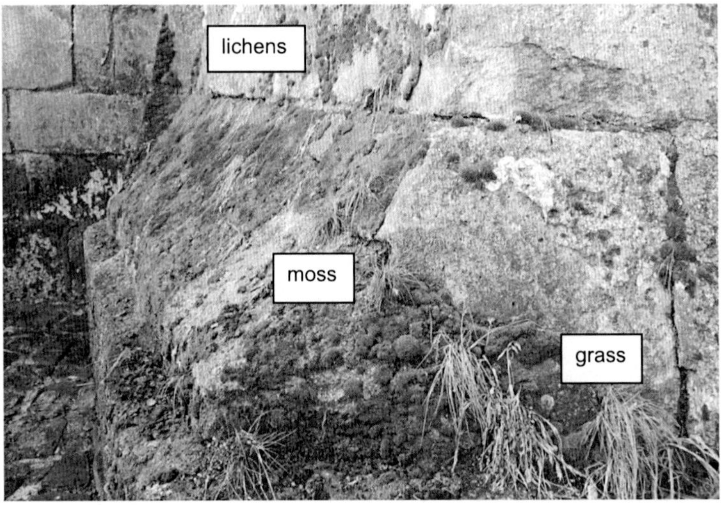

Fig. 1.10 Thirteenth-century church wall (Ile de France) made of limestone where lichens, moss, and eventually grasses fix clays and begin the development of soil formation

In some instances these contacts produce new silicate minerals (Adamo and Violante 2000), but most often they show a layer of calcium oxalate without the production of new silicate minerals. (Arocena, Univ. North BC personal communication). At water–rock interfaces, within the rock itself, alteration occurs when water, under-saturated with elements in the rock minerals, interacts dissolving the solids present. One can find congruent dissolution, total assimilation of the solids into the water, such as is the case for water–carbonate interaction, or incongruent dissolution where more soluble elements are taken into solution and less soluble elements reorganize to form new mineral phases.

1.1.4.2 Physical–Chemical Interactions

The interaction of water and rock follows physical pathways in the rock, cracks and fissures, opened by release of pressure (geological forces due to burial and asymmetric tectonic pressures) that occurs as the rocks are moved to the surface by tectonic forces. Several types of passageways can be described, according to size and resulting passage of water (Fig. 1.11).

Major tectonic movement produces fractures which transect mineral grains and which leave a passageway of several tens of microns to millimeters and more. Here the water moves rapidly, and in many cases it reaches the water table, zone of saturation of the pores, which then moves laterally toward a stream or river. The residence time of this aqueous fluid in contact with the rock is variable depending upon fracture size, but it can be relatively short in wide cracks and the interaction of rock minerals and water is limited. The water that remains which is far from

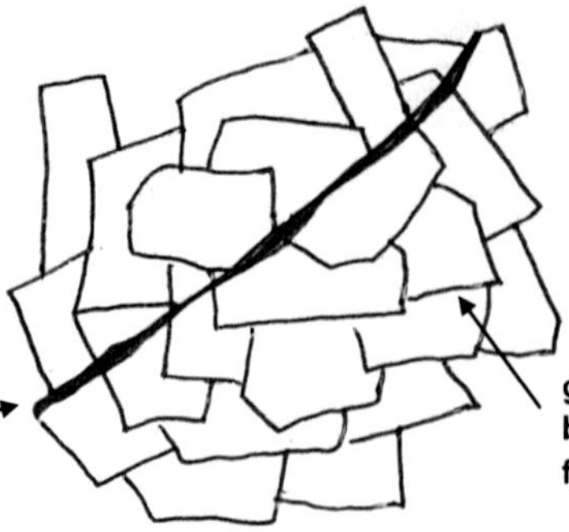

Fig. 1.11 Illustration of small infiltration passageways in a rock along grain to grain boundaries or along fractures induced by previous tectonic forces which transverse mineral grains in a rock

saturation with the minerals in the rock integrally dissolves the minerals on the edges of its passage widening them for increased water flow (see Meunier and Velde 2008).

Another important cause of small pores and cracks in rocks forming passageways for water flow is due to differential thermal expansion and contraction. Most minerals in rocks have a strong anisotropic thermal expansion coefficient following the different crystallographic directions of the crystal. This creates strong contrasts at grain boundaries, causing cracks to form between phases where water can pass and alter the adjoining mineral grains. If no water is present, in a desert for example, the grains will nevertheless be disjointed and the rock will gradually become unstable as a unit and become a mass of small grains (Fig. 1.12).

Slower movement, along fractures and cracks of less width, allows aqueous fluids to interact more with the minerals and produce more dissolution leaving alteration minerals in the rocks while transporting the more soluble elements from the high temperature minerals. However the fluids are far from chemical equilibrium with the rock and the minerals formed by incongruent dissolution are composed of the most insoluble elements combined with oxygen. As the pores are smaller, and the pathways more tortuous, the alterite minerals are less depleted in soluble elements forming more complex minerals, which eventually represent the phases, which would be in equilibrium with the rock under static hydrous conditions. At times material from one destabilized mineral can be incorporated into another mineral, which is forming at the same time. Hence the minerals formed in a rock under the initial stages of alteration will be heterogeneous in the sense that different minerals represent different types of interaction, or stages of disequilibrium between rock minerals and aqueous solutions. The rock–water interaction zone produces minerals that are not all in chemical thermodynamic equilibrium with each other. Although they are all stable under surface conditions as isolated phases, i.e., aqueous solution in an oxidizing atmosphere, they are not necessarily stable as an assemblage in aqueous solution, some containing elements that would otherwise be present in another phase which are released in solution to be incorporated into other phases [see Brantley et al. (2008), for a discussion of such kinetic effects].

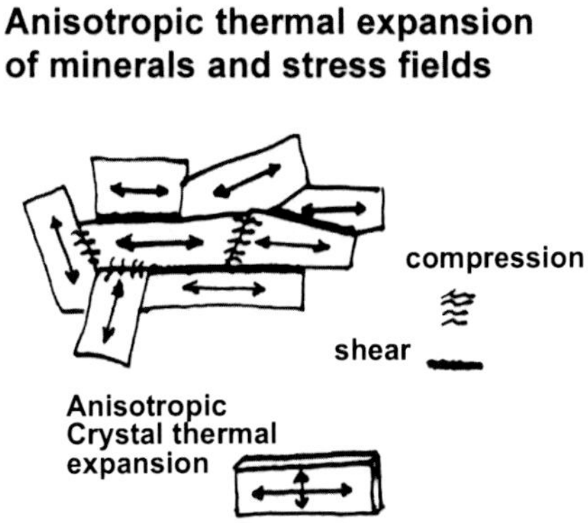

Fig. 1.12 Illustration of the effect thermal expansion on minerals in a rock. Most minerals have anisotropic expansion, which creates strong interface tensions that create grain contact fractures. At some interfaces compressive forces occur while at others extensive (shearing) forces occur

The amount of alteration of a rock can be roughly gauged by considering the clay content (the fine-grained material present), which represents the new material produced by the interaction. Most of the new minerals are hydrous, i.e., containing hydrogen ions in the form of H_2O but more prevalently OH units. This is the trademark of surface alteration. Overall the rate of interaction is not very great, at least on the scale of human life. Egli et al. (2001) indicate less than 5 % clay for alpine soils on granitic materials of 700 years exposure. Oh and Richter (2005) report clay content for granite and phyllite (low-grade argillaceous metamorphic rock) as being < 20 % for what they describe materials from the oldest soils in the world. The climate for these series of soils is humid to moderate in the South-eastern United States. Lowe (1986) indicates that the time necessary to totally transform volcanic tephra (highly unstable materials) in subtropical climates is near 800,000 years. It is clear that the transformation of unstable high temperature minerals into new surface-stable clay-sized particles is a slow process.

The result of chemical alteration then is that in the initial stages the rock is transformed along cracks and fissures, weakening it and allowing it to lose some of its initial structure through the dislocation and loss of material. This produces what is called saprock. With a complete loss of physical competence, collapse of dissolved zones, and displacement of clays, the material becomes incoherent and is called a saprolite. The extent of development of these zones depends upon the alterability of the rock and the length of time that alteration has occurred. Of course the major variable in such interactions is the amount of rainwater, which moves through the rock. Alteration in tropical zones will be more intense than that on alpine meadows. The end result is that the greatest part of the alterite material (partially transformed rocks) is a mixture of reacted and un-reacted minerals. Frequently portions of un-altered or little altered rock can be found in a matrix of altered material, ranging in size and concentration depending upon the stage of alteration of the materials. Throughout this zone, which can be several tens of centimeters in thickness to tens of meters, the interaction of moving water and unstable minerals occurs to produce new minerals or integrally dissolving preexisting phases. The more rainwater and the higher the temperature (wet tropical climate) the more water flow and the more dissolution, and a greater proportion of the new, alteration phases will be formed.

1.1.5 Alteration Profile

Looking in more detail at the structure of alteration at the surface, i.e., the interaction of rocks with the surface chemical environment, one can define two major critical interaction zones responsible for the transformation of rocks into surface minerals: one is that of water–rock interaction, which usually begins with a gradual disaggregation of rocks through alteration processes or physical fragmentation and the development of new low temperature, hydrous minerals and the plant–soil interaction zone (Fig. 1.13) where biological interaction with the alterite minerals occurs.

1.1.5.1 Water–Rock Interface

The water–rock zone is the first major step in chemical change where rocks of high temperature origin come toward chemical equilibrium with surface chemical conditions. The ambient hydrous state and oxygen-rich atmosphere effect hydration and oxidation of minerals in the rocks of higher temperature origin. New minerals form along fractures in the rock where water can pass and evacuate dissolved material. The relations of chemical potential and the consequent production of mineral phases in these stages of alteration are detailed in Velde and Meunier (2008, Chap. 2). The fundamental point to realize is that the initial stages of alteration of a rock will give heterogeneous results, following planes of physical weakness, and will contain new minerals of different type and composition. The mineralogical heterogeneity remains through most stages of alteration until the forces of hydration and dissolution reduce the initial materials to more simple chemical configurations, essentially hydrous oxide materials with a more simple chemistry.

1.1.5.2 Alterite Zone

As the rocks are altered to a greater degree, a zone of alteration material is formed, thickening with time of exposure. Here one finds new phases and un-reacted older mineral grains or rock fragments. Accumulation of this material can be called the alterite zone. The chemical forces operating on this material tend to be more homogeneous in that the rate of passage of rainwater is regulated by a more homogeneous material with fewer differences in flow rate due to a more homogeneous pore system. In general there is a re-adjustment of the minerals present which become more homogeneous in their composition and structures.

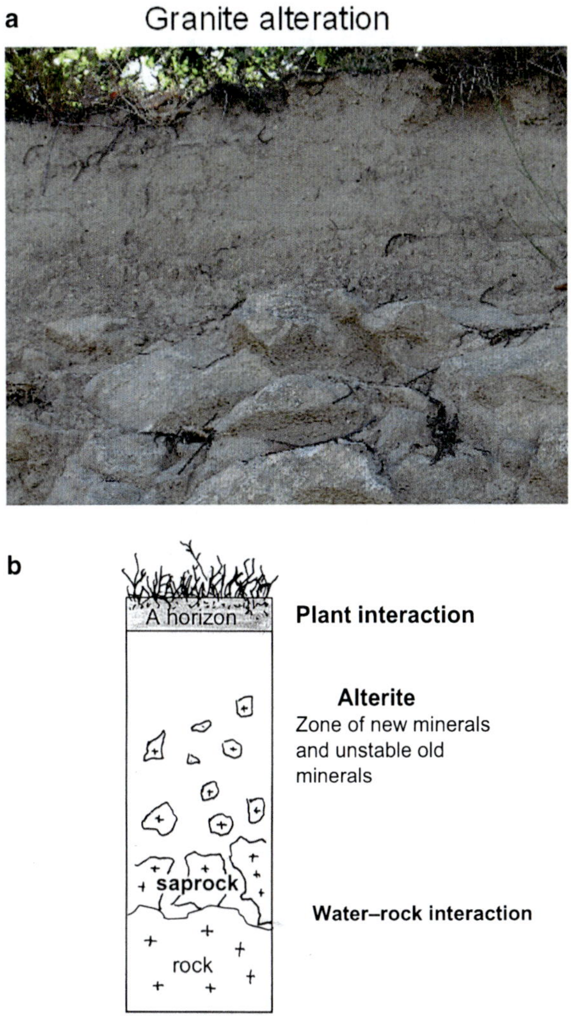

Fig. 1.13 (**a**) Alteration of granite near LaMastre (Ardeche, France). Granite alterite is seen in the large boulders at the *bottom* of the photograph, alterite material above, and a thin soil zone developed by grasses at the *top*. Larger roots penetrate into the alterite zone. (**b**) Illustration of alteration zones in a profile where the rock is altered at the water–rock interface producing new fine-grained minerals and rock and grain fragments which form the alterite. This material (alterite) is continually transformed by water interaction forming more fine-grained material. At the surface one finds the plant interaction zone (soil) where fine-grained material is to a large extent conserved by the action of plant roots, plant exudates, and deposition of mineral elements and formation of the clay-organic aggregate material

Nevertheless the mineralogy of this part of the alteration profiles remains multiphase in most cases.

The amount of new mineral material in the alterite zone depends upon several factors. Initially the amount of transformed material is a function of the amount of water–rock interaction. This is obviously a function of time, the older the alteration profile the more mineral transformation will have occurred, and the amount of water that interacts with the system. Climates with high rainfall will show more alteration in the alterite zone. The clay or alterite content can vary from several percent of the material to well above 60 % in the tropics. This indicates that the alterite is very rarely uniquely composed of new, surface-equilibrated minerals. This is because some of the minerals in the initial rocks are only very slowly reacted and remain in a state of metastable equilibrium. One notable case is quartz, which is not stable under weathering conditions but remains present metastably for very long periods of time being very un-reactive in its chemical context. The alterite zone is typically composed of old minerals and portions of un-reacted rock and new surface formed minerals, usually called clays due to their fine particle size.

The alterite zone usually shows a chemical progression from the bottom, rock–water contact zone, to the soil area of plant—alterite zone where the amount of interaction with the percolating rainwater interacts as a function of its degree of saturation with rock materials. The rainwater closer to the surface is less saturated with dissolved elements, especially of the unstable high temperature rock minerals and as a consequence it will transform the relict minerals present rapidly. Lower in the profile the percolating solutions, on average, are more saturated with the unstable mineral components due to dissolution–precipitation reactions and the reaction rate is lower. This general schema is of course conditioned by the rate of flow of water. The larger pathways, cracks and fractures, in the alterite will allow water to pass more rapidly and this water will be less saturated with dissolving elements. In this way one observes that the alterite zone is heterogeneous in the amount of new material present with high rates or reaction cole to pathways of rapid percolation and lower rates of reaction in zones of small passageways such as soil micropores.

1.1.5.3 Soil Zone

At the atmosphere–solid interface one usually finds plant life, as well as that of animals and different types of micro-organisms. This is the soil part of the alteration profile, marked strongly by organic matter and plant–soil interactions. Frequently the fluid chemistry in the soil zone is different from that lying below, due to plant induced uplift of mineral elements such as potassium, calcium, magnesium, phosphorous and silica (see Velde and Barré 2010). Plants also move minor elements and thus change the chemistry for major and minor elements in the soil zone compared to that in the alterite zone.

A new variable occurs in the soil zone, that of pH. Decaying organic matter or lack thereof can vary the pH significantly. This is largely determined by the type of plants present at the surface. Forests tend to form more acidic soils, with conifers producing the lowest pH. The acidity is variable depending upon the climate (rainfall, rainfall frequency, and temperature) and plant species present. These variables govern the biological activity in the soil. Prairies tend to form more basic soil contexts, usually several pH units higher than conifer forest soils. Deciduous forest soils are usually of intermediate pH values. Changes in pH can change the mineral stability forming new types, either oxides or silicate clay minerals.

An important vector of movement of fine material is within the soil profile itself. Macropores and cracks move rainwater to the groundwater level rapidly and in doing so can move clays, the fine particles of the soil, with the movement of water.

Such material accumulates in the subsoil horizon (named the B horizon), but it can also move outward to the groundwater table and eventually into stream or rivers. The movement is particularly important on slopes as one might suspect. An example is given in Fig. 1.13 where measurements of fine material, clay content, show an accumulation on the upland (some 3 m above the lake level, a loss of clay and no accumulation on the slope, and an accumulation near the foot of the slope.

In Fig. 1.14 one finds the clay content of profiles on uplands, a difference of tens of meters from the lowest zone, on the slope and at the bottom of the slope, near an irrigation rice paddy. In the upland profile under forest cover one sees a zone of clay accumulation below the major root zone (40–50 cm in depth). Clay movement is more or less vertical in the upper parts of the profile. On the slope there is no accumulation of clays, only a gradual increase in clay content into the alterite zone. At the foot of the slope, where water flow intensity decreases near the rice paddy, the clay content is greater at depth (60–80 cm in depth) than in the upland zone. This is a demonstration of the lateral movement of clays and their accumulation in the soils of the foot slope. Such movement has been described in a general sense, on the scale of a landscape in Millot (1970).

Such a vector of movement of fine particles is very important to consider in that often in areas covered with vegetation such as prairie grass, very little if any movement of particles comes from the plant-covered surface by erosion, but one finds that the streams draining the area have significant suspended clays in the waters during the rainy season due to transport within the soil zone. Soil clay particle movement is well documented from the soil horizon to the area just below it, forming the B horizon of clay enrichment. It is also the case that clays can move laterally, down slope into a zone of deposition or into ground water flow and hence into streams and rivers, thus exiting the local soil zones to be transported out of the immediate area.

1.1.5.4 Consequences of Alteration: Physical and Chemical Aspects

Smaller minerals, in the clay fraction, change the coherence of the rocks, as one would expect. The material becomes plastic and can be displaced by local differential gravitational forces. This produces slumps and landslides at the surface that are much more rare for rocks. One of the major reasons for this plasticity is the increased water content, water in the sense of H_2O. On any particle surface water adheres in the amount of one or several layers of molecules. This surface coating decreases the particle-to-particle friction factor and promotes plastic deformation in that the particles slide one over the other. Thus alteration promotes physical instability of surface materials.

More important to geochemical considerations is the relative surface area of particles, which increases with the decrease in grain size. In Fig. 1.15 one sees that the total surface area produced by alteration increases as particle size (indicated by the dimension of the edge of a cube) decreases. As the dimensions of the particles

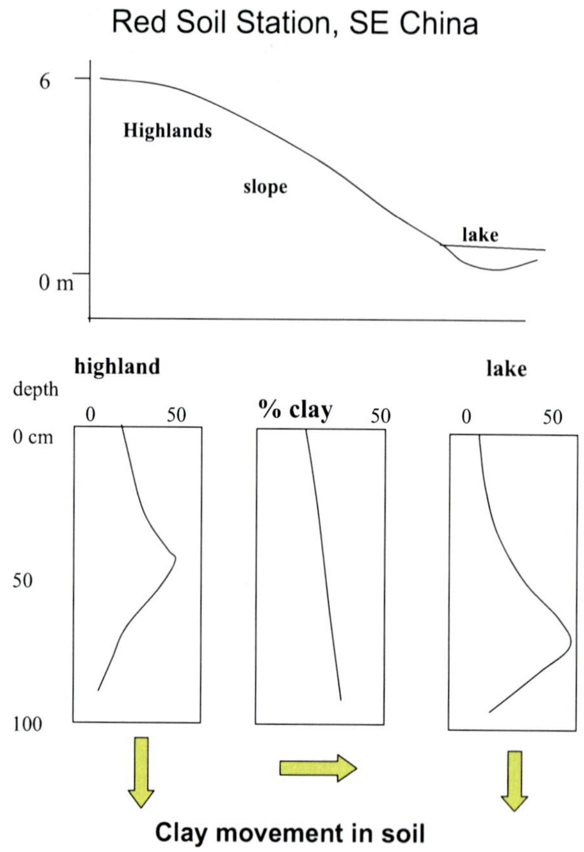

Fig. 1.14 Example of lateral soil clay transport within the soil zone down a slope. The site is at the Red Soil Station (Jiangxi Province, China; data kindly provided by Professor Bin Zhang, University Nanjing) where upland soil clay is transferred down slope to the edge of a rice paddy

decrease, the surface area compared to the volume of the particle increases very rapidly in a nonlinear fashion. This is important in that particle surfaces coated with a water layer permit the transfer and fixation of hydrated ions in the aqueous solutions that impregnate the soil and alterite zones (see Chap. 2). The solid–liquid interface is the zone of ionic contact and one of chemical importance. Hence the smaller the grain size the more chemically active surface areas are present concerning ions in solution. In contrast to rocks, alterite materials then have an enormous chemical activity by surface transfer and catalytic properties on these surfaces. For surface geochemists this is the primary factor in any study of the migration and fixation of elements under conditions of alteration and transport.

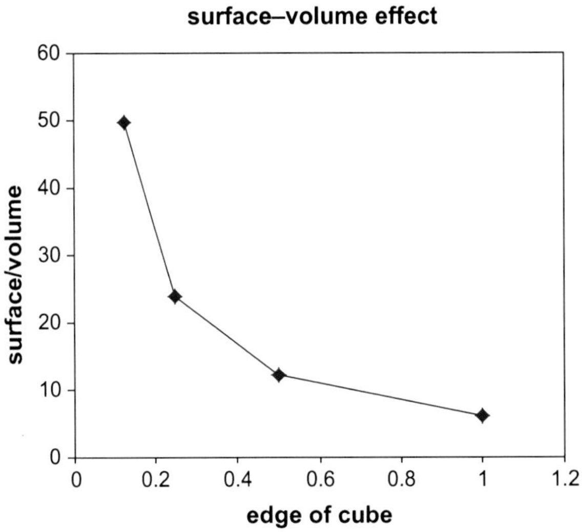

Fig. 1.15 Illustration of the relationship of surface to volume as a function of relative grain edge size. The smaller the edge or the grain size, the higher the surface area compared to the volume of the grain. Clays have a flat shape and are of small grain size, which produces a very high surface area compared to the volume of the particle

1.1.6 Plant and Soil

In the structure of alteration, there is a moderating factor at the top of the sequences. This is the biologic influence on the altered material issuing from water–rock interaction. Basically, plants need clay minerals for their biological cycle. The clays are storehouses of mineral elements, which can be extracted and exchanged at will (see Velde and Barré 2010). They form the aggregate structure of clay mineral particles fixed by organic matter into loose microporous volumes in soils that are the storehouse of water (in capillary sites). Clay minerals being vital to plant biotic processes, the plants attempt to keep them present in the root zone (that of plant–mineral interaction) so that the biological cycle is maintained. This is done by root action, which retains the clays and protects them from erosive processes. Typically clay concentrations are high in the upper parts of alteration profiles. One should note that if there are no plants present, there are no clays and there is no alteration profile. Without plants alterite material developed by physical fracturing (differential thermal expansion) and subsequent disaggregation of the rock is quickly eroded and the rock is constantly at the surface without its soil mantle.

In data of Huang et al. (2004) for Heinan (China) soils developed on basalt flows of different ages under subtropical conditions it is clear that the clay content in the profile increases toward the surface of the profiles and it increases in proportion as a function of time, but here again it is far from being a complete transformation despite the climate of humid tropical type. We can assume that in most cases soils and alterite will contain some minerals that have not yet transformed into surface materials. However, under tropical rainfall conditions, after very long periods of time, alteration can be almost complete with only the most refractory minerals resisting chemical alteration, such as quartz grains.

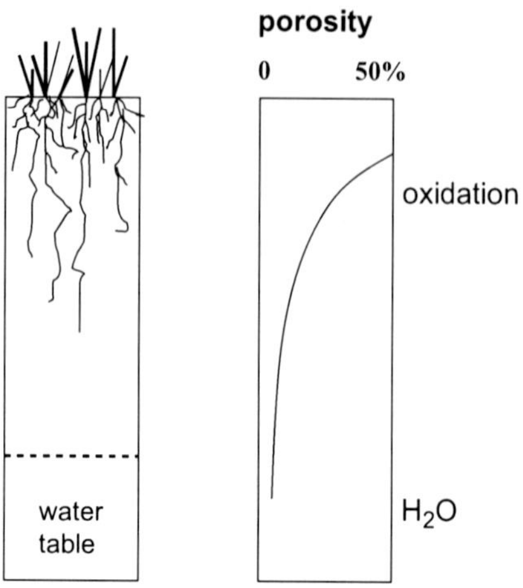

Fig. 1.16 Schematic illustration of the relationship of porosity and water content in an alteration profile where plant activity determines the porosity of the profile at the surface. In most soils, air is present in the surface porosity and oxidation potential is high. Most plants need oxygen for the soil system to function. Thus the soil root zone is one of a higher oxidation potential for alterite minerals

At the surface, biologic layers are conditioned by plants to be open to air infiltration in that most plants need oxygen to function in the root zone. Figure 1.16 indicates the structure of porosity and the effects on the infiltrations engendered by plant roots. As a result there is a stronger tendency to oxidation of minerals and materials in the root zones of alteration profiles than at greater depth. Thus the biologic factors in general complete and reinforce the chemical action of water and atmosphere as they affect the rock minerals of alteration.

The higher porosity in the soil surface zone (that where most of plants roots reside) is where there is a greater movement of fluids, keeping the pores open to air (oxygen) and where water can move downward easily. Such a physical structure has consequences on the movement of solids, clays, within the soil zone. In Fig. 1.17 it is apparent that the clay content increases from the bottom of the soil profiles, reaches a maximum at several tens of centimeters from the surface, and decreases into the soil zone. This is due to clay transport in the more porous zone, maintained by the plant regimes (in these cases forest and prairie) where fine particles are moved by moving aqueous fluids to a more less porous zone (the B horizon in terms of pedology) where clays accumulate.

Clay percent (amount of alteration reaction) increases with time and can be different depending upon the rock type. Also the biome can be important to soil clay movement. In Fig. 1.17 three rock types altered under similar conditions indicate the differences that one can find under oak–hickory forest cover in North Carolina [data from OH and Richter (2005)]. The topmost horizon, that of leaf litter and high organic content, is somewhat depleted of clays, which are found in high concentrations in the lower, B horizon. These are very old soils, and even so, they have not completely transformed to soil clay minerals. Some of the material is present in metastable form such as quartz, which would be found in the phyllite (shale in the figure) and granite based profiles, but in the basalt profile there would be little, if any, quartz and still one finds that the clay content is far from 100 %. In Fig. 1.17 for three rock types submitted to the same alteration conditions, the phyllite, low-grade

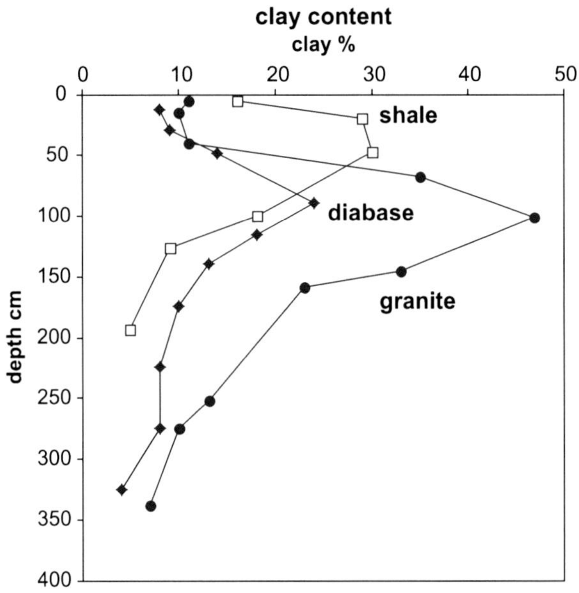

Fig. 1.17 Clay content as a function of depth for soils on different bedrocks in a sector of old soils [data from Oh and Richter (2005)]. Depending upon rock type the soil clay content varies and the transport of clays to moderate depth is more or less important

metamorphic rock which has a mineralogy similar to that of alterite products, the lower part of the rock is the least affected by water–rock alteration indicating that the minerals it contains are near chemical equilibrium with the altering fluids. Clay minerals are often recycled from one environment to another. As the product of alteration, clay minerals derived from a rock frequently pass through a soil with little or no obvious modification by soil process (Hillier 1995). Others are only partly modified (transformed). The changes for granite and basalt are greater than for phyllite. However, the increase to make clay-sized particles in the phyllite is accelerated in the plant–alterite zone where plants become the major chemical agent of change. This underlines the fact that the alteration profiles have two major parts. The water–rock interaction zone, which transforms solid material into lower density, has less structured material with a high clay size fraction. In the plant interaction zone the chemical forces are not the same as those in the rock–water interaction zone due to chemical control by plants and biological agents. New phases are formed in response to these solicitations.

1.1.6.1 Erosion, the Movement of Matter and the Role of Plants

One usually thinks of movement of solids at the surface as being due to erosion, which is generally considered to occur during periods of intense rainfall removing solids from the surface, top of the soil zone. However, one of the major functions of plants is to retain soil materials which are basic to their survival. In fact the greatest mass of materials displaced at the surface by erosion occurs through the massive failure of alterite materials on slopes through landslides and mass movement by creep mechanisms (see Fig. 1.2).

Probably the most important proportion of mountain "un-building" occurs this way. However when the material slides down slope its journey is not finished. Subsequent plant–soil interaction occurs, but in the initial stages the bare surface material is displaced by rain and subsequent by river movement. Here erosion of the finer grained material is important and the debris is moved further toward the sea, its final repository.

In agricultural areas, especially those where the soil is barren during the winter such as the North American continent, the erosion of fine-grained material by runoff of surface water into streams and eventually rivers is very important. Such runoff does not occur to a significant extent in areas of vegetal cover. The evolution of grasses during the Cretaceous or the development of agriculture had substantial impact on soil development and erosion. Plants use root structures at the surface to retain such clay material and preserve this as a soil base. If one looks at mountain streams there is still less material in suspension; clear water is more the rule than the exception. Hence the vegetal cover protects the soils from surface erosion, but massive displacement is more probable in areas of steeper slopes.

Basically plant retention of fine material at the surface in a root zone strengthens the surface material and rigidifies it to a certain extent. It is more resistant to lateral, and downhill movement. However, as alteration proceeds along the alterite–rock interface, the thickness of alterite increases. This material is more and more fine grained as water–rock alteration proceeds. Hence the alterite becomes less resistant to downhill movement through its increase in fine material while the soil plant–root zone becomes more rigid due to root retention and resistance to deformation. This system can eventually work to the disadvantage of plants in that by strengthening the surface and creating a contrast in mechanical resistance at the surface the interface between the alterite and the plant soil horizon is more and more contrasted increasing the likelihood of failure.

Overall the alteration process of unstable rock material forms small grains through chemical interaction of water and unstable minerals forming new small minerals grains. These grains are moved by various means (massive slope failures in mountains, stream, and river transport) under decreasing energy gradients where larger grains are deposited, and eventual deposition of the fine, clay-sized material, in lakes or more often the sea. The geographic slope gradient decides how far and where the different types of material are moved. The smallest goes furthest. This principle is extremely important for geochemical study in that the smallest grains have the greatest surface area per mass and are by this fact more chemically interactive, attracting different ions in solution to their surface. This activity of attraction is one of the key factors of movement of minor elements at the surface of the earth. Minor elements are those that do not make up a phase of their own and need to find a residence on another particle. Clays, fine-grained materials, are carriers of many elements, attracting them to their surfaces under certain chemical conditions and releasing them under others. This is the fundamental interplay of the geochemistry of the surface. We will pay particular attention to the components of the clay fraction, clay minerals (silicates), and oxides which are the fundamental parameters of surface geochemistry governing the retention or loss and transport of minor elements which are important elements in considerations of environmental problems and hazards.

1.1.6.2 Chemical Conditions of the Rock–Alteration Transition

Rocks are composed of crystalline minerals or at times amorphous materials such as volcanic glass. Rocks are hard and dense, not those materials one finds at the surface of the earth in equilibrium with the atmospheric chemical forces. The contrast is striking when walking outside or digging in one's garden. Earth (dirt) is soft and mobile, changing its texture and structure with the amount of water present while rocks are immobile (on the scale of ones walk in the woods or digging a patch in the garden). The contrast is one of the physical conditions of the formation of rocks compared to soil or surface alteration products. Rocks have been formed under conditions of relatively high temperatures, those with enough thermal energy to change and recrystallize the mineral assemblage present compared to those present in the initial state of sedimentation. Magmatic rocks show high temperature minerals formed below the melting point of their mass; some show relicts of the molten state. Metamorphic rocks show the effects of temperature and tectonic forces, which changed the mineral content, creating new phases and changing their shapes. Sedimentary rocks retain for the most part relicts of the materials that were sediments, i.e., materials formed in equilibrium with surface chemistry. However, in most cases the minerals have been transformed chemically to a certain extent and the phases have been recrystallized to from a more rigid mass. In all cases, the minerals in rocks are unstable at the conditions of earth surface chemical forces. They have an origin of higher temperatures and pressures. They have chemical characteristics that are not those of minerals at the surface. For this reason they react with the atmospheric conditions of the surface, water, and an oxygen-rich atmosphere, which destabilize the initial phases to create new ones.

In contrast to the formation and re-formulation of minerals in sedimentary, metamorphic, and igneous rocks, the transformation of materials at the surface includes the separation of some of the initial material from the solids, forming a new material of a new composition. Rocks are formed in essentially closed systems, chemically conservative, or nearly closed systems, whereas the formation of surface materials is accompanied by a loss of material from the solids through the interaction with surface waters. In the geological cycle, this "lost" material is often re-deposited or concentrated in sedimentary deposits when the water of dissolution is re-concentrated through evaporation or biological action. The weathering process (interaction of rocks with surface chemical materials) is one of segregation and re-combination according to the chemical properties of the elements in the minerals of the rocks that interact. Some elements are relatively easily dissolved, calcium for instance, which is transported out of the surface systems in part, often to be re-combined with elements in the air (CO_2) through the action of animal life to form carbonate material which in turn forms a new rock. On the other hand, quartz (SiO_2) is a mineral that remains metastably present, often concentrated through its grain size in contrast with smaller minerals of alteration origin, becoming sand deposits. Dissolved mineral elements that are not re-concentrated by evaporation of surface waters nor by organic interaction are transferred to the sea.

The rate at which the rocks interact with the surface chemical forces depends upon the difference in the conditions at which they were formed and those of the surface. Igneous rocks react relatively rapidly whereas sedimentary rocks interact relatively slowly. Generally speaking, sedimentary rocks will disaggregate into fine material through the action of physical forces, temperature differentials, but the minerals present will remain much the same. They are already the products of alteration.

There is a general hierarchy of instability for rock minerals, which is roughly related to the conditions of formation (temperature and pressure) and their chemistry. Concerning minerals in rocks at the surface, one can measure the rate of dissolution, which is the major factor in mineral reaction. The most soluble are the carbonates, not minerals formed at high temperature but those most affected by the exchange of a hydrogen ion for a cation in the structure. Calcite ($CaCO_3$) and dolomite ($CaMg(CO_3)_2$) show high rates of dissolution. These two minerals make up the greatest part of carbonate rocks, formed under sedimentary conditions. Next in order of instability are minerals from high temperature rocks, basalts for example, such as foresterite, an iron-rich olivine. Minerals from more acidic rocks, formed at somewhat lower temperatures, are less soluble (e.g., the plagioclase minerals anorthite and albite). Minerals found to form at the surface have lower dissolution rates, alumina, and iron oxyhydroxide. The high temperature mineral, muscovite, is the most stable of the composite minerals. It has a composition and structure very similar to one of the most common clay minerals formed at the surface, illite. The least soluble is quartz, which remains metastably present and is always a major component in surface materials.

It is evident that a rock will be composed of several minerals, which will have different rates of dissolution. This is a strong factor in the disintegration mechanism in that one phase dissolving differentially will create passageways in the rock that will favor differential fluid flow and subsequent dissolution due to the different amounts of rock–fluid interaction. The faster the flow rate, the more unsaturated the fluid and the greater will be the tendency to dissolve the surface of the minerals along the surface of the fracture pathway. Slower fluid movement will favor incongruent dissolution, i.e., reactions closer to chemical equilibrium which produce new phases and induct dissolved elements into solution. These reactions form clay minerals and oxides.

Another factor is the accessibility of water to the grain surfaces. In rocks fractured by thermal effects, those with heterogeneous grain types, water can infiltrate and remain present in the small capillary pores created in the rock. Here dissolution and mineral reaction will occur. In rocks that are more anisotropic, the thermal cracking factor is less important and less water penetrates the rock resulting in less water–rock chemical interaction.

The effects of differential dissolution rates on different minerals, the effects on desegregation, and enhanced alteration due to mineral heterogeneity and overall rock compositions determine the input of dissolved material into streams and rivers, which carry the geological elemental cortege to the sea. In doing so this material traverses the continents with interactions, some deposition, or halts in the movement of material (stream deposition, lake beds, etc.), which affect the chemistry of the surface. *Interaction of the alteration products with the changing chemistry of their transportation vectors is the basic interest of this book.* We wish to investigate not only the transformation of rocks to alteration minerals but also the effects of mixtures of this material as it moves from alteration site to final deposition site. Surface geochemistry is not only the formation of new minerals but also the effects of this initial interaction with changing chemical conditions during the transportation phase. For example, lakes can have different chemical constraints (pH, Eh, biological activity) than the soils that formed the clays and oxides of alteration, which are moved into lakes. The groundwater flowing from the rock–alterite zone can have different chemical constraints than those of the soils found above them. Streams and rivers will incorporate different materials from different setting of alteration and plant interaction as they flow to the sea and hence will have different chemical constraints imposed upon the suspended and dissolved material as it moves ocean-ward. These are the dynamics of surface geochemistry.

As has been mentioned above, geology is movement. The earth is in a constant state of change, albeit on a scale difficult for humans to comprehend. Nevertheless, the earth surface processes are such that they can be easily seen (landslide, muddy streams, mud flats, and sand beaches, etc.) as indicators of land surface change. These movements change the place of materials, in the dissolved state or in the particulate state. The new surface materials such as clays and oxides are moved in preference and carry dissolved elements on or with them in the process of adjustment to gravitational forces. This movement is the basic reality of surface geochemistry. Small particles move in aqueous solutions, and carry with them hydrated ionic materials, which can be easily exchanged with other elements when the chemistry of the solutions changes. Our task is to understand when the different ions released during alteration enter stably into newly formed alteration products or when they are fixed temporarily on the surface of a mineral due to inner-sphere or outer-sphere complexation or when the surface fixation forces change to release these elements into aqueous solution to be moved to a different site. Surface geochemistry is a study of the ephemeral.

1.2 Chemical Elements and Associations in Surface Environments

1.2.1 Affinities of the Major Elements and Surface Geochemistry

This book is not designed to give an exhaustive list of elements and their presence at the surface of the earth. The major reason for this is a lack of data for many of the less abundant elements found in the Mendeleyev table compilation. There are chemical rules and deductions from these rules, which allow one to predict the behavior of elements, but often there is not much data available to test these hypotheses. We will thus focus our objectives on the better-known elements and their reactions to changing chemical conditions of the surface chemical environment. This approach is fraught with problems in that the influences of agents of chemical change are multiple in natural environments at the surface of the earth. Major factors are the chemistry of solutions, driven by variations in pH and Eh. The presence of different types of material with strongly differing chemical properties, silicates, oxides, and organic molecules, creates further difficulties in that each phase or chemical type of material exerts different attractions on the diverse chemical elements present. Our approach is to gather the available information on elemental abundance and association in order to describe the behavior of elements at the surface of the earth. The chemical differentiation and segregation at the surface occurs in the context of geological forces, which determine what we can call geochemistry. However there is always an underlying chemical reason for the dispersion and affinities of the elements in the geological environment.

Initially, we consider the relative abundance of elements. The classification of major and minor element used here is rather classical, and consequently not very precise. A major element is one that contributes more than several tenths of a percent (oxide weight) in a given sample. Minor elements are normally considered in the several tenths to one-tenth of a percent and trace elements are of lower abundance. This definition is of course subject to variation depending upon the geologic context. Ore bodies are typically composed of minor to trace elements, unusual concentrations, but nevertheless real and realizable in the geological context. In some rocks major elements might be rare. In ultrabasic rocks sodium and potassium become of minor abundance. However, if we consider the concentrations of elements in common, abundant rock types the major, minor, and trace element categories are valid. In problems of surface geochemistry one may find, through anthropogenic action, that a trace or minor element is present in major element proportions. Thus, one has to be careful to stipulate the context and occurrence when considering the nomenclature of the different elements found at the surface of the earth.

Classification of the elements present in the earth's crust has been made numerous times over the past decades. McQueen (2008) gives a summary of the different

Table 1.1 Classification of chemical elements in the surface environment

I	II		III		IV				V		VI		
					Covalent with oxygen								
Ionic charge			Anions		oxyanion complex formers				Covalent				
(+1)	(+2)		(−1)		(+3)	(+4)			anions		Gasses		
H													
Li	Be		F		B	C	N		O		He		
Na	Mg		Cl		Al	Si	P		S		Ne		
K	Ca		Br						Sc		A		
Rb	Sr		I						Te		Kr		
Cs	Ba										Xe		
VII													
Moderately covalent cations (forming oxides or oxyanion complexes)													
Sc	Ti	V	Cr	Mn	Fe	Co	Ni	Cu	Zn	Ga	Ge	As	Sc
Y	Zr	Cb	Mo	Te	Ru	Rh	Pd	Ag	Cd	In	Sn	Sb	Te
								Au	Hg	Tl	Pb	Bi	
REE oxides and phosphates													
trans uranides													

The groups are based upon their behavior under surface conditions, the major differences being the valence of ions in aqueous solution. Cations are of positive valence, and anions are of negative valence, i.e., the relationship of completely filled electron orbitals compared to the ionized state where electrons are in lower or greater abundance than the normal equilibrium state. Association of elements with oxygen changes their behavior in many instances. Anions with a strong tendency to formal strong electronic bonds with other elements, covalent ions, and those forming strong bonds with oxygen are extremely important to surface geochemistry [table based upon, data in Krauskopf (1967), Mason (1966), Pauling (1947) among others]

types of categories of elements as they are incorporated into mineral phases. General considerations range from whether an element is found in the atmosphere, in silicate minerals, in oxide minerals, or in sulfide minerals. These are the major criteria of Goldschmidt (1954), which have been useful over a long period of time. Our interest here is that these classification schemes are applicable to the rock as it is presented at the surface to be transformed by interaction with slightly acidic, oxygenated water. Here the criteria of elemental distribution in rocks and minerals are for the most part no longer applicable. Some minerals of the high temperature assemblages are still present but they are not stable, and given enough time will be transformed into different surface minerals. We will not use the terminology common to geochemistry texts in that it is not all that applicable to our problem of surface geochemistry (Table 1.1).

The elements of major abundance and many elements of lower abundance can be classified as to their behavior under aqueous conditions and different redox conditions. At either end of the periodic table one finds the elements that form gases, Ne, Xe, and so forth. These elements are not active at the surface of the earth and will therefore be ignored totally. We will start with the ions, which are found in solids and aqueous solution, the major solid phases, which are present, and the major components that affect the geochemistry of the surface. Our objective is to discuss the reactions due to aqueous conditions experienced at the interface of the continents. Water is the major chemical agent and at the same time the physical agent causing movement of materials. Our discussion is oriented toward the reactions and associations of minerals and elements to aqueous chemical interaction at moderate temperature. Virtually all of the surface geochemical reactions pass through an aqueous stage. Ionic diffusion in solids is very slow at low temperatures. Thus in the absence of liquid water rock materials do not react or change significantly. The chemistry of elements in aqueous solutions is the key to understanding surface geochemistry.

1.2.2 Agents of Change

1.2.2.1 Oxygen

Oxygen is ubiquitous in the zone of surface interaction. This element is the predominant one present in geologic surface materials, as in most of the earth, at least from the crust outward. Oxygen commonly represents half of the atoms present in a rock or altered rock. This is in contrast to biologically related materials where oxygen usually represents $< 20\ \%$ of the atoms in a molecule. It is the one of two elements that dominates aqueous solution and it is very present in the atmosphere. Oxygen is an anion of various properties, most important of which is the ability (or prevalence) to share electrons with other ions, which is the basis for covalent ionic associations. The relations of other elements with the divalent negatively charged oxygen ion are primordial to the structure and coherence of materials. It is the background in which geochemistry evolves. However, the tendency of oxygen bonding in solids is quite different from that of its bonding in fluids and in the air. In air oxygen tends to form its own gases (O_2, O_3^{2-}), or be combined with carbon or nitrogen to form other gases. These molecules are quite stable especially CO_2. In water oxygen is associated with hydrogen, either as H_2O, H_3O^+, or OH^- ions. Some free oxygen (O_2) is dissolved and is a highly reactive substance. Oxygen in the air tends to combine with certain cations as they are released from rock minerals forming oxides. The activity of oxygen in the interaction with rock minerals in aqueous solutions is critical to mineral stability. When a constituent element (such as Fe) changes oxidation state, there is a change in the electronic balance of the original mineral which often causes the dissociation of the constituent elements resulting in the formation of new minerals. This action is frequently the initiation of alteration processes.

Oxygen is the primary oxidant for the degradation of organic matter in soils; in the absence of oxygen, however, other species oxidize organic matter following a thermodynamically predicated sequence of oxidants (oxygen $\rightarrow$ manganese oxides ~ nitrate $\rightarrow$ iron oxyhydroxides $\rightarrow$ sulfate (Froelich et al. 1979).

1.2.2.2 Water

Probably the most important relation in surface geochemistry resides in the very unusual properties of water, H_2O. It is one of the most singular liquids we know, with a negative thermal expansion in the solid state, a capacity to accept cations, anions, and a significant array of charged ionic complexes in solution, inorganic as well as organic. It is composed of oxygen and hydrogen. In the water molecular structure hydrogen ions are present at 120° one from the other which results in an asymmetric molecule, with positive ions on one side and the underlying oxygen ion (negative) on the other side. The asymmetry leaves a portion of the oxygen without electronic compensation such that the water molecule has a tendency to attract cations to the "vacant" side of the molecule. This slightly ionic (polar) character gives water the capacity to attract and retain ionic species. Thus one can consider water to be a solvent, bringing other chemical materials into a stable state in the liquid. Also water is relatively easily dissociated in two components, H^+ ions and OH^-. Both are chemically active creating soluble units from solid materials. Both H^+ and OH^- units can associate with elements to form complex associations. The prevalence of one or the other depends upon the relative amount of H^+ or OH^- (pH), which is critical to most of aqueous solution chemistry.

Water is at the same time the prime altering agent by its chemical activity and thus the prime agent of movement of material in solution and of course it is the major medium of movement of particulate (non-dissolved and dissolved) material. Water is a chemical and a physical agent. The importance of water cannot be underestimated in surface geochemistry. Further it is the basic substance necessary for life on the surface, acting as a relay for dissolved substances to be moved into a plant and as a major part of the chemical functioning of plants. The "blue planet" is essentially governed chemically by the activity of water at the surface.

1.2.3 Bonding Between Elements

Rocks and alterite minerals, the products of incongruent mineral reaction in aqueous media, are largely found as silicates, which are covalent associations of oxygen with Si ions which form an oxyanion structure, such as carbonates, and oxides or hydroxy-oxides. The cation–oxygen associations are the primordial key to understanding the structure and chemistry of surface minerals.

The elements of earth material at the surface are for the most part linked by either covalent or ionic bonding forces or intermediate types. Ionic bonds and covalent bonds differ in their structure and properties. The exact definition of covalent and ionic bonding is difficult to determine in that there are different grades of intensity of the two opposing bonding types.

Ionic bonding is the electrostatic force of attraction between positively and negatively charged ions. These ions have been produced as a result of a transfer of electrons between two atoms with a large difference in electronegativities. If a compound is, for example, made from a metal and a nonmetal, its bonding will be ionic. For example, sodium and chloride form an ionic bond, to make NaCl, or table salt. For covalent bonding we distinguish between two types. The first type is nonpolar bonding with an equal sharing of electrons. But usually, an electron is more attracted to one atom than to another, forming a polar covalent bond. For example, the atoms in water, H_2O, are held together by polar covalent bonds. If a compound is made from two nonmetals, its bonding will be covalent. Whether two atoms can form a covalent bond depends upon their electronegativity, i.e., the power of an atom in a molecule to attract electrons to itself. If two atoms differ considerably in their electronegativity (measure of the attraction of an atom for the electrons in a chemical bond)—as sodium and chloride do—then one of the atoms will lose its electron to the other atom. This results in a positively charged ion (cation) and negatively charged ion (anion). The bond between these two ions is called an ionic bond.

The electron sharing links the atoms into a very stable chemical structure. It is difficult to extract one element from the others in covalently bonded compounds. A demonstration of this fact is the low solubility of covalently bonded structures in water. Ionically bonded atoms do not share electrons and are loosely held together. Materials composed of these bonded atoms are easily dissolved in water, such as is NaCl.

By contrast, in surface geo-materials, oxygen is the major anion present and bonding to oxygen is primordial. Two types of bonded materials are present. The first type is an oxyanion, where oxygen is bonded to a cation. The bonding is in a largely covalent mode, with a residual negative charge on the oxygen satisfied by bonding to a cation in a more ionically bonded form which is strongly covalent, silicates for example. Oxoanions are formed by a large majority of the chemical elements. The other type is an oxide where oxygen is bonded to cations directly to form a neutral compound. The equilibrated associations of quadrivalent cations occur when the cation is joined with two oxygen atoms (SiO_2, CO_2, SO_2).

Krauskopf (1967, p. 136) gives relations of bonding types for most of the common (major and some minor) elements found in surface geochemical situation. One can summarize this information as follows:

Type I: Ionic character of $> 70\ \%$ with oxygen

$$\mathrm{Ca} > \mathrm{K} > \mathrm{Na} > \mathrm{Li} > \mathrm{Ba} > \mathrm{Ca} > \mathrm{Mg}$$

Type II: Intermediate bonding character with oxygen 60–50 % ionic

$$\mathrm{Be} > \mathrm{Al} > \mathrm{B} > \mathrm{Mn} > \mathrm{Zn} > \mathrm{Sn} > \mathrm{Pb} > \mathrm{Fe} > \mathrm{Cu}$$

Type III: Covalent bonding ($< 50\ \%$ ionic character)

$$\mathrm{Si} < \mathrm{C} < \mathrm{P} < \mathrm{N}$$

Classically these values are based upon the electronegativity of the element ionic state, which can be found in tables of chemical character in many textbooks and reference works. The covalent bonds between Si, C, and P (type IV ions) with oxygen to form an oxoanion are very strong where the charged complex ion can be satisfied electronically by association with an ion of high ionic character (type I) or can be associated with ions in the intermediate category either in covalent or partly ionic bonding (type II). In some cases, oxoanion bonding with two types of cations in ionic or covalent bonds occurs (types I, II, and III) which is typical of silicate minerals. Hence within the same chemical compound ions can be present in different states of bonding. In such complex structures there is room for a significant amount of ionic substitution of one element for another. This "solid solution" is again typical of silicates.

Each ion has a general sphere of electron influence, which can be described by an ionic radius, in angstrom (Å) dimensions. The radius essentially determines the number of oxygens, which can be closely associated with a cation in a crystalline structure. This number of ions is often referred to as a coordination number ranging from 12 to 4. Anions of course have the same ionic radii zones of electron influence. Figure 1.18 indicates the number of oxygens in association with different cations, which can change when the ion changes oxydation state such as is the case for iron. The most frequently observed ionic substitutions in a crystal occur when the ionic charge is the same and the ionic radius similar.

An important effect on bonding type is the oxidation state of the element concerned. In general monovalent cations are highly ionic in character. As oxidation state of an element increases the tendency to form oxyanions is greater. If an element can have several oxidation states under surface conditions, it can change its ionic form going from cation to an oxoanion. Arsenic is an example of oxygen–oxidation state interactions. (Arsenite ($As[III]O_3^{3-}$), Arsenate ($As[V]O_4^{3-}$). In aqueous environment the inorganic arsenic species arsenite (As^{3+}) and arsenate (As^{5+}) are the most abundant species. Some elements considered as metallic in type, chromium for example, can change ionic association in aqueous solution from cation to oxoanion. Chromium, named for its many-colored compounds, exists in the oxidation states of −2 to +6, inclusively. The existence of a particular oxidation state is dependent on many factors including pH, redox potentials, and kinetics. The +3 and +6 oxidation states are the most common ones found in aqueous solution. Change in oxidation state can change the affinity of an element in its chemical associations greatly in the surface environment.

Fig. 1.18 Examples of some elements and their ionic characteristics with the coordination characteristics with oxygen in the case of cations [data from Krauskopf (1967)]

Cation	Charge	Radius (Å)	Coordination
Na	1	0.97	6, 8
K	1	1.33	8, 12
Mg	2	0.66	6
Ca	2	0.99	6, 8
Sr	2	1.22	
Fe	2	0.77	6
Fe	3	0.63	
Mn	2	0.74	6
Mn	4	0.53	
Ti	4	0.68	6
Al	3	0.51	4, 6
Si	4	0.42	4

Anion	Charge	Radius (Å)
S	2	1.85
O	2	1.4
OH	1	1.4
F	1	1.36

1.2.4 Cation Substitutions

The elements, which do not form strong covalent bonds form largely ionic bonds, tend to have a lack of an electron or more in their electronic orbital structure. Ionic bonds are less strong than covalent bonds. Li, Na, K, and Ca like cations which have a high ionic bonding tendency are very important to surface geochemistry in that they are easily displaced from a mineral under surface alteration conditions. Such of alteration is for the most part, concerning cation movement, initiated by the substitution of a hydrogen ion (H^+) for a cation in a structure. This ion exchange of the cation results in its migration toward the ambient aqueous fluid from which the hydrogen ion came. This ion exchange phenomenon is electro-neutral. The driving force is the low concentration of the cation in the aqueous solution relative to that in the solid phase (in this order). In order to attain an equilibrium concentration of cations in solution in the presence of a solid phase, hydrogen ions migrate into the solid and displace cations in the structure. This cation exchange is a vital initial starting point in attaining chemical equilibrium between unstable minerals, in rocks, in contact with aqueous solutions, in particular surface waters, which are unsaturated in cations due to their atmospheric origin (rain water).

An interesting example of relative bonding stability in a complex silicate material is shown in Fig. 1.19. The elemental concentrations of cations in a thirteenth-century cathedral window from Angers (France) have been affected by hydrogen substitution and cross diffusion in materials altered by rainwater for six

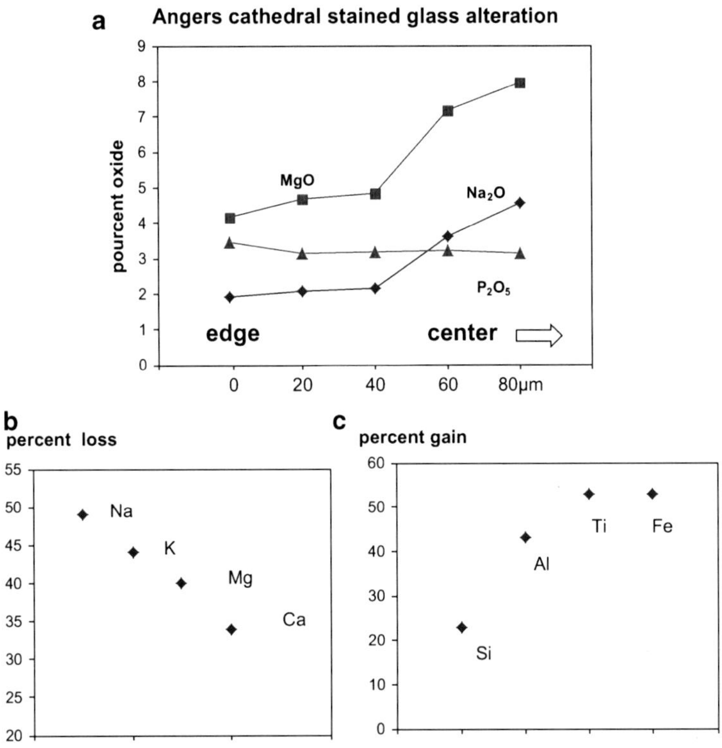

Fig. 1.19 Analyses of progressive stages of alteration through the surface layers to the unaltered glass on a thirteenth-century stained glass fragment from the Angers cathedral France (data from F Pivet). (**a**) Sequential analyses from the exterior to the un-altered portion of the glass in relative distance, approximately 20 μm between measurements. (**b**) Relative loss of Na, K, Mg, Ca. (**c**) Relative loss of Si, Al, Ti, Fe

centuries. The initial material is glass, an amorphous silicate formed at high temperature and highly unstable chemically under surface conditions. The chemical organization of the glass is essentially the oxyanion complex (Si–O, Al–O) compensated by more ionically bonded cations. In the glass there is a progressive loss of alkali and alkaline earth ions from the amorphous solid material (glass), which is reflected by a relative increase in silica content. Taking the most external point of the measurement zone, it is possible to classify the elements by relative percent loss compared to the initial glass material. The sequence Na > K > Mg > Ca is quite clear. The most easily moved ion is sodium, while the most difficult to remove in the group is calcium. The difference in mobility is obviously due to the charge on sodium ions, 1+ while that of calcium is 2+ which produces a charge density twice that of sodium (Lide 2000). Less mobile ions such as Si, Al, or P change in relative abundance (Fig. 1.17c) indicating that even the less mobile, more covalently bonded ions can be removed during interaction of water with a silicate material. In the case of window glass, the removal of alkali and alkaline earth cations by hydrogen cations does not result in a stable structure. The hydrogen ions react to form water molecules leaving a highly hydrated silica mass on the surface of the glass after the ionically bonded cations have been removed.

In this example the cations that are less strongly bonded (ionic bonding) are removed, being replaced by hydrogen ions which become water molecules. The remaining material represents more strongly bonded cations where the bonding type is largely covalent. One important way to characterize ions is by their ionic potential. Ionic potential is an ion's charge divided by its radius, and it is thus a measure of density of charge. Ionic potential gives a sense of how strongly or weakly the ion will be electrostatically attracted to ions of opposite charge, and to what extent the ion will repel other ions of like charge. There is a reasonable correlation between the observations in Fig. 1.19 and the data in Fig. 1.18.

1.2.5 Chemical Types of Atoms and Multi-element Units

1.2.5.1 Anions and Cations

The corresponding low bonding energy ions with negative charge (Cl, Br) are frequently associated with the cations mentioned above, forming salts, i.e., high solubility materials formed at the surface of the earth as a result of strong water loss and subsequent accumulation of dissolved elements into residual minerals upon evaporation.

These anion and cation elemental combinations are the most soluble elements found at the surface.

1.2.5.2 Cations and Oxygen Ions (Oxoanions)

An oxyanion or oxoanion is a chemical compound with the generic formula $A_xO_y^{z-}$ (where A represents a chemical element in a cationic form and O represents a negatively charged oxygen atom). Some elements are able to form more than one oxyanion, each containing a different number of oxygen atoms. For example, chlorine can combine with oxygen in four ways to form four different oxyanions: ClO_4^-, ClO_3^-, ClO_2^-, and ClO^-. Here the charge remains the same; only the number of oxygen atoms varies. Oxoanions are formed by a large majority of the chemical elements.

Associations of cations with oxygen through largely covalent bonds are a fundamental characteristic of surface geochemistry. The cation–oxygen units can form an anionic unit, which is completed by ionic bonds to cations when forming a solid phase. The most strongly bonded associations are in the order boron, carbon, silicon, aluminum, phosphorous, sulphur, arsenic, uranium, and oxygen. The configuration of these associations is that of an oxyanion complex that has an overall charge which is satisfied by another cation forming a solid phase. Thus the covalently bonded cation is part of a tripartite structure where it satisfies part of the negative charge of the oxygen anion. The tendency is to form a strong, partly covalent bond, which determines the stability (solubility and resistance to cation exchange) of the phase. Usually the cations associated with a given oxyanion structure are multiple and can be exchanged under strong chemical potentials. The relative solubilites of these associations depend upon the cation forming the phase. The solubilities of oxyanion complexes can be considered as a measure of their stability under conditions of alteration. In general sulfates are the most soluble with Ksp (solubility products) of 10^{-5}–10^{-10} (barium sulphate). Carbonates are less soluble with Ksp values ranging from 10^{-9} to 10^{-17} and the least soluble (most stable) oxyanion complexes are phosphates with Ksp values of 10^{-24} to 10^{-37} [data from Lide (2000)]. One can expect to see few sulfates in surface alteration environments, i.e., those where flowing water is the alteration agent, and phosphates should remain present.

1.2.6 Reduction of Oxoanions

Among these oxyanions some are chemically stable, at one specific oxidation state while others are susceptible to redox reactions. Sulfur and arsenic are typical elements of this type. This tendency to respond to changes in Eh is crucial to understanding surface geochemistry in that the environments of surface materials are often subject to changes in redox potentials generated by biologic activity.

Some oxoanions change oxidation state under surface conditions and are no longer associated with oxygen, such that they form cation–anion associations. Such is the case of sulfur and arsenic. Sulfur forms sulfide phases associating with cations. These states are highly dependent on specific Eh and pH conditions. The change in oxidation state and charge on the ion complex under surface environments is extremely important to an understanding of movement of materials under surface geochemical conditions.

Sulfur is a very important case with respect to redox sensitivity in that it is often used as an energy source in bacterial activity, either in the oxidation of sulfides or the reduction of sulfates. These changes can be very important for minor elements, which can be incorporated in one or the other of the sulfur phases. Under surface conditions it appears that the appearance of sulfides is largely due to biological action, especially bacteria where reduction of sulfate ions occurs (Pósfai and Dunin-Borkowsky 2006).

Sulfides are of highly variable solubility, but they are susceptible to oxidation and hence dissolution by surface chemical effects which leaves their presence dependent upon Eh factors more than water concentrations that promote dissolution.

1.2.6.1 Redox Cations (Transition Metals)

The transition elements are group of industrially important metals mainly due to their strong inter-atomic metallic bonding giving them generally high melting/ boiling points and high tensile strength. The transition elements have low ionization energies. They exhibit a wide range of oxidation states or positively charged forms. The positive oxidation states allow transition elements to form many different ionic and partially ionic compounds. Some typical characteristics of transition elements are formation of colored compounds, and paramagnetism, variable oxidation states, and tendency to form complexes.

Redox reactions are reactions in which there is a simultaneous transfer of electrons from one chemical species to another. *Redox reactions* are really composed of two different reactions: *oxidation* (a loss of electrons) and *reduction* (a gain of electrons). A relatively large number of elements are generally associated with what are called metals. They can be reduced to the pure state by reduction of the electronic charge (copper is perhaps an example, Manceau et al. 2008). They are present in surface minerals or associated at their surfaces as cations. Some of these elements have the particular property of changing oxidation state under conditions of surface geochemistry. For example Plutonium can, for example, have four oxidation states (+3 to +6) under natural conditions. They also have a higher electronic affinity for other elements, especially oxygen. Transition metals may form oxides–cation or oxygen bonded units or they may enter into the oxyanion complexes. Very rarely they are present as a single phase due to strong reductive conditions. Normally elements in the metallic state become oxidized at the surface to form a metal oxide phase or hydroxy-oxide phase. The most common type of such elements in alterites is iron due to its high abundance, followed by manganese, and eventually other elements such as Ti, Cr, Ni, Co Cu, Zn, V which are considered in most geological contexts as trace elements due to their low abundance and incorporated into other phases as minor substitutions.

Redox cations are then found as oxides or in oxyanion complexes under surface chemical constraints. The possibility to change chemical association is very important in following the movement and fixation of these elements in surface environments. Scott and Pain (2008) indicate that in general a change in oxidation state to the more oxidized form reduces solubility in aqueous solution and hence a tendency to form an oxide or oxyhydroxide. In the reverse, a change in redox conditions to a more reduced state can move an ion from an oxide state into solution. This is a very important concept for the movement of materials at the surface under conditions of biological activity engendered by the plants. Biological action frequently affects the redox potential of surface environments and changes oxide materials into solutes.

1.2.6.2 Heavy Metal Elements

Over the past two decades, the term "heavy metals" has been widely used. It is often used as a group name for metals and semimetals (metalloids) that have been associated with contamination and potential toxicity or eco-toxicity. There is no authoritative definition. Normally the term heavy metal refers to any metallic chemical element that has a relatively high density and is toxic or poisonous at low concentrations [for example mercury (Hg), cadmium (Cd), or thallium (Tl)]. Human activities have drastically altered the balance and biochemical and geochemical cycles of some heavy metals. Therefore, the concentration of heavy metals in soils has been an issue of great interest in the past few years not only to ecologists, biologists, and farmers but also environmentalists. Heavy metals are dangerous because they tend to bioaccumulate. In small quantities, certain heavy metals are nutritionally essential for a healthy life. Heavy metals are natural constituents of the Earth's crust. They can be associated with sulfur, carbonate, or oxygen. Among the most prominent in surface materials (present in ppm or ppb quantities) are Sn, Sb, Cd, Pb, As, Bi, Hg.

A somewhat extraneous pair of elements, Ga and Ge, is chemically closely related to Al and Si, respectively, and can be seen to follow these major elements closely in relative abundance suggesting a constant substitution in mineral structures for the two major surface mineral-forming elements.

1.2.7 Metals

Some elements are most often found as metals such as gold, platinum, silver, and at times copper for example. Their abundance in ambient surface conditions is very low in general. However, they do form a pure mono-element phase.

1.2.7.1 Oxyhydroxide

The group of oxides, hydroxides, and oxyhydroxides comprises several cations such as Fe, Al, Mn, or Ti, where the anionic part is an oxygen (−O), a hydroxyl group (−OH), or an oxygen and a hydroxyl (−OOH). In alterites, oxides (hematite α-Fe_2O_3, magnetite $Fe^{2+}Fe^{3+}{}_2O_4$, maghemite γ-Fe_2O_3, wuestite FeO, ilmenite $FeTiO_3$), and oxyhydroxides (goethite α-FeOOH, lepidocrocite γ-FeOOH, ferrihydrite ~FeOOH) do occur. This category of elemental combination is rather difficult to define since the material is rarely well crystallized and often of varying composition. The main cation components of these phases are iron or manganese. However, the surfaces of these materials attract many types of hydrous cations, which can be incorporated in the host mineral structure. In fact the attractive power of manganese oxides to incorporate transition metal and heavy metal elements is many times greater than that of other phases in surface deposits (Manceau et al. 2000, 2007). Their attractive properties are perhaps due to their incomplete structures. At the edges of these minerals one finds many free or unsatisfied bonds. At these points, an unsatisfied Fe or O ion can attract ions to form a bond. The many uncompleted bonds at the edges of the grains create a capacity for fixing ions from aqueous solutions. In some instances, the alteration of vitreous volcanic rocks, one finds the same situation except that the cations are silicon and aluminum (Andosol materials, from Japanese *an* meaning dark and *do* soil), where the major elements of the un-crystallized silicate material form incomplete structures which fix and attract ions from solution that would be otherwise lost to the aqueous phase. In fact these poorly crystallized materials in soils are very reactive and can often determine the geochemical fate of trace elements in aqueous solution.

1.2.8 Special Elemental Groups

1.2.8.1 Rare Earth Elements (Lanthanides)

Rare earths are an interesting group of metals that have recently become quite useful in high tech, and today they are strategic materials in the world economy. Rare earth elements are most often treated as a group. They occur together either when present in concentrates or dispersed as trace elements in a silicate matrix. They have a general capacity to change oxidation state under surface conditions and can be selectively mobilized, at least in part, according to these electronic states. Most often they are associated with oxygen or oxyanions such as in zirconates or phosphates. There is some association with surface alteration minerals, probably silicates, but little is known about the crystallo-chemical attractions and incorporation. It appears that rare element transport predominantly occurs via preexisting high temperature minerals which are enriched in these elements (McLennan 1989).

1.2.8.2 Transuranic Radionuclides

Taking their name from being trans- or beyond uranium, transuranic radionuclides have atomic numbers greater than that of uranium, which is 92. All transuranic isotopes are radioactive. Transuranic radionuclides do not occur naturally in the environment. Very minor amounts can be present with some uranium ores. In Gabon, Africa, in a natural underground nuclear reactor sustained nuclear reactions estimated to have occurred about 1.9 billion years ago. Some transuranic radionuclides, including neptunium and plutonium, were produced. Like other radioactive isotopes such as those of naturally occurring uranium, radium, thorium the transuranic radionuclides undergo radioactive decay to create typically long chains of decay products.

Plutonium was dispersed worldwide from atmospheric testing of nuclear weapons conducted during the 1950s and 1960s. The fallout from these tests left very low concentrations of plutonium in soils around the world. Normally trans-uranic radionuclides are rare, but due to anthropogenic contaminations they might become important in the future in terms of surface geochemistry. The Fukushima Daiichi nuclear power plant (DNPP) accident caused massive releases of radio-activity into the environment. The released highly volatile fission products, such as ^{129m}Te, ^{131}I, ^{134}Cs, ^{136}Cs, and ^{137}Cs, were found to be widely distributed in Fukushima and its adjacent prefectures in eastern Japan. Zheng et al. (2012) reported the isotopic evidence for the release of Pu into the atmosphere and deposition on the ground in northwest and south of the Fukushima DNPP in the 20–30 km zones. These elements are typically more soluble in their reduced states, under certain conditions of redox potential, which change rapidly under surface conditions. Under other conditions they tend to form oxyhydroxide phases.

Uranium is a component of practically all rocks and therefore it is classified as a lithophilic element. Its relative abundance compares to silver, gold, and the light rare earths elements and it is more common than tin, mercury, and lead. It occurs in numerous minerals and is also found in lignite, monazite sands, phosphate rock, and phosphate fertilizers, in which the uranium concentration may reach as much as 200 mg kg^{-1}. Typical concentrations in phosphate fertilizer are 4 Bq (=0.32 mg) Uranium-238 and 1 Bq Radium-226 per g P_2O_5. Uranium is usually present in minerals either as a major or as a minor component. Sometimes they are altered to form the bright-colored secondary uranium minerals (complex oxides, silicates, phosphates, vanadates). Also specific micas contain uranium in the form of sulfates, phosphates, carbonates, and arsenates, which are products of the weathering of original uranium ores. Uranium is found at an average concentration of ~0.0003 % in the Earth's crust.

Rarely, both ions are oxygenated, having both an oxycation and an oxoanion. One of the better-known examples of this is uranyl nitrate $(UO_2)(NO_3)_2$. They form oxy-cation units in many cases (yl-ions, plutonyl for example). The uranyl ion is an oxycation of uranium in the oxidation state +6, with the chemical formula $[UO_2]^{2+}$.

1.2.9 Association of the Elements in Phases (Minerals) at the Surface

A very general classification of mineral phases can be made into a few groups, which dominate the minerals of the surface:

1. **Aluminosilicates** dominated by oxyanions with Si–O as the major component. These minerals have a large variety of cation substitutions as solid solutions in different crystal structures, with major elements, i.e., several percent of the element, or minor elements with less that 0.5 % of the element present substi-tuting one for the other.

2. **Phosphates** are moderately stable structures involving the oxyanion P_4O^{3-} with low solubility at intermediate pH values but which are strongly affected by acid conditions. Usually Ca is the dominant cation present in common surface phosphate minerals, which are for the most part of biological origin. However, there are 391 different phosphate minerals listed in Huminicki and Hawthorne (2002), which suggest very complex substitutions and variations in crystal structures and chemistry in particular in the higher temperature phases.
3. **Oxides** where oxygen is the anion associated with various cations. These phases are usually composed of moderately covalent ions, which can change oxidation state relatively easily. Metals (i.e., cations of intermediate covalent tendency such as Fe, Ni, Cr, Mn) are oxidized from the divalent state prevalent in high temperature rocks and then they form stable oxide phases which are typically very insoluble. Metal cations which change oxidation state only in extreme cases (Cu, Ni, Zn, for example) tend to be dissolved into the aqueous solution but can be fixed temporarily on mineral surfaces (see Chap. 3) and are thus mobile but often follow minerals which are stable such as oxides of iron or manganese.
4. **Carbonates** where the oxyanion is CO_3^{2-} coupled to divalent cations (Ca, Mg, Fe for the major part). These minerals are special in the sense that the source of carbon is essentially atmospheric and hence the incorporation into a carbonate is through biological concentrations, either shell producing animals or through the decay of plant material by bacteria, which releases CO_2 in soils. This CO_2 can be combined with other elements present through further biological action. The solubility product constant (measure of relative stability under near-neutral pH conditions) indicates a wide range from 10^{-8} to 10^{-14} mol/l (Lide 2000, p. 174). The relative stability for several phases is

$$\mathrm{Mg} < \mathrm{Ca} < \mathrm{Ba}, \quad \mathrm{Sr} < \mathrm{Zn} < \mathrm{Fe}, \quad \mathrm{Mn} < \mathrm{Pb} < \mathrm{Hg}$$

   Basically the most common carbonates, calcite and magnesite, are less stable than the phases with higher degrees of covalent bonding character.
5. **Sulfates** where the oxoanion SO_4^{2-} is joined by divalent cations (Ca, Mg, etc.). Sodium, potassium, and magnesium sulfates are all soluble in water, whereas calcium and barium sulfates and the heavy metal sulfates are not. Dissolved sulfate may be reduced to sulfide, volatilized to the air as hydrogen sulfide, precipitated as an insoluble salt or incorporated in living organisms. Seawater contains about 2,700 mg sulfate per liter (Hitchcock 1975). Because sulfate is highly soluble and relatively stable in water, sulfate minerals are generally formed by concentration of surface waters or seawater in evaporitic basins. Atmospheric sulfur dioxide (SO_2), formed by the combustion of fossil fuels and by the metallurgical roasting process, may also contribute to the sulfate content of surface waters. It has frequently been observed that the levels of sulfate in surface water correlate with the levels of sulfur dioxide in emissions from anthropogenic sources.
6. **Halogenides** where the anions Cl, Br, I are associated with cations such as Na, K, Mg, Sr, among others. Again these minerals are evaporitic concentrates

of elements dissolved in surface water. The stability of these phases is low in the presence of water. They occur in situations where free water is scarce such as in evaporate deposits. The most soluble elements, the anions Cl, Br, I, will be removed from a solid rock most easily from an alteration site and moved over the longest flow paths.

7. **Alkali and alkaline earths**. Cations of low covalent tendency (Na, Ca, K for example) will initially be removed differentially into solution leaving less soluble cation–oxygen associations in the solid form, as silicates, oxides, and hydroxides. These elements can be associated with silicates in loosely held associations (K, Ca, and Mg), but they are in the long run removed from the alteration zone under conditions of strong water–solid interaction.

1.2.9.1 Oxoanions

Many different oxoanions are found in the soil environment, and the chemistry of these oxoanions is quite varied. Some oxoanions such as phosphate and sulfate are essential nutrients for plant growth and are found in relatively high concentrations in soils. Other oxoanions such as borate are micronutrients. They are essential for plant growth at low concentrations but become toxic at higher concentrations.

Another group of oxoanions, such as arsenate, arsenite, selenate, selenite, and chromate are frequently studied because they have little agronomic use and are instead detrimental to human health. These elements are both metals and metalloids and include antimony, arsenic, chromium, molybdenum, selenium, tungsten and vanadium. The oxyanions are all present in low concentrations in crustal rocks and can be concentrated in certain systems as a result of weathering reactions if geochemical conditions are favourable. Because most of the elements form anionic species at neutral to alkaline pH values, their adsorption behaviour onto oxide surfaces is different from that of trace metals such as Cd, Cu, Pb, Zn, Ni, Zn.

We may compare the solubility of minerals in the same groups, i.e., carbonates, sulphates, phosphates by their solubility products (Lide 2000), which indicate the effects of different ionic substitution on mineral stability. In general sulphates are the most soluble (and hence least stable under conditions of surface alteration) having a Ksp (solubility product constant) $< 10^{-10}$ for the elements Ca, Ba, Hg, Pb, Cd, Sr, while carbonates (Ca, Ba, Fe, Mg, Mn, Hg, Zn, Pb, Cd, Sr) are less soluble with Ksp values of 10^{-14}–10^{-17} and phosphates are even less soluble with Ksp constants of $< 10^{-24}$. The stability sequence would then be, in general, sulfate < carbonate < phosphate. Using Ca compounds as an example, the sequence is sulfate (10^{-5}), carbonate (10^{-9}), and phosphate (10^{-33}).

1.2.10 Elements in Surface Phases

Geochemistry is the study of the distribution and stability of elements in the environment that we live in. Surface geochemistry is especially concerned with the distribution of elements in solids, on solids, and in aqueous phases. Solids tend to move slowly and liquids move more rapidly. However, solids can be moved in liquids, in suspension, if they have a sufficient size. Aqueous phases provide readily available elements for biological activity, for the better or the worse. The distribution of elements dissolved in aqueous solution is more or less straightforward concerning their chemical activity, which follows concentration and the chemical parameters of interaction with the water molecules that surround them (see Chap. 2). The distribution of elements in solids is more complicated.

In the formation of phases, there is the necessity of having sufficient amounts of the proper components present to provoke the formation of a phase: one reaches the saturation limit of its presence in the medium of transport. However, in each phase, there is the possibility of substituting one element (atom by atom) by another within the structure. Also there is the possibility of attracting elements from solution, as hydrated ions, to the surface of a solid. Thus, several possibilities are possible for different elements to be present within or on a mineral phase.

Generally speaking the elements of major abundance form phases that correspond to their relative abundance in the chemical–geological system in which they form (not necessarily corresponding to their solubility product). This results in a limited number of species for a given chemical environment, usually three or four. Minor elements have lower elemental abundance and are not sufficiently chemically active to form a specific phase. Hence, they either find a space in a major element phase, in substitution at low concentration levels, or they can be attracted to mineral surfaces. If none of this is possible, the element remains in aqueous solution and is transported out of the system. Of course this situation pertains for some major elements, which are moved out of the system of solid phases in solutions. Sodium is a classical example.

1.2.10.1 Atomic Substitutions within Crystals

Solid phases are for the most part composed of what can be called major elements, major in the sense that their concentrations are sufficiently large to form a specific mineral phase. Mono-elemental phases are extremely rare in nature, gold being one. Consequently, the distribution of major elements is interdependent upon the presence of other elements. The most abundant element in surface minerals is oxygen, making up around half of the atoms present in the phases. There is no restriction to its presence, since it is available in the atmosphere and it is dissolved in surface waters. Thus surface minerals are essentially oxygen-rich. Other elements are combined with oxygen forming minerals depending upon the type of ions that are involved with the oxygen and its relative ionic concentration.

1.2.11 Silicates

The most common minerals at the surface are called silicates, where the silicon–oxygen covalent structure dominates the chemistry and structure of the minerals. Silicates can be defined as compounds containing $[SiO_4]^{4-}$ anions. The main structural unit of silicates is a tetrahedral cluster containing one silicon atom and four oxygen atoms. This means that silicon is a dominant cation numerically and that the highly covalent Si–O bond determines the mineral chemistry. Frequently aluminum is associated with the silicon–oxygen bonding structures, as a substituting ion. These covalent ion structures need more cationic compensation to form a complete structure, except in the case of the pure mineral SiO_2 quartz. This mineral is special in that there is no substitution of other cations in the structure. It is a pure phase, a rare occurrence in surface mineralogy. Curiously it has a very great stability under surface conditions, often persisting beyond its thermodynamic limits to remain present when other minerals have been dissolved or transformed. This is why white beaches over the globe are most often formed of quartz grains, which are metastably present but very evident.

The tetrahedral clusters can be polymerized, i.e., linked to each other through the bridging oxygen atoms. They are able to form polymers by means of linkage with one, two, three, or four neighboring tetrahedra, forming siloxane Si–O–Si bonds. Other ions can be located in the silicate lattices. Usually these ions are "major elements" Na, Ca, K, Fe, Mg. Hence most silicates are composed of several of the ten major elements with minor proportions of other elements in the minerals.

One feature of great importance in silicate minerals is the sharing of electrons by oxygens with more than one cation. In fact a strongly attached network, in two or three dimensions, is formed through this inter-linkage and electron sharing. The dominant feature of silicates is the arrangement of oxygen ions around silicon cations; for the most part four oxygen ions are linked to a silicon ion forming a tetrahedral structure. These same silicon linked oxygen ions can be linked with other cations forming yet other geometric structures around the other cations, such as an octahedron when six oxygen ions are linked to the cation. Ionic substitutions are regulated by charge and more importantly relative ionic diameter since the oxygen polyhedra have a given space within the polyhedral structure that they form. Substitutions of one cation for another in silicates are to a large extent a function of charge and ionic size.

Elements of minor abundance substitute for major elements in crystal structures following the same chemical criteria. In general alkali elements substitute for other alkalies, transition metals for transition metals, and so forth. However there are some surprises, which will be discussed in further chapters. One is lead and another is potassium, lead being a large and heavy divalent ion, while potassium is smaller and monovalent. Nothing could be more dissimilar at first glance than this ion pair. Yet lead is often found to substitute in high temperature silicate mineral structures for potassium (Taylor and Eggleton 2001, p. 142). It is probable that this relationship continues at the surface, if one considers the data for elemental distribution in materials where mineral structures are such that lead could substitute for potassium (micas and smectites). Minor element substitutions do not follow the rules in all cases.

1.2.12 Oxides and Hydoxides

A second possibility is the formation of what are called oxides and hydroxides, i.e., minerals with oxygen but not containing silicon. The oxides and hydroxides present in soils reflect the environmental conditions of soil formation. They are variable in structure, composition, and degree of crystallinity. The formation of different types of oxides and hydroxides is controlled by temperature, moisture, organic material, pH, and Eh control. The oxides are most commonly dominated by Fe or Mn cations. The oxides may also occur in a more complex manner as oxyhydroxides. Oxyhydroxides are chemical compounds that commonly form in aqueous environments with different content in cations (e.g., Fe^{2+} and Fe^{3+}), oxygen, hydroxyl, water, and some amounts of SO_4^{2-}, CO_3^{2-}, and Cl^-.

1.2.13 Carbonates

Other surface minerals, frequently encountered, are carbonates. Carbonate minerals are the most stable minerals of carbonic acid. Most of them are simple salts (e.g., Calcite $CaCO_3$), others contain additional anions (e.g., malachite $Cu_2CO_3(OH)_2$). Carbonic acid is relatively weak and prefers to bond with elements of low ionic potentials (sodium, potassium, strontium, calcium). Cations with high ionic potentials such as Cu^{2+} or rare earth elements form only if F^-, OH^- groups or O^{2-} are present to weaken the CO_3^{2-} complex. An example is azurite $Cu_3(CO_3)_2(OH)_2$. About 170 different carbonate minerals are known. In nature carbonates are mainly composed of calcium carbonate as calcite and rarely as aragonite, of magnesian calcite, of dolomite, and occasionally of iron-rich carbonate such as ferroan calcite, siderite and ankerite.

1.2.14 Phosphates

Phosphate is a salt of phosphoric acid. The fundamental building block is the PO_4^{3-} tetrahedron. Soils generally contain 500–1,000 parts per million (ppm) of total phosphorus (inorganic and organic), but most of this is in a "fixed" form that is unavailable for plant use.

The two main categories of phosphorus (P) in soils are organic and inorganic. Organic forms of P are found in humus and other organic material. A mineralization process involving soil organisms releases phosphorus in organic materials. Inorganic phosphorus is negatively charged in most soils. Because of its particular chemistry, phosphorus reacts readily with positively charged iron (Fe), aluminum (Al), and calcium (Ca) ions to form relatively insoluble substances. The solubility of the various inorganic phosphorus compounds directly affects the availability of phosphorus for plant growth. The solubility is influenced by the soil pH. Soil phosphorus is most available for plant use at pH values of 6–7. When pH is less than 6, plant available phosphorus becomes increasingly tied up in aluminum phosphates. As soils become more acidic (pH below 5), phosphorus is fixed in iron phosphates.

Apatite is the most common phosphate mineral. The bones and teeth of most animals, including humans, are composed of calcium phosphate, which is the same material as Apatite. Apatite is named from the Greek word *apate* (in Greek mythology, Apate was the personification of deceit and was one of the evil spirits released from Pandora's box) since Apatite has a similar appearance to so many minerals.

For the most part the formation of phosphate sedimentary material is due to precipitation from aqueous solution. Many freshwater deposits of fish remains become phosphate repositories.

1.2.15 Sulfates

Sulfates are salts of sulfuric acid, containing the group SO_4. Sulfates occur naturally in numerous minerals, including barite ($BaSO_4$), epsomite ($MgSO_4{\cdot}7H_2O$), and gypsum ($CaSO_4{\cdot}2H_2O$). Sodium, potassium, and magnesium sulfates are all highly soluble in water, whereas calcium and barium sulfates and many heavy metal sulfates are less soluble. Atmospheric sulfur dioxide, formed by the combustion of fossil fuels and in metallurgical roasting processes, may contribute to the sulfate content of surface waters.

Sulfate occurs naturally in the aquatic environment or it can have an anthropogenic origin. When sulfate naturally occurs in aquatic environments, it can be the result of the decomposition of leaves, atmospheric deposition, or the weathering of certain geologic formations including pyrite (iron disulfide) and gypsum (calcium sulfate).

1.2.16 Substitutions of Ions in Mineral Structures

The major phases, composed of major elements, are rarely pure in the sense that only one or two cations are present. However in some minerals the cations can be of several types. The mixture of major elements in the same mineral in varying proportions is called solid solution (the minerals being in the solid state as opposed to liquids). Depending upon the mineral and the temperature at which it has formed,

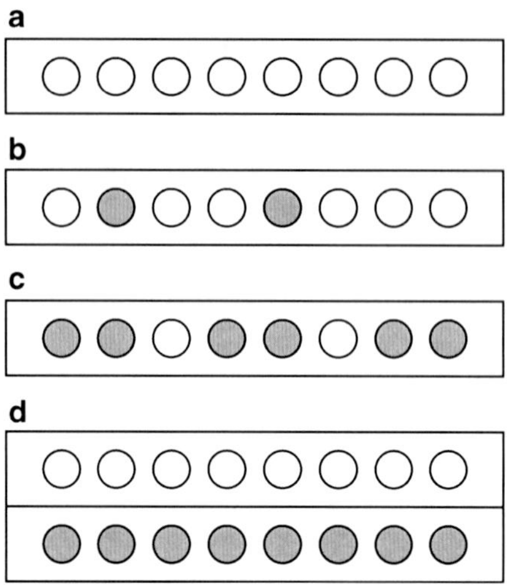

Types of chemical substitutions in
a mineral structure

Fig. 1.20 Substitutions of ions within a fixed crystal structure. This can occur in a random manner, or in a more ordered and regular manner in specific crystallographic sites. The phyllosilicate structure is based on a network of polymerised SiO_4^- ions extending essentially in a two-dimensional structure, represented by rectangular forms in the diagram. The substitutions are in identical crystallographic sites where elements of the same ionic charge are distributed in different manners. (**a**) No substitution in a given crystallographic site. (**b**) Random substitution in a given crystallographic site. (**c**) Regular substitutions in a given site. (**d**) Segregation of ion types layer by layer in a given crystallographic site

minerals can accommodate more or less substitution of one element for another. Surface minerals tend to show a range of substitution for some minerals and no substitution for others. Figure 1.20 shows several types of substitutions found in surface minerals.

One may have a pure phase (case A), a substitution in a random manner, i.e., no regular succession of elements in substitution sites (case B), regular substitution (case C), or segregated substitutions (case D). If the segregated substitutions form large segments of the crystal structure, one will distinguish two phases by X-ray diffraction identification methods and even though they form one contiguous crystal, one may consider them as two phase systems. The segregated substitutions can be in a random order within the crystal, or with a regular repetition case as shown in example D in Fig. 1.20.

Elements of minor abundance (<0.1 wt%) can be found in various types of substitutions, but are rarely identified as such. Usually one assumes random substitution in a crystal for elements of minor abundance. These minor element substitutions within the crystal are stable, i.e., the elements will not be released unless the crystal becomes unstable and is transformed into another or dissolved. The amount of substitution of different ions in a crystal is to a certain extent controlled by temperature. It is well known that at high temperature (approaching the temperature of melting), more elements can be substituted in a given structure for a silicate mineral. However one finds also that substitutions are found in phases formed at very low temperatures. Surface minerals often contain significant amounts of various elements in substitution in mineral structures when formed under surface environments. This phenomenon can be explained as follows: as temperature is high, thermal agitation loosens the mineral structure allowing different elements to enter a structure even though they have slightly differing ionic sizes. At lower temperatures these elements diffuse out of the structure to form a specific phase of their own. At very low temperatures the thermal energy is very low. Here there is a disorder in elemental occupancy of crystallographic sites as minerals are formed because there is not enough energy to allow misplaced ions to diffuse out of the structure to form another specific phase. Disorder occurs for lack of thermal agitation energy at low temperature whereas disorder is created at high temperatures because there is too much energy present. This is illustrated in Fig. 1.21 where element Y is substituted in phase A in varying amounts according to temperature conditions.

Ionic substitutions follow, generally, a pattern of constraints based upon ionic charge and ionic size. In Fig. 1.22 several elements are shown as a function of ionic charge and ionic diameter.

The elements grouped in this space can often be found in different proportions in the same mineral type. Those found at some distance tend not to be associated. This is especially true for ions of high charge. However in more complex situations one may find multiple substitutions where an ion of higher charge substitutes for an ion via a coupled substitution of an element of lower charge substituting in another site. One example is

$$Al^{3+} = Si^{4+} \text{coupled with } Al^{3+} = Mg^{2+}$$

This occurs in the same crystal but different crystallographic sites. In one site the ionic charge is increased and in the other it is reduced by the substitution keeping overall electronic balance, which is necessary for mineral stability. Multiple site substitutions are especially common in silicate minerals, especially those of the clay mineral group.

Substitutions of elements, one for another in a mineral structure, can then be rather complex. The simple case is exchange of ions of the same valence, for example, Na for K. Here the limiting factor is the size of the ion within the mineral structure. The greater the difference in size, the less substitution will occur in the same mineral. However, in a crystal which has a large number of oxygen ions present, the positions of cations is largely determined by the arrangement of oxygen ions. If there are four contiguous with a cation the space is smaller than if there are six coordinated to the ion. Data in Steinberg et al. (1979, p. 316) indicate that the space left for an atom in fourfold coordination is roughly 20–30 % less than when in sixfold coordination. Oxidation state of an ion changes its volume and hence

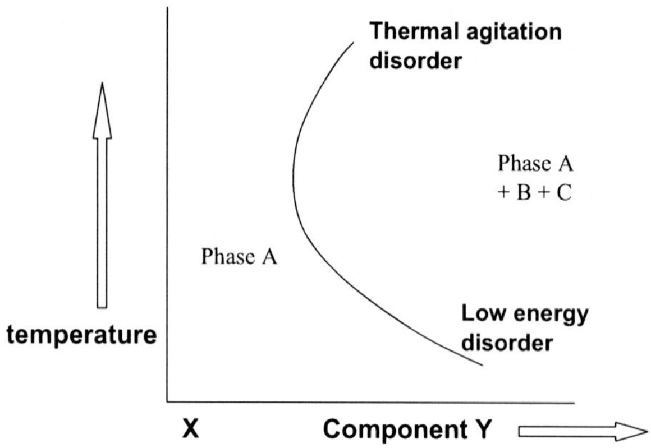

Fig. 1.21 Illustration of the effect of temperature on the substitution of element Y in phase A. More Y is substituted at low and high temperatures

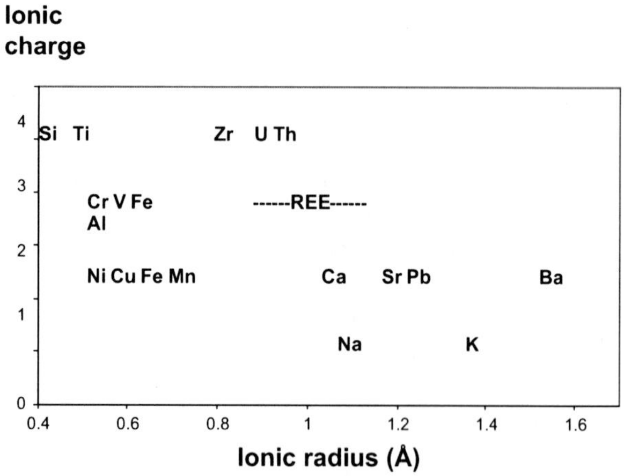

Fig. 1.22 Charge–ionic size relations after Brookins (1989)

influences its acceptation in a crystal structure (see Fig. 1.18). Hence the oxidation state of an ion plays a role in substitutions where it must follow charge balance criteria as well as ionic dimension characteristics which can determine its presence in a given mineral structure.

If we consider the alkali feldspars, for example, there is a high degree of substitution in the structure at high, magmatic temperatures, which reduces as the ambient temperature of the rock or crystallization of new minerals occurs. From an almost complete solid solution in high temperature magmas the limit decreases with temperature until at near surface conditions one finds only almost pure Na or K minerals forming low albite and adularia. One would expect then almost pure mineral compositions in surface minerals, however this is far from being the case. In fact it appears that those minerals formed at surface conditions have a greater capacity for solid solution (elemental substitution) than those minerals formed at slightly higher temperature, 120 °C, or so under diagenetic mineral conditions of formation.

In Fig. 1.21 the substitution of element X for element Y is seen to occur at high temperature due to thermal agitation as well as at low temperatures where there is not enough energy to form a well-ordered mineral with the constituent elements in chemical equilibrium. An example of this sort of initial wide range of substitution is given by Velde et al. (1991) where diagenetic chlorites, formed in sedimentary rocks at temperatures between 50 and 120 °C, have a relatively wide range of composition from one grain to another in the same sample at lower temperatures which decreases to a small range of compositional variation in rocks formed at higher temperatures.

The surface minerals with the highest degree of solid solution or elemental substitution are the clay minerals. These minerals have high degrees of substitutions in certain parts of their structure while in others there is relatively little substitution. This particularity has led to a great amount of mineral nomenclature and perhaps confusion as to what the true functions and range of compositions are in these minerals.

1.2.17 Mineral Surface Reactions

Some elements can be attracted to mineral surfaces due to charge imbalances on crystal faces in particular edge faces where the crystal ionic linkages are not compensated, such as the oxygen–cation compensations, and either cations or anions are present without a compensating ion to equilibrate electronic charge on the ion. To these locations charged elements in solution can be attracted by the electrostatic imbalance and fixed, temporarily on the crystal surface. The ions in solution are always hydrated and have some water molecules associated with them.

Since they are not integrated into the overall structural electronic compensation structure, their presence is ephemeral, and another ion can replace the surface ions due to a higher activity in solution and/or higher activity for the surface. This phenomenon, cation exchange, is extremely important for the movement of minor elements at the surface of the earth. Since they are not present in high enough concentration to form a phase or mineral corresponding to their chemistry, they are attached in very small quantities to sites on other crystals. Surface mineral phase can carry these minor elements, but their presence is determined by the chemical activity or their concentration in the aqueous solution of their environment. This action is quite important when it comes to understanding the pathways of minor elements at thc surface of the earth. The phenomenon of surface attachment is treated in detail in Chap. 2.

A second site of temporary ion retention is within the clay structure when a charge imbalance due to ionic substitution occurs in the silicate structure. Temporary ionic substitution occurs between the charged clay layers. These are called "interlayer" substitutions, where the presence of an ion is due to its relative concentration in solution and the affinity the mineral has for it as a function of its charge and ionic dimensions. Again this phenomenon, extremely important for surface geochemistry is treated in detail in Chap. 2.

The surface attraction phenomenon is driven by the surface availability, or the amount of surface compared to the volume of the minerals. The smaller the crystal, the greater the surface area and hence the greater will be the possibility of attracting elements from the aqueous solutions onto the mineral surfaces for a given mass of solid. Since the underlying principle of surface phase transformation of rocks into alterite and eventually soil materials is the formation of small crystals [see Chap. 4 in Velde and Meunier (2008)] the materials formed at the surface become very "surface reactive" with substantial crystal surface area being present. This property is very important when one wishes to follow the fate of elements at the surface of the earth. An enormous amount of transportation and displacement of minor elements, i.e., those not involved in the formation of mineral phases, is accomplished on mineral grain surfaces.

1.2.18 Summary

The major chemical influences on surface chemistry are water and oxidation. In aqueous solutions pH is a major factor for dissolution of minerals. The stability of a mineral can be measured relatively in terms of its solubility as a function of pH. Low pH increases the dissolution of carbonate and silicate minerals. Intermediate pH values, those of soils in the range pH 4 to pH 8, tend to favor mineral stability. Carbonates are the least stable (most soluble) followed by minerals from basic magmatic rocks (olivine, pyroxene) and then by minerals from acid eruptive rocks (feldspars, muscovite). Minerals with the lowest solubility are those formed at the surface of the earth such as iron oxides, clays, and eventually the least soluble, quartz.

The effect of pH can be correlated with rainfall, the more rain the lower the pH. This control is largely one of biome types, which follow rainfall abundance. In fact evergreen forests, either in the tropics or at altitude in mountain areas (conifer forests), produce acidic soils whereas deciduous forest and prairies produce soils of higher pH. The biome effect can change pH by three units in well-drained soils. The biological factor for pH control is very important to the stability of surface minerals, especially those with cations susceptible to changes in oxidation state. Not only does biology control pH it also controls Eh conditions. Thus iron minerals and manganese minerals will be susceptible to the biocontrols of Eh and their dissolution or precipitation can be induced by biologic action.

The same occurs exists for sulfur-bearing minerals. The more reduced form of sulfur (S^{2-}) where sulfur is an anion as in pyrite, FeS, can be changed into an oxyanion upon oxidation. The reverse can be true, when bacterial action reduces the sulfur oxyanion to form the sulfur anion. This example would occur in a reversible reaction of weathering, oxidation, and dissolution of a sulfur-bearing mineral which is re-generated due to biological action. Sulfur is an element of minor abundance in surface environments, but it is extremely important. In forming an iron sulfide, for example, it will incorporate a large range of trace elements in the structure in stable form as long as the redox conditions persist. Thus the immobilization or release of a range of trace elements depends upon the action of bacteria, which use the change in oxidation state of sulfur as a source of energy. In doing so they can change the balance of bioavailable elements in a surface environment.

In the literature and our text, one speaks of soluble elements and insoluble elements. This of course indicates the propensity of an element to be found in ionic form in aqueous solution. Cations (such as Na, K, Ca, and Mg) are easily dissolved as are anions (such as Cl, F, Br). The elements that are bound to oxygen, such as Si and Al, are less likely to be found in solution and are called insoluble elements. This means in fact that the solubility of the oxides is low compared to that of an ionic species. The type of chemical bonding of an element relative to oxygen is primordial to its behavior at the surface of the earth.

1.3 Useful Source Books

Gill R (1989) Chemical fundamentals of geology. Chapman & Hall, London, p. 290

Goldschmidt V (1954) Geochemistry. Clarendon Press, Oxford, p. 730

Holland H, Turkian K (eds) (2004) Treatise on geochemistry, vol. 5. Elsevier, Amsterdam

Krauskopf K (1967) Introduction to geochemistry. McGraw Hill, New York, p. 721

Mason B (1958) Principles of geochemistry. Wiley, New York, 329 pp

McQueen K (2008) Regolith geochemistry, Ch 5. In: Scott K, Pain C (eds) Regolith Science. Springer, Heidelberg, p. 461

Wedephol H (1969) Handbook of geochemistry, vol. I. Springer, Heidelberg

Chapter 2
Elements in Solution

2.1 Ions and Water

Growth and dissolution of minerals in contact with fluids in the crust or at the surface determines the elements and their concentrations in surface waters. Also anthropogenic activities can result in locally high concentrations of trace elements. In polluted areas near waste disposals, chemical factories, smelters, and mining tailings for example. In viticulture, grapevines are plagued yearly by "Downy Mildew," Plasmopara viticola, triggering the use of copper fungicide by farmers. Over time copper has accumulated in the topsoil, reaching toxic levels, causing plant stress and reducing soil fertility (McBride 1981). These (bio)geochemical enrichments generally persist as a result of element immobilization by inorganic solids. But how are these elements immobilized? Are they fixed by sorption or do they become oversaturated and precipitate as a phase of their own?

There are many factors that govern the rate of growth and dissolution of minerals in contact with fluids in the crust or at the surface. The alteration of minerals on the surface of the earth, especially the dissolution and precipitation processes, is under the influence of chemical, physical, and biological factors. Sorption processes often control the speciation of metal ions in low temperature geochemical environments and thus influence mobility, toxicity, and bioavailability (Hayes and Traina 1998).

Elements in aqueous solution are the basis of surface geochemistry. Water is the agent of change concerning the materials brought to the surface by geological actions. Rocks containing minerals of high temperature origin interact with water because the constituent rock minerals are unstable under conditions of aqueous abundance where the water contains no more than minor amounts of solutes and is in equilibrium with atmospheric oxygen and CO_2. The dissolution of an element into aqueous solutions is the major result of the interaction of surface water and the geologic materials found at the surface (rocks), which are for the most part composed of silicate minerals that have formed at significantly higher temperatures and these minerals are no longer stable under the current surface conditions. The chemical weathering of rocks results from the differences in thermodynamic

A. Bauer and B.D. Velde, *Geochemistry at the Earth's Surface*,
DOI 10.1007/978-3-642-31359-2_2, © Springer-Verlag Berlin Heidelberg 2014

conditions that existed at the time of mineral formation and that of current ambient conditions at the earth surface. The surface environment is one of high water content, and hence a tendency to dissolve minerals, to form hydrous minerals of low thermal energy, and hence a tendency to form minerals of low atomic ordering.

In the last years major advances in understanding the fundamental chemical processes which control mineral weathering and in the efforts to quantify weathering rates on both laboratory and field scales have been made. A very good review can be found in White and Brantley (1995).

2.1.1 Ions

The key to chemical interaction in such states is the intensity of the chemical forces between the different major elements (cations) and oxygen atoms. If O-cation bonds are weak, the element will be present in the solution. If O-cation bonds are strong, the element will tend to form a solid phase and remain out of the aqueous solution. Typical strong bonds occur between Si–O, Al–O, and Fe–O. These bonds tend to be strongly covalent, whereas the less strongly bonded elements have more ionic bonding tendencies in solids. The more chemical reactions occur at the surface, the greater is the tendency to retain materials with covalent bonds which is important to the geochemistry of the surface. Initially the covalent Si and Al bonded ions form silicate mineral phases at the surface, usually called clay minerals. As weathering intensity increases, and solutions contain less and less dissolved materials due to abundant and frequent rainfall, eventually the Si ions go into solution with Al and Fe remaining as solids in the oxide—oxyhydroxide state. Thus the tendency of ions to form covalent bonds with oxygen as silicates and oxides determines their initial presence as solids in alterite materials. Eventually only oxides and oxyhydroxides remain (Fe and Al).

Bond strength between different elements and oxygen atoms (Lide 2000) could be used as a guide for estimating the stability of different materials. In such tables we can see general trends, where alkali and alkaline earth ions tend to be less strongly bonded to oxygen, and hence one can assume them to be more likely to enter into aqueous solution from a complex mineral phase. The elements Si, Ti, C, Zr, C, and Al have strong element—oxygen bonds and are most often found to be difficult to dissolve under natural conditions remaining as solids in alterite materials. However the elements B, P, N, and S for example even though of high bonding energy especially for oxygen tend to be dissolved, albeit as oxyanions, under conditions of surface alteration.

An interesting sequence of element—oxygen bond energies is seen in the transition metal sequence (Fig. 2.1) where significant differences are seen for elements lighter and heavier than iron, but values for Mn, Fe, Co, and Ni, elements similar to iron, are quite similar in bond energies. As it turns out, as will be seen in further chapters, Co and Ni are usually very strongly related to Fe–Mn abundance in alterite materials whereas Zn, Cu, and Cr are often but not as closely as Ni and

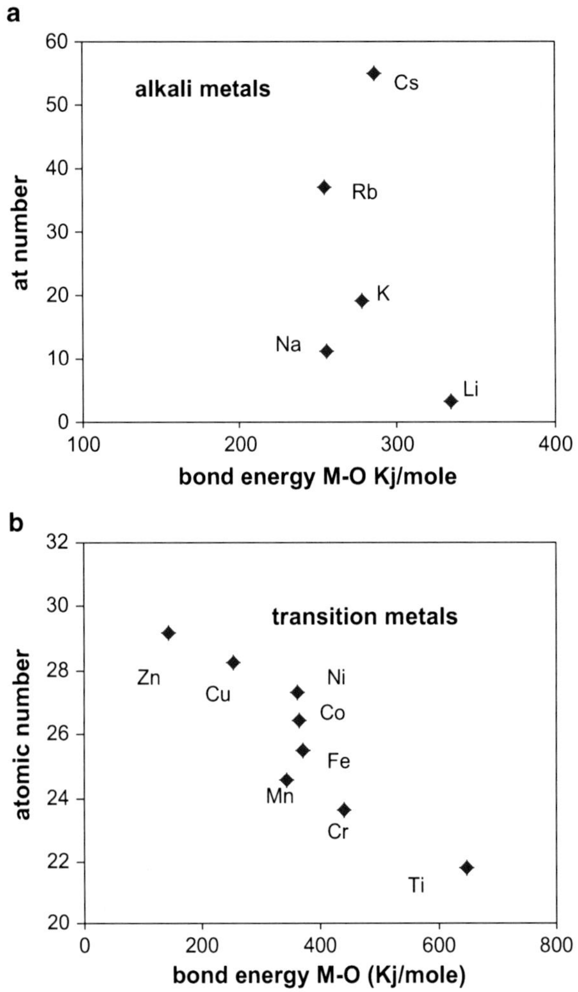

Fig. 2.1 Plot of data from Lide (2000) for bond energies between various alkali metal elements and oxygen atoms (**a**) and transition metal elements (**b**)

Co. However Cr can be strongly bonded to Fe in laboratory experiments. Bond energies of alkali metal ions with oxygen are similar, but Li shows a stronger affinity for oxygen. This is reflected in the tendency of Li to be found in solid alterite materials (clays) whereas the other alkali metals are more often in solution and remain in solution until the last stages of evaporation.

The energy released in the process of hydration is known as hydration energy. The hydration energy can be plotted as a function of ionic radius for alkali elements with a nice relationship due to the charge density, where the highest hydration energy occurs when charge density is highest (Li compared to Cs). Remember that the atoms get bigger as you go down groups where atoms are present with more electrons. These elements are more active in water solutions as cations than are

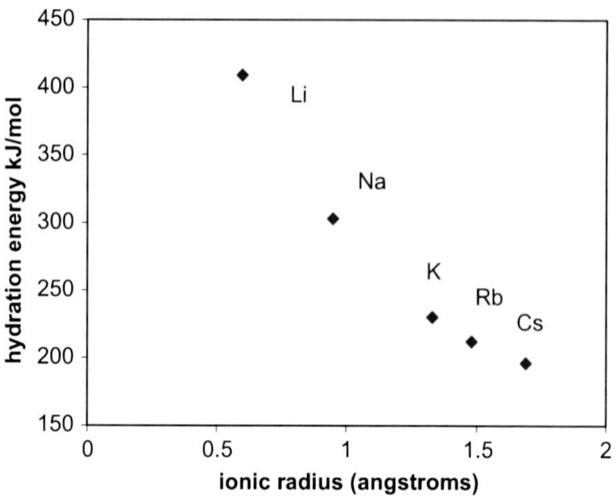

Fig. 2.2 Hydration energy (kJ/mol) and ionic radius, angströms (data from Lide 2000) for alkali metal ions

transition metals. Such chemical tables are frequently used as the basis for modeling the behavior of elements in surface environments (Fig. 2.2).

However using the measure of covalency tendencies of elements alone is not sufficient. For example using tables of electronegativity (Lide 2000) does not always give relations consistent with observations in natural geochemical systems concerning relative attraction of elements in the ionic form to charged surfaces or to other ions.

2.1.2 *Ions in Water*

The dominant feature of aqueous chemistry is of course the chemistry of the water molecule. The structure of the water molecule is simple. It is composed of one oxygen atom and two hydrogen atoms. Each hydrogen atom is covalently bonded to the oxygen via a shared pair of electrons. Oxygen has six electrons in its outermost orbit. Out of these six electrons, one electron each is shared with two hydrogen atoms, which have one electron in the outermost shell. Through covalent bonding, oxygen shares two pairs of electrons with hydrogen atoms to form the water molecule. Besides these two shared pairs of electrons, water has two pairs of unshared electrons that repulse each other, leading to distortion in its shape. The water molecules themselves can decompose into two other types of atomic arrangements: The anionic hydroxide ion, OH^-, form and the cationic form, H^+ or H_3O^+, both of which have a charge of one, negative for the hydroxide and positive for the proton or hydronium ion. For different chemical reasons, one can at times find that there are more H^+ than OH^- ions in a given solution at the surface of the earth, and hence the solution is called acidic. If more OH^- units are present than H^+ ions the solution is basic. The designation for the different proportions of H^+ and OH^- ions present is expressed on a logarithmic scale, and called pH.

2.1.3 Inner Sphere: Outer-Sphere Attractions

The polarity of the water molecule arises from its greater electronegativity, or electron loving nature, than hydrogen. Even though the pairs of electrons are shared with the hydrogen atoms they are not shared equally as oxygen pulls the electrons more toward itself. This gives rise to a greater negative charge on one side of the oxygen atom and a positive charge on the other with hydrogen atoms. That means, water molecules become polar because of the greater ability of oxygen to attract electrons toward itself. This polar nature of water molecules gives rise to an electric dipole moment and the ability of ions and other molecules to dissolve in water. These attractions in the presence of water molecules create what is called the inner sphere of hydration on cations in solution. A sort of coating of water molecules occurs around the ions in solution, especially the cations. The water molecules are relatively strongly attached to the ions. More diffuse charge compensations beyond this inner layer of water ions, which are in direct contact with the cation, create a second, outer sphere of water ions attracted to the cation. Beyond this second layer water molecules can be attracted to a much lesser extent and in a disordered manner [see Tournassat et al. (2009) for example. Hence ions in water are intimately associated with water molecules themselves].

2.1.4 Attraction of Ions to Solids: Absorption–Adsorption

Sorption is arguably the most important chemical process governing the retention of nutrients, pollutants, and other chemicals in soils. The term sorption (Lat. sorbere, sorptio = swallowing, absorption) includes a variety of possible processes and does not make any distinction with regard to the underlying mechanism. In this section, we deal with metal ion reactions at mineral surfaces involving various surface phenomena such as outer-sphere attachment and inner-sphere surface complexation of ionic species.

We will use the case of a 2:1 clay mineral to demonstrate the relations between molecules in water and surface charge sites on sorbers. The same scheme can be used for oxides. The situation is more complex for organic material in that the functional groups on such surfaces are quite different and vary depending upon the initial material and its state of degradation by bacteria in surface environments.

Two major types of chemical interaction are found between ions in solution and solids at their surface. They can be described as absorption, i.e., introduction of an ion within the mineral structure (usually a partially hydrated cation), or adsorption, i.e., interaction at the surface of a mineral, within the aqueous medium, of a cation or an anion. The bonding type is due to very different causes of residual electro-negativity on the solid phase. Absorption is due to internal electronic disequilibrium due to substitution of ions of different charge within the mineral and adsorption is due to the existence of incomplete bonding of atoms on the edges of the structure where local charge imbalance is present.

Absorption, introduction of a hydrated ion in a mineral structure with electrostatic attractions of one sort or another, is called outer-sphere attraction while adsorption refers to a more strong chemical bond called inner-sphere adsorption. In the case of outer sphere surface complex of water molecules the cation remains bonded to the hydration shell; thus there is no direct chemical bond to the surface of the solid particle. The attraction is purely electrostatic and the complexation is a rapid process that is reversible, and adsorption occurs only on surfaces that are of opposite charge to the adsorbate.

Inner-sphere surface complexation involves chemical bonding of aqueous ionic species in direct contact with mineral surface functional groups. However, depending on geochemical conditions in the aqueous solution and the type of mineral concerned, reactions other than mere sorption, such as surface-induced redox reactions and cation incorporation into the solid matrix, are possible. The formation of strong cation metal–surface oxygen atom bonds with significant ionic or covalent contribution results in inner-sphere surface complexes where metals bind directly to the surface with no intervening water molecules. Water molecules are present on the non-bonded portions of the ion in question. This strong binding to the mineral surface has a more pronounced effect on reversibility and desorption kinetics compared to outer-sphere sorption (Maher et al. 2012; Tinnacher et al. 2011). Inner-sphere and outer-sphere complex formation can occur simultaneously.

2.2 Absorption (Outer-Sphere Attraction and Incorporation Within the Mineral Structures)

A special case of chemical interaction between solids and ions in solution is to a large extent particular to silicate clay minerals, where an interlayer site of surface charge attraction can occur in the interior of the clay mineral crystal. In clays, where the structure is lamellar in type, the simple ideal formulation of the completed structure is electronically neutral at the surface where oxygen ions are present. However if ionic substitution within the structure occurs, where ions of different charge are substituted, there will be a permanent, pH independent residual charge on the surface of the clay layer. In clays, and some other silicates, ionic substitutions of different types occur within the structure where one ion is substituted for another but does not have the same electronic charge. For example Al in AlO_6 octahedral sheets is substituted by divalent cations (Mg/Fe(II)) and/or Si in tetrahedral SiO_4 sheets by trivalent cations (Al/Fe(III)). There are many variations on this theme. In these instances the overall structure lacks an electronic charge which is expressed as an attraction on the surface of the crystal.

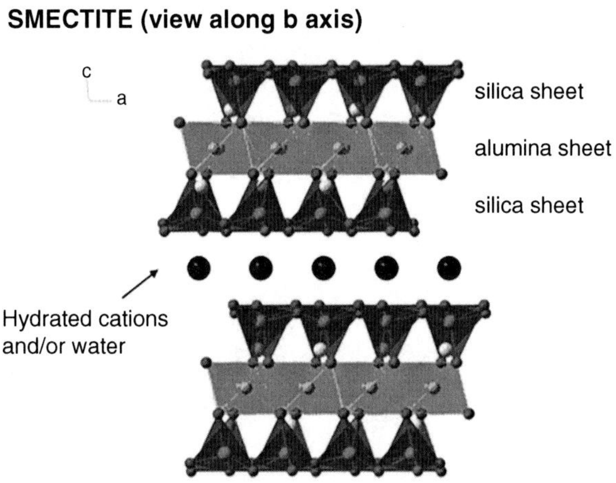

Fig. 2.3 Schematic representation of the smectite structure where silicon and aluminum cations are coordinated to oxygen anions. Elemental substitutions in either of the cation sites can lead to an overall charge deficiency on the surface of the crystal, which in turn attracts hydrated cations as absorbed ions in the structure

The maximum layer charge for the 2:1 layer silicates is −1 based on M_4O_{10}. This is half of the unit cell of a layer silicate. The layer charge is often used as a criterion for the classification of the 2:1 clay minerals.

The substitution results in a charge imbalance on the overall structure in that the charge is not compensated on the substitution site itself but diffused over the oxygen surface of the clay layer. This charge imbalance is independent of pH. Hydrated cations in solution are attracted to these diffuse interlayer charged sites by purely electrostatic attraction. When the absorption of an ion is dominated by electrostatic forces, the resulting attachment is weak electrostatic interaction. Neutrality is restored by having hydrated cations between the layers, called the interlayer space. The end result is a permanent charge diffused on the surface of the clay layer structure over a number of oxygen atoms (Fig. 2.3).

The weak electrostatic nature of cation attachment to permanently charged surfaces makes this interaction reversible and strong competition with other cations in the surrounding aqueous solution takes place. For example in order to become available to a plant, a cation adsorbed on a clay particle must be replaced by a cation present in the soil solution. Plant roots facilitate this process by excreting a hydrogen ion (H^+) into the soil solution in order to exchange this for a cation (e.g., K^+).

The most common exchangeable soil cations in the interlayer charged sites are: calcium (Ca^{2+}), magnesium (Mg^{2+}), potassium (K^+), ammonium (NH_4^+), hydrogen (H^+), and sodium (Na^+). They form an outer-sphere attachment with the charged surfaces in which waters of hydration exist between the charged ion and the oppositely charged mineral surface.

The composition of the exchangeable ions in the interlayer is a function of their concentration in the soil solution and the affinity of an ion for the exchange site.

The ion affinity is a function of the charge and hydrated radius of the ion with small, highly charged ions being preferred over large ions with low charge. Exchangeable ions are easily displaced into the soil solution making these ions

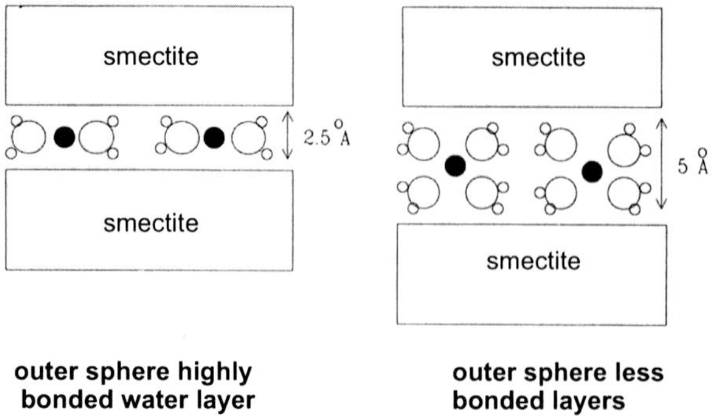

Fig. 2.4 Representation of water–cation (*shaded circle*) interlayer structure in a clay mineral for strongly attracted water molecules with a spacing of 2.5 Å and less strong attractions with a spacing of 5 Å

readily available for plant and microorganism utilization. Summarising important characteristics of the outer-sphere attachment:

1. Non-localized electrostatic charge neutralization (surface charge).
2. It is developed by isomorphic substitution within the crystal.
3. It is pH independent.
4. It is weak electrostatic interaction.
5. The exchangeable ions are fully solvated and are readily available for plant and microorganism utilization (Fig. 2.4).

2.2.1 Dynamics of Interlayer Absorption of Hydrated Cations in Clay Minerals

The source of charge on a clay layer through isomorphous substitution results in a permanent charge on the surface of most layer silicates. The expression of this charge on the clay surface is to a large extent independent of pH and responsible for the absorption of hydrated cations in the interlayer or the adsorption on the surface of clay minerals. The case of hydrated cation absorption within clay minerals is perhaps the most studied cation retention on solids in an aqueous environment. The reason is that cations fixed on clays are a basic source of mineral element nutrition for plants and hence agricultural scientists have investigated this aspect of soil science for quite some time. Numerous texts have been written on the subject. Several aspects of cation retention (base saturation since most of the elements considered to be associated with clay minerals of agricultural interest are alkali or alkaline earth elements) are important to note. A key point is the selectivity of one element over another in the removal from solution. Given equal amounts of two dissolved elements in solution, clay minerals usually are found to attract more of one element than the other (Fig. 2.5a). The chemical affinities of different ions in solution have been studied in order to determine which ion is favored to be included into the clay structure by cation exchange. The relations are not simple since charge

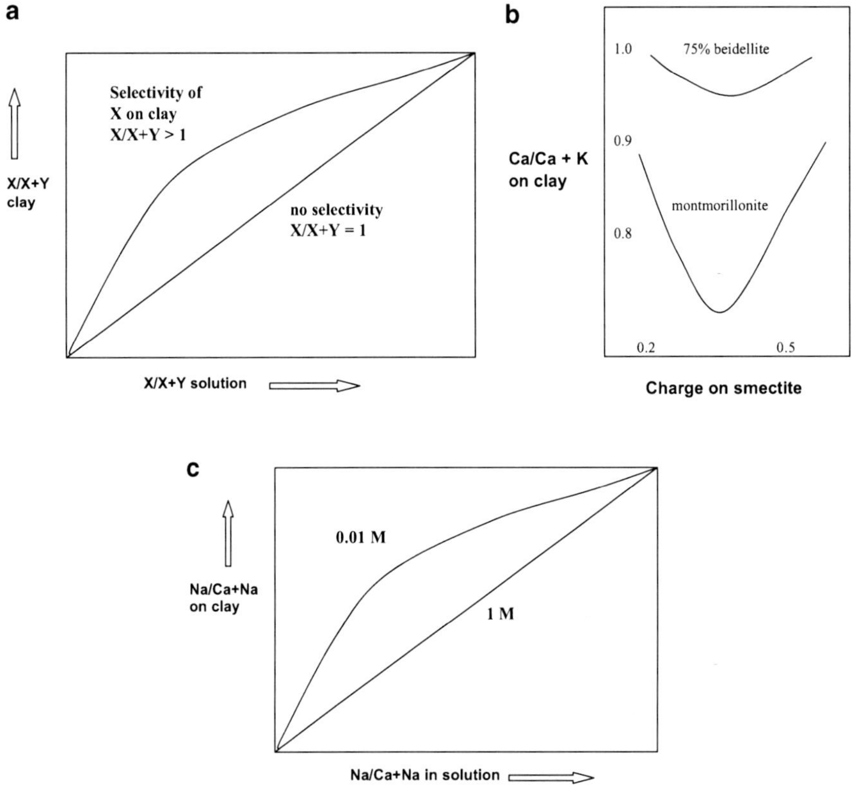

Fig. 2.5 (**a**) Selectivity of one ion compared to another by comparing the relative concentrations (X/X+Y) in solution and that found on the clay (see Stumm 1992). (**b**) Selectivity of K over Ca in smectite interlayer sites for clays of different composition. Mo = montmorillonite, for which ionic substitution creating charge deficiency is in the central cation site (octahedrally coordinated) and B (Beidellite) for which the charge is generated in the more superficial silica tetrahedral sites. Charge here actually indicates the charge deficit on the structure as a function of 22 negative charges (after Velde 1985). (**c**) Cation selectivity as a function of ionic concentration of the aqueous solution (after Stumm 1992)

on the structure and the origin of the charge can be factors in the selectivity process [see Tournassat et al. (2007, 2009) for instance].

It is important to consider that the solution concentration of dissolved ions can change over a season in soils or alteration sites on rocks near the surface. The relative concentration of elements can be affected by other factors such as input by plants at the surface. Hence the absorption of cations can change over short periods of time at the surface, but it will be more stable at depth where fluctuations will be less important due to the more regular, slow flow of fluids down toward the water table.

2.2.1.1 Selectivity of Cations

The extent of this charge on the clay surface is independent of pH and responsible for the retention of hydrated cations in the interlayer or on the surface of clay minerals. A key point in surface geochemical problems is the selectivity of one element over another as hydrated ions enter a clay structure. Given equal amounts of two dissolved elements in solution, clay minerals usually are found to attract one of them more strongly than the other (Fig. 2.5a). The chemical affinities of different ions toward clays have been studied in order to determine which ion is favored to be included into the clay interlayer by cation exchange.

One aspect that is frequently overlooked concerns the amount of charge on the clay structure due to ion substitution and the site from which it is generated within the clay structure. We will use the example of smectite in the presence of K–Ca solutions to demonstrate these effects. These ions are interesting in that they have roughly the same ionic radii while one is monovalent and the other divalent (hence twice the charge density of the former). Figure 2.5b shows the relations of the proportions of K and Ca fixed on a clay from a solution containing equal amounts of both ions in solution. The differences in the clay structures are in the amount of charge generated by internal substitution and the distance from the surface where the charge is generated. Montmorillonite has a structure where the charge is generated in the internal octahedral site (Al) of a three-sheet structure of Si–Al–Si oxygen coordinated layers. Beidellite substitutions are in the surface Si site near to the zone of hydrated cation attachment. Beidellitic surface charge types appear to give more or less homogeneous results over the range of charge variation showing a clear preference for Ca cations. Montmorillonite in turn exhibits significant selectivity for K under conditions of intermediate and especially high charge where more K is selected on the clays relative to Ca or Mg cations. One would expect that Ca would be preferred over K since its charge density is twice as high. Thus the amount of charge and the proximity of the charge substitution to the surface layer of hydrated ion attachment can affect the ions fixed on a clay. Figure 2.5c shows that solution concentration also changes ion selectivity.

2.2.1.2 Hydration and Ion Exchange

Hydration of an ion depends on the electrostatic attraction of water molecules to that ion. In moving down a given group of the periodic table, the atomic and ionic radii increase (Fig. 2.2) while the hydrated radii of the cations decrease. The extent of hydration of the cation depends on their size, which reflects the average charge per surface area. The smaller the size of the ion the greater will be the charge density and, hence, the greater will be the ability to draw electrons from molecules and the larger will be the hydration sphere. Therefore, smaller cations (and thus ions of greater ionic potential) attract more water molecules. The result is the inverse relationship between the hydrated and the non-hydrated radius of the ions.

When the valence of the cations is equal (i.e., both +1 charge) the cation with the smallest hydrated radius is more strongly absorbed. In the case of the monovalent cations of potassium and sodium, the potassium ion is more strongly adsorbed since it has a smaller hydrated radius (the radii of the unhydrated ion is the first value and hydrated ions represent the second value behind the element, values in pm) and hence is more strongly absorbed to the site of the negative charge. So for the alkaline metals the elements with the larger ionic radii and the lower hydration energy are preferred over smaller cations with a higher hydration enthalpy.

$$Cs^+(186,329) > Rb^+(163,329) > K^+(149,331) > Na^+(117,358) > Li^+(94,381)$$

A similar behavior can be observed for the divalent cations of calcium and magnesium. The hydrated magnesium ion is larger than that of calcium; the magnesium ion is held more weakly and behaves in some instances in soil (i.e., when calcium is low) like sodium.

$$Ba^{2+} > Sr^{2+} > Ca^{2+}(100,412) > Mg^{2+}(72,428)$$

In case of vermiculites (high charged clay minerals) there are no differences among the heavier alkaline earth metals.

$$Mg^{2+} > Sr^{2+} \approx Ca^{2+} \approx Ba^{2+}$$

A competitor for the interlayer sites can be the hydrogen ion (H^+) at high concentrations, i.e., low pH. Ferrage et al. (2005) showed that at low pH solution values the hydration state of Ca-Montmorilonite can be decreased significantly ($pH < 2$) and that H^+ ions would replace the hydrated Ca^{2+} in the interlayer sites of smectites. Tertre et al. (2009) show that the pH effect on Na^+ hydration state is such that it begins to reduce the amount of bound Na^+ at pH values below 5. The selectivity coefficients are affected by pH also, as K, Ca, and Mg are also affected by pH but not in the same way. Uptake curves for different minor elements in solutions as a function of the mineral substrate (Arnfalk et al. 1996) indicate that retention of cations commonly occurs at values above pH 2 on different types of smectites. A general observation from the literature is that in metals with higher valencies, heavy metals, and actinides, outer-sphere sorption to clay-type minerals becomes prevalent at low pH ($pH < 5$) and low ionic strength.

Bonneau and Souchier (1979) and more recently Tertre et al. (2011) demonstrate the importance of the concentration of dissolved elements in solution concerning the selectivity of one ion over another. This factor is important in considering soil chemistry in that soil solutions change concentration with season and rainfall, which could very well change the selectivity of a given ion over another over a relatively short period of time.

Bergseth (1980) determined the selectivity sequencies for Illite, Vermiculite, and Ca-smectite in solutions containing two heavy metals. The selectivity is increasing in the following order:

$$Cu^{2+} > Pb^{2+} > Zn^{2+} > Cd^{2+} > Mn^{2+}$$

Here the hydration enthalpy is decreasing from Cu^{2+} (−2,102 kJ mol^{-1}) to Mn^{2+} (−1,834 kJ mol^{-1}). This means that in this case the more hydrated cation is taken up preferentially. So an interpretation only based on the hydration enthalpy is not possible.

In summary, the charge of the cation and the size of the hydrated cation essentially govern preferences in cation exchange equilibria. Highly charged cations tend to be held more tightly than cations with less charge and, secondly, cations with a small hydrated radius are bound more tightly and are less likely to be removed from the exchange site.

2.2.1.3 Hydration State of Ions in Clay Minerals

One of the characteristic features of certain clay minerals, the expanding types, is the ability to incorporate cation–water complexes into the mineral structure. This occurs in certain sites of the layered structure where permanent charge is present on oxygen ions of the sheet structure surface.

In aqueous solutions where clays are present cations enter the interlayer sites with water molecules around them. Out of solution, in dry clays, when the cations are in the interlayer they will be regularly arranged with the water molecules. The water molecules around the cations form planes of strict geometry and dimensions between the clay layer surfaces. They are included within the water–cation configuration such that the water molecules form geometric, quasi-crystalline layers of a given dimension, 2.5 Å thick per layer. This can be seen by X-ray diffraction in the stepwise expansion of the 00*l* basal reflections with an increasing amount of water layers in the interlayer of the clays. The different steps correspond to the intercalation of 0, 1, 2, or 3 planes of H_2O molecules in the interlayer. Moore and Hower (1986) and Sato et al. (1992) found that up to three H_2O layers are stable in dioctahedral smectites, depending on the relative humidity. Different layer types were thus defined for smectite: dehydrated (0 W, d001) 9.7–10.2 Å, mono-hydrated (1 W, d001) 11.6–12.9 Å, bi-hydrated (2 W, d001) 14.9–15.7 Å), and tri-hydrated (3 W, d001) 18–19 Å) layers. Apparent basal spacing (or apparent layer-to-layer distance) does not vary in a stepwise function, because of the ordered interstratification of two appropriate hydrates.

Factors controlling the hydration state of smecite include composition (total layer charge and charge location), interlayer cation (type, valency, and hydration energy), and environments (humidity or vapor pressure, temperature, and H_2O pressure). Kawano and Tomita (1991) emphasized the importance of cation radius and the location of layer charge on the rehydration properties of smectite.

2.2.1.4 Dehydration and Layer Collapse

As already described the effect of cation hydration is important for the structure and properties of hydrated clays, especially regarding sorption by compensation of structural charge unbalance. Cations with low hydration energy, such as K^+, NH_4^+, Rb^+, and Cs^+, produce interlayer dehydration and layer collapse, and are therefore fixed in the interlayer positions. Conversely, cations with high hydration energy such as Ca^{2+}, Mg^{2+}, and Sr^{2+} produce expanded layers and are more readily exchanged (Sawhney 1972).

Sometimes in the literature concerning the pH independent dehydration and retention the term inner-sphere complexation is used. This term is ambiguous because it leads to confusion with the same term used in the context of surface geochemistry. In the context of surface geochemistry the pH-dependent formation of strong metal ion–surface oxygen atom bonds results in inner-sphere surface complexes where a part of the metal ion's hydration sphere is removed (see next chapter).

One aspect largely overlooked is the amount of charge on the clay structure and the site from which it is generated. Layer charge arises from substitutions in either the octahedral sheet (typically from the substitution of low charge species such as Mg^{2+}, Fe^{2+}, or Mn^{2+} for Al^{3+} in dioctahedral species, typical for Montmorillonites) or the tetrahedral sheet (where Al^{3+} or occasionally Fe^{3+} substitutes for Si^{4+}, typical for Beidellites), producing a negative charge for each such substitution. If the substitution occurs in the octahedral layer the charge deficiency is distributed on ten surface oxygen atoms of the tetrahedral sheet, belonging to four tetrahedra linked to the octahedral site. This charge is diffuse. It leads to the formation of an outer-sphere complex on crystal surfaces. A substitution directly in the tetrahedral sheet, results in a localized charge deficiency. That enhances the dehydration and layer collapse. Since the radius of K^+ is very close to the diameter of the hexagonal siloxane cavity, it may form a very stable pH independent surface.

2.3 Adsorption (Inner-Sphere Surface Chemical Bonding)

2.3.1 Edge Surface Sites and their Interactions with Cations and Anions

Most solid phases found at the surface of the earth are oxygen–cation minerals, i.e., a combination of oxygen atoms and cations of different types. The types of bonding between the oxygen and cations are generally of covalent nature, and often a complex series of covalent sharing of electrons between oxygen atoms and several cations occurs, as in silicate minerals. In silicates for example it is common for one oxygen atom to be electronically related to two or more cations. This being the case, at the edges of crystals where the structure is not fully compensated electronically, an oxygen or cation can be present with a residual charge present due to incomplete linkage. Usually oxygen atoms are present on edge sites either linked or not to hydrogen ions.

Edge sites on silicates can then be considered as covalently bonded oxygen anions with a residual, unsatisfied charge. The same situation occurs for other minerals, such as oxides for example, which have uncompensated charges on their surfaces. Yet another type of material has unsatisfied bonds on its surface, organic material, but here the types of bonding are even more complex than in the case of silicates or oxides. In all three types of materials, silicates, metal oxides, and organic matter, the surfaces attract charged ions to compensate their unsatisfied charges. The attraction and chemical bonding, largely ionic in nature, of adsorbed ions is one that is strongly influenced by the chemistry of the aqueous solutions, i.e., the concentration of dissolved ions in solution and the relative amount of the anionic OH^- molecule and the cationic H^+ ion.

Edge site attraction can be found on clay minerals, on oxides and hydroxides in alterites and soil, and on organic molecules in soils. In fact most materials, except those that are well crystallized with a well-defined crystal structure expressed in crystalline form, attract ions from solution to their surfaces which are bound chemically. Surface site adsorption can occur on clay minerals (silicate minerals with highly covalent oxygen–cation bonding), oxides, and organic molecules, which expose a wide variety of edge sites.

The major solid phase materials (sorbents) in soils are clays, metal-oxyhydroxides, and soil organic matter (SOM). Most clays are negatively charged. In many soils, they represent the largest source of negative charge. While formation of outer-sphere bound ionic species at ion exchange sites of clay minerals is pH independent, sorption by inner-sphere complexation is considered to take place on amphoteric surface hydroxyl groups and thus varies with their pH-dependent protonation/deprotonation (Geckeis et al. 2013). This is the other source of charge in soil minerals. Oxygens on metal-oxyhydroxides surfaces in contact with water are bonded to protons. Unsatisfied bonds at the terminal ends of minerals and organic matter result in surface charge. These ions are called surface sites and act as an amphoteric substance. At acid pHs there are doubly protonated surface oxygens (water molecules) and singly (hydroxo groups) while non-protonated surface oxygens (oxogroups) are present at basic pH (Schindler et al. 1976). Thus metal-oxide surfaces can undergo acid–base reactions and behave amphoterically as a function of pH. The formation of strong metal ion–surface oxygen atom bonds with significant ionic or covalent contribution results in inner-sphere surface complexes where part of the metal ion's hydration sphere is removed. This strong binding to the mineral surface has a more pronounced effect on reversibility and desorption kinetics compared to outer-sphere sorption (Geckeis and Rabung 2002).

A few important points to remember about the characteristics of variable charged surfaces:

1. Adsorption process can be written as a chemical reaction between a surface and an ion.
2. The reactions take place on amphoteric surface hydroxyl groups.
3. They are pH dependent.

4. The hydroxyl groups develop at unsatisfied bonds at the terminal ends of minerals.
5. The surface reactions result in very strong association.
6. They are in principle reversible.

2.3.2 *Origin of the Surface Charge of Soil Minerals*

The charge results from chemical reactions at the surface. Many solid surfaces in soils contain ionizable functional groups, for example, −OH, –COOH. Soil minerals are variably charged because their surfaces become hydroxylated when exposed to water and assume anionic, neutral, or cationic forms based on the degree of protonation ($\equiv$M–O$^-$, $\equiv$M–OH, or $\equiv$M–OH$_2^+$, where $\equiv$M represents a metal at the edge of a crystal structure), which varies as a function of solution pH. The variably charged surfaces are chemically active and the charges are localized at specific sites. Adjacent sites may even have opposing charges!

In general, variably charged minerals may adopt a net positive surface charge at low pH and a net negative surface charge at high pH. The solution pH can have a strong effect on the surface charge of the minerals and hence on sorption of nutrients or contaminants. As the solution pH increases, hydroxyl groups depronate. This deprontonation is decreasing the positive or increasing the negative charge density on the mineral, thus facilitating cation adsorption. At the same time the ability for anion adsorption is decreasing. The ability of a mineral or a soil to retain cations increases with increasing pH. Parameters affecting inner-sphere sorption are ionic strength, electrolyte composition as well as metal ion concentration, presence of competing cations, complexing ligands and temperature, etc. (Geckeis and Rabung 2002; Hayes and Katz 1996).

The importance of edge site charges can be seen in the data given by Grim (1953) shown in Fig. 2.6 for kaolinite (a non-permanent charge clay mineral) where the reduction in grain size is seen to generate a higher cation exchange capacity (CEC) indicating the presence of more charged edge sites per unit volume. The smaller the crystallite size, the greater the crystal edge surface. Thus it is evident that the surface edge sites of clays can be an important source of cation retention.

We will use the case of a silicate clay mineral to demonstrate the relations between molecules in water and surface sites on adsorbers. The same scheme can be used for oxides, but the situation is more complex for organic materials in that the functional groups on the latter are quite different and vary depending on the initial material and its state of degradation by bacteria in surface environments.

On edge sites the covalently bonded silicate structure is not electrochemically complete in that many of the oxygen atoms in the covalently bonded structure are not bonded to cations (Bickmore et al. 2003). For example, the structural unit of silicon, which is tetrahedrally coordinated in a silicate mineral, would be Si–O$_4$ where the oxygen atoms are coordinated with silicon atom. There are four oxygen atoms and one Si atom. Since Si has a charge of +4 and oxygen of −2, there is a

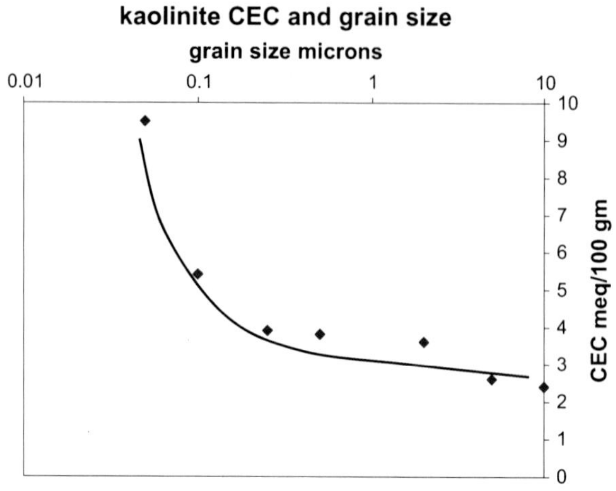

Fig. 2.6 Relations of grain size and cation exchange capacity (CEC) in millequivalents of charge per 100 g of clay. Data from Grim (1953)

surplus of charge as the unit is written. In a completed structure the oxygen atoms are shared in a complex manner with other cations so that the excess charge is shared also with the different cations. Each oxygen atom shares a charge of +1 from the initial silicon atom and must share a residual charge of plus 1 with another cation. In the case of Al ions in a silicate clay structure, the Al cation is coordinated to six oxygen atoms, and again there is an excess of charge on each oxygen cation which is shared with other cations in a covalent coordination complex. Each oxygen atom shares a half charge with the initial Al cation. In the complex covalent structures of silicates, oxygen atoms are coordinated to two or more cations, sharing electrons in covalent bonding with different cations. For example the Al ion is normally hexa-coordinated to oxygens, which are bound to Al but also Si ions. Since Al has a charge of +3, each oxygen represents one-half of a charge in compensation on the Al ion. If the structure is not complete, a cation can be missing, and each oxygen atom at the surface presents a residual negative charge of one in the case of Si–O complexes and +0.5 in the case of Al–O complexes.

The oxygen on the uncompensated site in an AlO_6 complex gives a residual charge of -1.5. If a proton is fixed on the site, the residual charge will be -0.5. If a second proton is added on the site, a residual charge of +0.5 will remain (Fig. 2.7). The case for ions on incomplete Si–O_4 complexes is such that if an oxygen atom is present, the charge will be -1, if a proton is present there will be no residual charge, and if a second proton is present the charge will be +1. The more hydrogen ions attracted to the surface sites, the higher the positive or the lower the negative charge on the surface (Fig. 2.7).

A change in pH from acidic to more basic values decreases the hydrogen activity and this gradually decreases the positive charge on the structure and eventually a series of negatively charged sites appear. This is the basic explanation for the retention or adsorption of oxoanions on solids in solution in acidic solutions and the adsorption of cations in basic solutions. It is clear that the edge site structure is a series of different oxygen sites where charge occurs depending on the nature of the

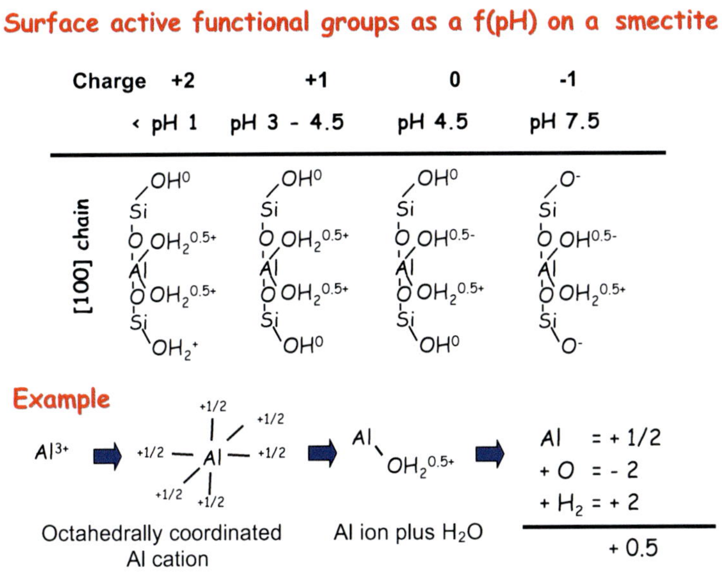

Fig. 2.7 Illustration of the possibilities of charge at the surface of a clay mineral where edge site cation ions (Al and Si) have a residual charge due to a finite crystalline structure that does not fulfill the shared charge of cations with oxygen ions. The charged ions Si and Al attract ionic components of water molecules, O^{2-}, OH^-, H^+, and water itself either neutralizing the residual charge on the cation or leaving a positive or negative charge on the clay edge surface. The case of an aluminum ion compensated by a water molecule is shown at the bottom of the figure (after Bickmore et al. (2003))

site(s) on the crystal and on the pH of the aqueous solution in contact with it. Acidic solutions produce edge surfaces of silicates that will attract anions and oxoanions from solution and basic solutions will produce surfaces that attract cations from solution. The balance of charge on the different sites of incomplete charge compensation of oxygen–cation complexes is shown in Fig. 2.7.

2.3.3 *Acid–Base Reactions at the Surface of Minerals and the Notions of Points of Zero Charge*

As explained above among the various reactions, which lead to charged interfaces, the adsorption of hydrogen and hydroxyl ions is of importance. In the case of mineral/water interfaces, the adsorption of H^+ and OH^- is related to the protolysis of surface OH^- groups. Such ≡M–OH groups exist on all soil minerals. The adsorption reaction of protons can be described the following equation: in

$$\equiv \text{M–OH} + \text{H}^+ \rightleftharpoons \equiv \text{M–OH}_2{}^+ \quad (2.1)$$

and the uptake of the hydroxyl ions can be presented as deprotonation

$$\equiv \text{M–OH} + \text{OH}^- \rightleftharpoons \equiv \text{M–O}^- + \text{H}_2\text{O} \quad (2.2)$$

A point of zero charge (PZC) can be seen as an intrinsic property of a mineral surface. There are different notions of points of zero charge and it is best to distinguish between them to avoid confusion. In the present case the point of zero charge is the pH at which the total concentration of surface deprotonated sites is equal to the total concentration of surface protonated sites. This PZC is best called the point of zero net proton charge (PZNPC), because the charging mechanism is related to the interaction of the surface with water ions (i.e., protons and hydroxide ions, which in turn are related to each other). According to our definition the PZNPC is found when

$$[\equiv \text{M–OH}_2{}^+] = [\equiv \text{M–O}^-] \quad (2.3)$$

The surface charge changes with pH not only in magnitude but also in sign. At low pH values the surface is positively charged; as the pH increases the positive charge is reversed to a negative charge, which increases with a further increasing pH. So there is a pH of a solution in equilibrium with a mineral, at which the total positive and negative charges are equal. At this point, the PZNPC, the net proton surface charge, is zero (Fig. 2.8). This does not mean the surface has no charge, but rather there are equal amounts of positive and negative charges. The PCZ is normally determined by potentiometric titration measuring the adsorption of H^+ and OH^- on amphoteric surfaces in solutions of varying ionic strength. An overview of different point of zero charge values for different minerals can be found in Kosmulski (2009).

A related concept in electrochemistry to the point of zero charge to represent the surface charge is the isoelectric point (IEP). The isoelectric point is determined by establishing the pH value at which soil particles do not move in an applied electric field. ("electrophoretic mobility measurement") or at which settling occurs in a suspension of soil particles ("flocculation measurement") (Sposito 1989). The reason that there is an IEP for the mineral surfaces is that the particle charge is determined by a competition between two reactions—one that makes the surface positive and one that makes it negative [see reactions (2.1) and (2.2)]. At some intermediate pH, the two reactions will be in balance and zeta potential will be zero (2.3). The isoelectric point is the pH at which a particular particle carries no net electrical charge. It is the pH value of the dispersion medium of a colloidal suspension at which the colloidal particles do not move in an electric field. The IEP has been related to the solubility minimum of the solid, i.e., minimum solubility in water or salt solutions at the pH, which corresponds to their IEP.

Although they should be similar, there is a huge discrepancy in reported points of zero charge for clay minerals based on the compilations by Kosmulski (2009).

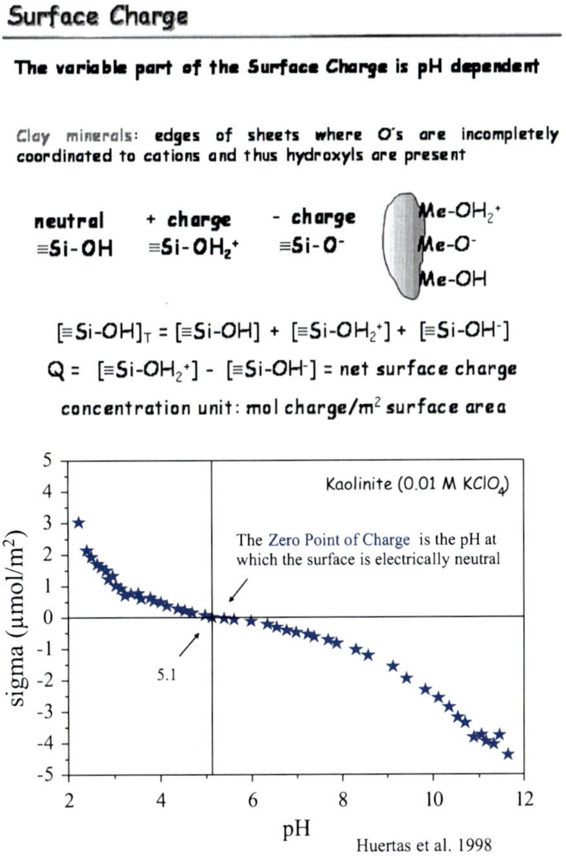

Fig. 2.8 Surface charge vs. pH diagram after Wanner et al. (1994) and an illustration of the origins of different surface charges on silicate structures due to incomplete bonding of surface cations. Plot in lower portion of the figure (data from Huertas et al. 1998) shows the overall charge on kaolinite (positive or negative) as a function of solution pH. The crossover from positive to negative residual charge is the "Zero point of charge"

This is partially due to the fact that PZNPCs determined from potentiometric titrations and IEPs from electrokinetic measurements measure different quantities. The isolectric point is the point at which the zeta potential (surface charge) is zero and the particles in the applied electric field don't move. The electrophoretic mobility is zero. Lützenkirchen et al. (2012) showed that the PZNPC varies as a function of ionic strength (Fig. 4), probably because the permanent charge of the clay minerals influences the overall electrostatics of the system.

While the potentiometric titrations measure the interactions of the surface with protons and hydroxide ions, the measurements of the IEP quantify the electrokinetics of a system. It is a measure of the net charge within the shear-plane of a particle and therefore results obtained by this method include contributions from permanent charge, charge due to proton ad/desorption and accumulation of counter-ions within the shear plane. On clays the potentiometric titrations therefore miss the negative permanent charge that appears to dominate the electrokinetic charge. For example in the case of illite the point of zero charge based on potentiometric titrations varies from pH < 3 up to 9.6, while the IEP is 3.2 or below. For

Table 2.1 The values of pH corresponding to the pH for the point of zero charge for kaolinite and smectite, reported in recent literature

Mineral	Background electrolyte	pH	Method	Sources
Na-Montmorrilonite	0.1 and 0.5 M $NaClO_4$	7.8 and 6.5	PCZ	Baeyens and Bradbury (1997)
Na-Montmorrilonite	0.006–0.088 M NaCl	8.5–8	PCZ	Avena and De Pauli (1998)
Montmorrilonite	0.005–0.5 $NaNO^3$	6.1–7	PCZ	Wanner et al. (1994)
MX-80 bentonite, purified	0.002–0.007 M	<3 if any	IEP	Garcia-Garcia et al. (2009)
K-montmorillonite	0.001–0.1 M KNO_3	7.6–8.1	PCZ	Rozalén et al. (2009)
Kaolinite	0.001–0.1 $KClO_4$	5.5	PCZ	Huertas et al. (1998)
Kaolinite	0.1 NaCl	3.9–4.3	PCZ	Brady et al. (1996)
Kaolinite	None	4.5	PCZ	Motta and Miranda (1989)
Kaolinite	None	<3 if any	IEP	

kaolinite/kaolin Kosmulski (2009) reports literature values from titrations that range from pH 2.2 to 7.5, while IEPs are systematically below pH 5. In the same compilation the points of zero charge from titration for montmorillonite range from pH 2.8 to 10.6, and the IEP was always lower than 5.6. Obviously, titration measurements (that include single potentiometric titrations, common intersection points, and mass titration endpoints) are higher than the IEP (Table 2.1).

2.3.4 *What Is the Significance of PZCs?*

Points of zero charge in general terms may indicate the pH beyond which minerals will exhibit either cation or anion exchange. Minerals with opposite surface charges (i.e., different sign of charge) will attract each other. Consequently, the respective PZCs play an important role in soil aggregate formation and define the ranges of pH in which stable colloidal suspensions can form. The notions of PZCs should be kept in mind. For aggregate formation it is generally assumed that the IEP is the relevant PZC for homogeneous particles. However, for hetereogenous particles like clays, the individual PZCs for edges (PZNPC) and of the planes are relevant. The individual determination of the IEP of the two basal planes of kaolinite has been recently achieved experimentally (Gupta and Miller 2010).

In Fig. 2.8 the overall charge on clay structures is shown as measured in an aqueous solution and the pH at which the proton-related charge goes from positive (low pH) to negative. This transition point is of great importance. At this point anions and oxoanions will no longer be favored and cations will be attracted to the clay or other mineral surfaces. Data for kaolinite (Fig. 2.8) indicate a crossover point of zero charge at near pH 5, while montmorillonite shows crossover points between 6.5 and 8 (Baeyens and Bradbury 1997; Avena and De Pauli 1998). Crossover points for oxides, Si, Al, Fe^{2+}, Fe^{3+} (Sverjensky 1994) range from

3, Si, to 11.6 for FeO. Thus cations can be attracted to mineral surfaces found in alteration environments between pH values of 3–12 depending upon the material present. If one considers the crossover points for Si and Al oxides at pH 2.9 and 8.5 and the values for kaolinite near pH 5 it would appear that the Si and Al ions in the structure contribute to the crossover point values giving the overall value in between those of the oxides themselves. This clearly indicates that it is the cation within the mineral structure at edge sites which determines the edge effect pH attractions.

As a result of these relations hydrogen ions are in competition with cations for charged surface sites on solids. The competition is a function of the intensity of ion uptake chemistry and thus will vary from one elemental species to another. Furthermore, the exact chemical configuration of the materials giving rise to surface charges will vary from mineral to mineral due to crystallographic structure differences and hence the selectivity between a given elemental cation and hydrogen ions will be different. Thus the observations in laboratory experiments can be used as a general schematic representation of ion uptake on pure minerals, but the more complex configurations of composite material in nature will not necessarily follow the predictions that one can make based on experimental determinations in the laboratory. Nevertheless some very useful general trends can be noted concerning ionic attraction to solids in surface environments.

The terminology for the bonding of ions in solution to charged surfaces is that when an ion is simply attracted to a charged surface and fixed by weak forces, such as Van der Waals attractions, there are two or perhaps more spheres of water molecules surrounding the ions. This is outer-sphere attraction. If the ions are linked by an oxygen, a hydroxyl group or an aquo group, it is termed inner-sphere attraction. Here the chemical forces are stronger involving ionic and covalent bonding. The stronger bonding forces will be more ion specific and selective concerning the electronic forces and configuration of the ion in solution.

2.3.5 Ions and Factors Affecting their Attraction to Solids

2.3.5.1 Cations

The attraction of a specific cation to highly charged, local sites related to a given cation in a mineral structure is also of very great importance. This type of attraction has been noted to be highly pH dependent. The term of variable charge site was applied to indicate that the attraction of cations from solution could be modified by the addition of hydrogen ions that "neutralized" the charge on the site by fulfilling the charge imbalance. As pH increases, the hydrogen ion saturation of the sites decreases and other cations find their place on the clay, oxide, or organic molecules in natural alterite materials. The influence of pH on cation uptake can be seen in the range of pH 4–8, which covers a large portion of the values found in surface environments. Thus this effect is very important to consider when looking at minor element retention or release in surface geochemical environments.

The importance of surface sites on clay edges and oxides can be illustrated by the effect of grain size (amount of edge sites) on the cation exchange capacity (CEC) of kaolinite which has no internally generated charge imbalance and hence no fixed charge on the crystal structure but only charge variable edge sites [Fig. 2.6, data from Grim (1953, p. 136)]. The values double as the grains become smaller. These are the edge sites, which attract ions from solution. However, the most common range of cation exchange capacity for smectites (clays with an interlayer site capable of exchanging cations in solution) is on the order of 110–130 milli-equivalents per 100 g. Thus the interlayer sites of clays have a potentially much greater capacity to fix cations. Laboratory measurements of cations fixed on crystal edge sites indicate a low capacity, but nevertheless this capacity to adsorb ions from solution can be important when weakly absorbing clay material (essentially smectites) is present.

One measure or estimation of the importance of surface sites (variable charge) which can be affected by natural environment changes in chemistry is that given by measurements of fixed cations as a function of pH on natural materials. A variety of studies have been made to determine these effects (Bourg 1983; Arnfalk et al. 1996; Hooda 2010; Ross 1994 to cite a few). Most often the studies deal with Pb, Cd, and transition metal ions. Edge site retention of cations on minerals such as kaolinite or illite is inhibited by pH values below about 4–5. Values of pH are higher for smectites, in the range of 6–8. Exotic elements such as the heavy metal element Hg are fixed at higher pH values, near 7. Thus the variable charge sites come into action at pH values in the lower range of natural environmental waters.

2.3.5.2 Oxoanions

A very important relation of the chemical attraction between ions in water is the tendency for cations of high charge, +3 and above, to be strongly associated with oxygen ions which changes the overall charge on the association and a cation can become an anion, or oxoanion. Sulfur is an example changing from S^{6+} to SO_4^{2-}, or chrome going from a Cr^{6+} cation to the oxoanion form of CrO_4^{2-}. Oxoanions behave in a very different way from the cation equivalent in solutions, especially as a function of pH. In fact oxoanions are fixed on absorber variable charge sites at low pH and released as pH increases usually above pH 6 (see Fig. 2.7). The opposite behavior of cation and anion uptake is very important considering that many of the oxoanions are formed from heavy metals, such as Bi, Se, As, Se [see Drever (1982) or Stumm (1992) for example]. Oxoanions tend to form inner-sphere surface complexes which gives them more access to charged sites on available surfaces. The oxyanions can eventually be precipitated from solution by combining with a cation of lower valence, usually +2, producing a solid phase.

If we consider the schema in Fig. 2.8, we see that the higher pH solutions will favor the existence of negatively charged sites on complex solids, while low pH solutions will give rise to positively charged surfaces which will attract negatively charged ions such as oxoanions. Hence the adsorption curves of ions in solution will show low values for cations and high values for anions and oxoanions at low pH. This contrasting behavior is potentially very important for the distribution and movement of dissolved ions in altering waters, transport waters, and those of sedimentation. The oxidation state of an ion is intimately connected with its response to solution pH.

As is the case with transition metal cations, the pH at which oxoanions are attracted to charged surfaces is variable, most likely depending upon the strength of bonding of the oxoanion with a given charged site on the mineral substrate [see Gaillardet et al. (2004), Drever (1982), Arai (2010) for examples]. Usually the elements involved in the oxidation—oxoanion change are heavy metals such as As, Se, or the transition metal Cr and S or P (see Stumm 1992).

The information one can find in the literature on bonding characteristics of ions in solution as a function of pH is derived from calculation of bonding based upon chemical experiments or thermodynamic values involving hydration energy and other parameters of water ion interaction. This data is of course very informative, but it does not give an insight into the preferences of different ions for adsorption sites, either those related to Si or Al–O surface functional groups for example, nor the effects of ion concentration on adsorption or ion selectivity by the different adsorption sites.

Thus one has two types of charged sites on clays; edge sites of more or less point charge associated with a specific oxygen ion or a basal surface charge spread over several oxygen sites. Since the bonding energies of the two types of chemical attractions are rather different, one essentially a point source atom to atom bonding and the other a diffuse surface attraction, the behavior and preferences of attraction for cations in solution are rather different.

Oxides in soils and weathered materials, especially Mn oxides, can incorporate cations also, often in a pure ionic non-hydrated form but also in a hydrated form. Lanson et al. (2002) and Manceau et al. (2007) indicate that heavy metal and transition metal elements can migrate into pre existing oxide structures forming substitutions within the crystal lattices, of course in a non-hydrated form.

2.3.5.3 CEC

Soils can be thought of as storehouses for plant nutrients. The cation exchange capacity (CEC) of a soil is defined as the total sum of exchangeable cations that can be adsorbed at a specific pH. It is a measure of the quantity of negatively charged sites on soil surfaces that can retain positively charged ions (cations) such as sodium (Na^+), calcium (Ca^{2+}), magnesium (Mg^{2+}), and potassium (K^+), hydrogen (H^+), and aluminum (Al^{3+}) by electrostatic forces. The capacity of the soil to bind these cations is called the cation exchange capacity (CEC, measured in milli-equivalents of charge per 100 g of clay).

Only a small portion of the plant nutrients is in the soil solution. The exchangeable cations, which are bound to the soil surfaces, are in equilibrium with soil solution. The CEC, therefore, provides a reservoir of nutrients to replenish those removed by plant uptake or leached out of the root zone.

Cation exchange sites are found primarily on clay minerals and organic matter (OM) surfaces. Soil OM and clay minerals will develop a greater CEC at near-neutral pH than under acidic conditions (pH-dependent CEC) The charge of the soil components that contribute to the CEC is affected by the soil pH. These components have OH groups on their edges. OH groups can release or bind protons. At high pH protons are released from these groups, their charge becomes negative, and, as a result, CEC of the soil increases.

The four most abundant exchangeable cations in soil of humid regions are hydrogen, calcium, magnesium, and potassium. Of these, all except hydrogen are absorbed by plants in large quantities. Soil located in less humid regions (i.e., arid and semiarid) generally contain little or no exchangeable hydrogen, and often contain large quantities of exchangeable sodium.

CEC may be the limiting factor for having a fertile soil. You can improve CEC in weathered soils by adding lime and raising the pH. Otherwise, adding organic matter is the most effective way of improving the CEC of soil. On the other hand, a soil's CEC can decrease with time as well, through natural or fertilizer-induced acidification and/or OM decomposition.

Mineral	Type	CEC (surface charge, $cmol_c/kg$)	Surface area (external, m^2/g)
Smectite	2:1 clay mineral	−80 to −150	80–150
Vermiculite	2:1 clay mineral	−100 to −200	70–120
Fine Mica	2:1 clay mineral	−10 to −40	70–175
Chlorite	2:1 clay mineral	−10 to −40	70–100
Kaolinite	1:1 clay mineral	−1 to −15	5–30
Gibbsite	Al-oxide	+10[a] to −5	80–200
Goethite	Fe-oxide	+20 to −5	100–300
Allophane	Amorphous	+10 to −150	100–1,000
Humus	Organic	−100 to −500	Variable

[a]Positive sign indicates that the minerals no longer exhibit a cation exchange capacity, but rather an anion exchange capacity

Adapted from Table 8.1, Brady and Weil (2002)

Soil CEC is normally expressed in one of two numerically equivalent sets of units: meq/100 g (milliequivalents of charge per 100 g of dry soil) or $cmol_c/kg$ (centimoles of charge per kg soil).

2.4 Eh–pH Relations: The Effects of Redox Reactions

Redox reactions are largely controlled by biologic activity at the surface in that the normal tendency of air saturated water is to oxidize most of the elements present that are susceptible to change in oxidation state. Microorganisms in organic systems tend to reduce the oxidation state of materials present in their use of change in oxidation state for their metabolic energy. This is especially true for organic matter where the end result is the production of gases. However minerals are often affected by biologic activity such as sulfides and oxides where both oxidation and reduction can occur. This produces ions of different oxidation state that may subsequently be released back into the ambient aqueous solutions.

The ions in solution are of different valences depending upon the electronic structure of the element forming the ion, and the ambient conditions of redox potential. Values for various reactions involving the oxidation of an ion or the reduction of an ion are given in Lide (1999, listed as electrochemical series).

Positive values indicate the energy necessary to lose an electron while negative values would indicate a driving force to gain an electron. In natural systems at the surface of the earth there is often a coupling of change in oxidation potential (oxidation or reduction) and pH conditions. This is illustrated by Brookins (1988) in his Fig. 1. Fields of ionic stability can generally be defined in zones of decreasing Eh (reduction of ionic oxidation state) as pH increases. In surface systems where pH and Eh can change largely due to biologic action, there is a tendency for lower pH to be accompanied by reducing conditions. In the various diagrams presented by Brookins, it is clear that the elements that can have variable oxidation states tend to be oxidized at higher pH values. Such variations in pH can change cations (positive charge) into anions (negative charge) through a combination of cations with oxygen ions which often gives an oxoanion, where the positive charge is compensated by the oxygen ion, but nevertheless the combined elements form an overall negatively charged unit in aqueous solutions. Some elements such as As have a dominant oxoanion structure (Arsenate: AsO_4^{3-}; Arsenite: AsO_3^{3-}) of various negative valences as a function of pH and Eh. Further, redox reactions can create oxides or hydroxides, which are no longer soluble in aqueous solution. Extreme cases of ionic reduction potential, Cu for example, give the formation of a metallic phase. The values of pH and Eh at which elements change oxidation state and hence functional state (negative or positive ions in solution, or phases of low solubility) depend on the electronic configuration of the element of course.

From a geochemical point of view, we are interested in major element phases and the elements that are less abundant in the various phases of the surface environment. Iron and manganese are two such elements, which have affinities for transition metal ions of minor abundance and some other heavy metal elements. The redox conditions of a soil are very important in the formation conditions of Fe and Mn phases. Fe and Mn oxyhydroxides tend to incorporate other elements, especially transition metals, into the solids. If these elements remain in solution they will be carried on colloidal mineral surfaces or as dissolved species in solution along with minor elements.

2.4.1 *Eh and pH in Weathering*

In general the chemical composition of a rock will determine the overall pH of the aqueous solutions interacting to produce new mineral phases in the rock–water interaction zone. Acidic rocks, such as granite, produce overall slightly acidic

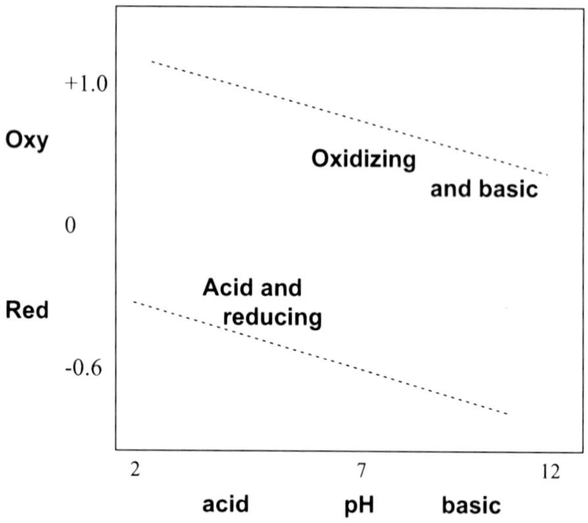

Fig. 2.9 Indications of the relations between redox and pH relative to natural surface environments. *Dashed lines* indicate the limits of the existence of water, which occurs between the lines (after Brookins 1988). Generally speaking, water rock interaction chemistry would occur in the upper part of the diagram where oxidation is prevalent and reducing solutions rare. Acid soil solutions are indicated in the diagram. The ranges of Eh–pH conditions, which produce cations or oxoanions, are of course quite different for each element due to its specific electronic configuration

solutions and basic rocks, such as basalts, produce overall basic solutions. When silica is in excess in the alteration reactions, silicic acid is produced in the solutions, and hence the term acidic rocks. Basic rocks give soluble magnesium and iron, which attract OH and the solutions become basic. The initial rainwater is only slightly acidic in character and hence strongly influenced in its chemistry by the rocks in which it comes in contact.

Soils, i.e., the bio-zone, are a different matter. Here the chemistry of the altering solutions is strongly influenced by the organic material present and the biological activity engendered by its decomposition. Acidic soils can be produced by types of vegetation, such as conifer forests, or special geomorphological situations, such as poorly drained areas. They produce bogs where organic matter is only slowly degraded and acidity is high. The effects of pH and Eh are coupled in affecting the stability of ions in solution. The cations are found in the more oxidizing and middle pH ranges while oxoanions, hydroxides, and oxides are produced in the more reducing and higher pH range. The relations found in soils are shown in Fig. 2.9.

Redox conditions are then very important in surface geochemistry in the plant-influenced areas. The various states of oxidation can occur in rock–water interaction zones, but the plant influenced zones can produce reduction of ions producing different phases either solids or ionic forms in solution. Plant action can also influence solution pH that in turn determines the fixing powers of solids concerning ions in solution. The formation of oxides or their dissolution and the change of ionic character from cation to oxoanion are fundamental concepts and mechanisms that should be considered in the study of surface geochemistry.

2.5 Observation of Absorption Phenomena for Some Specific Elements in Solution

2.5.1 Transuranium Elements

The adsorption or absorption of radionuclides is of particular interest in that these elements can at times become unwanted residents in the environment. Therefore quantifying the transport of radioactive contaminants within soils or the aquifers relies on a detailed knowledge of the geochemistry of the system and of the radionuclide retention mechanisms along the migration pathways (sorption, incorporation, precipitation). The retardation of these contaminants is strongly affected by their sorption behavior at the mineral water interface. The trivalent actinides are often replaced in the laboratory by lanthanides, which have very similar chemical properties.

The fact that the transuranium elements have a large number of electrons per atom helps in determining bonding properties by modern laboratory methods of detection of physical measurements such as X-ray absorption spectroscopy (EXAFS) or laser fluorescence spectroscopy (TRLFS).

Tan et al. (2010) summarize available studies on actinide/lanthanide interaction with mineral surfaces, including detailed descriptions of spectroscopic methods. The authors discuss examples of actinide/lanthanide interface reactions in order to elaborate mechanistic insight provided by individual spectroscopic techniques, such as TRLFS and EXAFS. Another review on actinide speciation was recently published by Maher et al. (2012). The review provides an overview of actinide reactions in environmental compartments including complexation with major relevant natural ligands abundant in groundwater and naturally occurring biological (microbial) and mineral surfaces. A very good overview of the mineral–water interface reactions of actinides is given by Geckeis et al. (2013).

It appears that the transuranium elements, trivalent for the most part, have a varied behavior concerning the type of attachment they effect on clay minerals with different charge sites. Some elements are absorbed as outer-sphere hydrates on the clay structure, while others are found as inner-sphere adsorbates depending upon the pH of the solution. One of the major questions in these studies is the hydration state of the ions in solution and the type of chemical attachment between ion and charged site on the minerals observed.

With help of EXAF measurements Stumpf et al. (2006) indicate that Am(III) is fixed by a double bond to surface oxygen atoms bound to Fe atoms on iron oxide. For clay minerals Stumpf et al. (2001) show that Cm(III) is sorbed onto kaolinite and smectite as outer-sphere complexes at low pH (ion exchange, we here sometimes use the notion outer-sphere complex in this sense, note also that Am(III) and Cm(III) have very similar beahvior). The number of water molecules was the same no matter if the Cm(III) was adsorbed or in solution. As pH increases (pH > 5.5) Cm(III) is bound as a inner-sphere surface complex. No incorporation of Cm was seen within the smectite structure; only edge sites are involved in Cm inner-sphere adsorption. The Al ions of the clay structure are those that react creating the hydrated inner-sphere complex that bind the Cm(III) ions (Stumpf et al. 2001).

The retention mechanisms of Am(III) onto smectite and kaolinite as a function of pH were studied by (Stumpf et al. 2004). Am(III) is absorbed as outer-sphere complex at pH < 5, and is becoming inner-sphere ion absorbates as pH increases. At pH 8 the coordination number of the Am(III) decreased which may be affected by the formation of a ternary OH-/Am/clay absorbed species. It appears that the site of inner-sphere surface attachment is associated with Al ions.

Uranyl complexes, studied experimentally by Greathouse et al. (2005), are sorbed at low pH in the interlayer sites of smectites as outer-sphere complexes giving a 14.58 Å layer spacing, somewhat lower than larger cation complexes such as Sr or Ca which give a 15.2 Å spacing.

Extended X-ray Absorption Fine Structure (EXAFS) spectroscopy has been used to investigate the sorption mechanisms of uranyl on different clay minerals under different conditions in the absence of carbonate (Chisholm-Brause et al. 1994; Sylwester et al. 2000; Hennig et al. 2002; Catalano and Brown 2005; Schlegel and Descostes 2009). Depending on pH and U(VI) loading, these studies were able to differentiate between outer-sphere complexation, i.e., cation exchange and inner-sphere complexation at the edge sites of clay minerals. Preferential sorption to Si tetrahedral or Al octahedral edge sites in these studies was not obvious. In presence of carbonate EXAFS measurements did not prove the formation of ternary complexes on the montmorillonite surfaces (Marques et al. 2012).

As the light actinide ions (Pa–Pu) tend to exist in various oxidation states in aqueous environment, surface-induced electron transfer processes trigger coupled sorption/redox reactions, which strongly affect their mobility (Bruggeman et al. 2012). Under naturally relevant conditions actinide ions can exist in various redox states (An(III), An(IV), An(V), An(VI)). The actinide redox state and speciation of the actinides change due to surface reactions and can form a broad variety of complexes with groundwater constituents, mainly OH^- and CO_3^{2-}. While the tri- and tetravalent actinide species form more or less spherical aquo-ions with 9–10 water molecules in the first coordination sphere, penta- and hexavalent cations form actinyl cations with covalently bound axial oxygen atoms and four to six fold coordination in the equatorial plane.

Redox reactions on mineral surfaces attract increasing attention since they may have a great impact on the mobility of the light actinides uranium, neptunium, and plutonium. It seems that electron transfer from Fe(II) solid phases is kinetically preferred for surface sorbed actinide ions over reduction of dissolved actinide species by Fe(II) in solution (Geckeis et al. 2013). In this context, natural Fe (II) containing minerals in the soil or in the aquifers represent huge electron donor reservoirs. Actinide species will therefore most likely be present in deep geological repositories in oxidation states III or IV, with low solubility and strong sorption as a consequence.

2.5.2 Lanthanides

Like trivalent actinides trivalent lanthanide ions exist in aqueous solution as hydrated species. In solid form those hydrates are known to form tricapped trigonal prisms and each metal ion is coordinated to 9 H_2O molecules. Coordination numbers of around 9 have been experimentally determined also in solution (Rizkalla and Choppin 1994; Stumpf et al. 2004; Lindqvist-Reis et al. 2005).

From a comparison between the data derived from Cm fluorescence spectra and batch data of Eu sorption on the sodium form of soil mica it is clear that outer-sphere surface sorption occurs at low pH in agreement with earlier studies (Stumpf et al. 2001). Under these conditions, the first hydration sphere of the metal ion remains unaffected and the sorption reaction is "invisible" to TRLFS. Outer-sphere complexation can be quantified by the difference between the distribution coefficient (Kd values) measured in batch sorption tests and those determined by TRLFS. The comparison of TRLFS with batch data of Cm and Eu sorption data respectively on Ca-montmorillonite and Na-illite shows that both elements behave similarly. Sorption edges obtained from wet chemistry experiments and spectroscopic information do not show evidence of any significant differences neither in the pH-dependent sorption behavior nor in the surface speciation.

Several studies have been carried out concerning the sorption of lanthanides on clays [see Bruque et al. (1980) or Coppin et al. (2002) for example]. It appears in the study of Coppin et al. (2002) that lanthanide sorption on kaolinite and Na-montmorillonite not only depends on the nature of the clay minerals but also on pH and ionic strength. At high ionic strength (0.5 M) both clay minerals exhibit the same pH-dependent lanthanide sorption edge. At low ionic strength (0.025 M) the permanent charge is compensated at low pH by physical sorption of the lanthanides on the basal planes for smectite. At pH values >5.5 lanthanide inner-sphere complexation takes place on amphoteric sites at the edge of particles to compensate for the variable charge. A fractionation is observed between the heavy rare earth elements (HREEs) and the light rare earth elements (LREEs) at high ionic strength, with the HREEs being more strongly sorbed than the LREEs. The fractionation observed at high ionic strength could be interpreted as either a consequence of a competition effect with sodium or the formation of inner-sphere complexes. This contrasting behavior should be reflected in rare earth content and elemental distribution in natural systems.

2.5.3 Transition and Other Metals

Uptake studies of transition and heavy metal ions on clays and the construction of sorption models describing this retention phenomenon have been ongoing for a certain range of elements on different types of materials likely to be found in alterites and soils such as iron oxide phases and clay minerals (Arnfalk et al. 1996; Evans et al. 2010; Tertre et al. 2009; Ross 1946; Gaillardet et al. 2004) and Al, Fe gels (Ross 1946). Studies from different laboritories on nominally identical systems usually yield similar results that differ in certain details though. In general the measurements of metal ion attraction to mineral substrates show that pH affects the accumulation, with a general trend, that can be given as follows. As pH increases from low values to about pH 10, some of the elements first extracted from solutions by solids are Pb^{2+}, Cr^{3+}, Cd^{2+} in the pH range of 3–4 where they are fixed in smectites, in the interstitial position where a double layer of water molecules is present around the ions. In general the adsorption of these elements is found to occur at higher pH values on oxides, gels, and kaolinite where the values of pH 4–6 prevail. The elements Zn, Cu, Ni, Co are found to be fixed on clays and iron oxyhydroxides minerals in the pH range of 5–6. One can consider that the complex, higher oxidation state cationic elements (+3) forming oxoanions tend to be attached to the mineral surfaces in preference to hydrogen ions. The accumulation of ions from solution and on the surfaces of solids of different types begins at pH 3–4. Below these values protons dominate on particle surfaces.

Several studies have been conducted to compare the affinity for sediments of ions in solution (Bourg 1983; Salomons and Förstner 1984; Gaillardet et al. 2004). The results are quite variable concerning the elements fixed, but the ranges of pH values of retention are quite similar. It is very striking to see the attraction of Cu, Zn, and Cd compared for two river systems, the Gironde and Rhone rivers in France. In the cases of these three elements the Rhone River fixes the elements at values between 5 and 6 while the Gironde River suspended matter attracts them at pH values of 3–4. This indicates a strong difference in the absorbing materials, most likely the Gironde River carries more smectite than does the Rhone River and hence the ions are absorbed in interlayer ion exchange sites which are less affected by pH. Meuse River bottom sediments indicate that Cu is absorbed on interlayer smectite mineral sites (pH 3–4) while Zn and Cd are adsorbed on variable charge sites at pH 5–6.

It seems clear that the pH of the solution determines which elements will be attracted to the solids as adsorbed or absorbed ions. One must also consider that the competition between hydrogen cations and other cations will determine the extent of adsorption on surface sites. We do not have a good idea of this dimension of cation retention yet. However it appears that there are two thresholds of cation retention concerning the solution pH: one being near 2–3 and the other near 5–6. These values are those found in acid soils under conifer forest or broad leaved evergreen (tropical plants). The value of pH 5–6 is closer to that of normal prairie soils or broad leaved deciduous forests. One would expect that many soils and rivers draining them will carry transition and heavy metals in northern latitudes and in tropical forest areas. Less transport in solutions as dissolved species is to be expected in temperate climates. However the reverse is true for oxoanions, such as As, Bi, Cr forms which are held on surfaces at low pH, acidic conditions with their concomitant positive charges on edge sites of crystals. Overall one would expect to find that transition metal and some heavy metal ions are lost from soils under conifer trees while oxoanions would be preferentially held on the positively charged mineral surfaces.

Cations can be absorbed in and adsorbed on minerals such as smectites with permanent charge sites expressed as diffuse charged sites on crystal surfaces or within the crystal or on variable charge punctual edge sites. The uptake site of an element on an edge or in the inner layer depends upon the characteristics of the attraction of the ion to water molecules and the pH of the solutions, and for example the extent to which protons compete for absorption or adsorption sites. As a result it is not a simple matter to estimate the amount of a cation that will be fixed on a clay mineral or oxide without taking into account the specific characteristics of the ion in question and the pH of the solution. Of course the oxidation state of an atom will change the characteristics of the water–ion relations and its attraction to charged sites on solids in contact with the aqueous solution where the ion is held.

2.5.4 *Oxides and Oxyhydroxides: Complex Cases*

The elements Fe and Mn form oxides and oxyhydroxides in alterites and soils which are preserved and at times added to different geological surface environments. The surfaces of these minerals are similar to the edge site surfaces in clays as far as charged ions are concerned. One can find such elements as Zn, Ba, As and various transition metals in coatings on oxides (Manceau et al. 2007). Lanson et al. (2002) showed that various cations can diffuse into the manganese oxide structure to form new structural molecules and units. Here the existence of a water layer around the ions is initially important to the attraction of ions on absorbers, but it seems to be of less importance as the cations enter into the mineral structure in an ionic form. This method of transfer and absorption is very important for the idea of surface phenomena, in that these elements enter into the mineral structure with neither the water nor hydroxide forms of an oxygen assemblage.

2.5.5 *Summary*

From the above summary of ionic relations in aqueous solutions, it seems that the role of pH is extremely important in determining the interaction between solids and ions in solution. Low pH values of 2 or less appear to exclude absorption (incorporation and solid state diffusion) and adsorption (retention at the surface) of cations

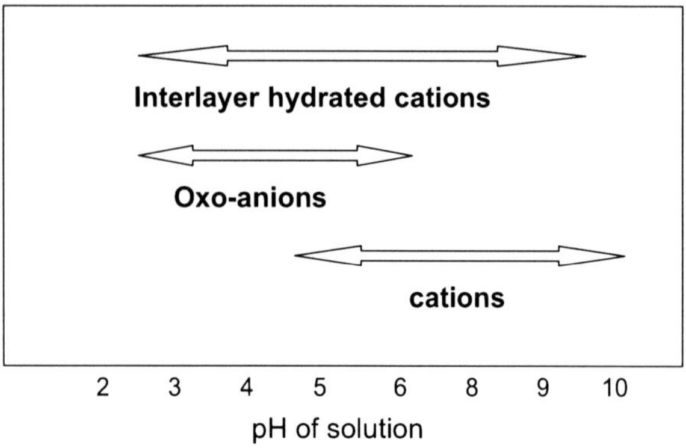

Fig. 2.10 Indication of the ranges of pH where the different types of ions can be fixed on (adsorbed) or within (absorbed) on particulate material as hydrated forms from aqueous solution

in clay interlayers and phases with variable charge sites of adsorption. Hydrated cations can enter clay structures (interlayer sites in smectites) at higher pH values. In the range of low pH between 2 and 5–6 oxoanions as well as anions are adsorbed on surface variable charge sites because the high hydrogen activity gives an overall positive charge to the materials, thus attracting anionic matter from solution. Above the pH range near 5–6 the overall charge balance on the charge variable surfaces becomes overall negative and cations are fixed on surface sites.

These relations are governed by the fact that the chemistry of ions in solution is strongly governed by oxygen–hydrogen balances which govern the electronic charge on ions at the surface of minerals. The balance of hydrogen and OH determines the charges on the surfaces of alterite minerals and thus determines which elements will be transported to solids and which will remain in aqueous solution. This is the basis of the movement of elements in aqueous solutions which is in reality the foundation of surface geochemistry. In general cation attraction to solids (minerals) is determined by pH where negative and positive charge sites are related to solution pH. A summary of the ionic forms which are fixed on particulate matter from aqueous solution is given in Fig. 2.10.

2.5.6 Soils and Cation Retention: Clays Minerals Versus Organic Material

Soils contain the different types of cation fixing materials, clays minerals, and oxides as indicated above, with variable charge edge sites on silicate clays and oxides and permanent charge interlayer sites on silicate clay minerals. Another, well-known component of soils is organic matter, either above surface detritus or root exudate and decayed roots below the surface. These materials are strongly affected by bacterial action and during this process the surface chemistry of organic material changes. Incomplete organic molecular chains, surface functional units, and others give this material a strong potential for interaction with ions in aqueous solution. Further as soils are the contact zone between the atmosphere and alterite

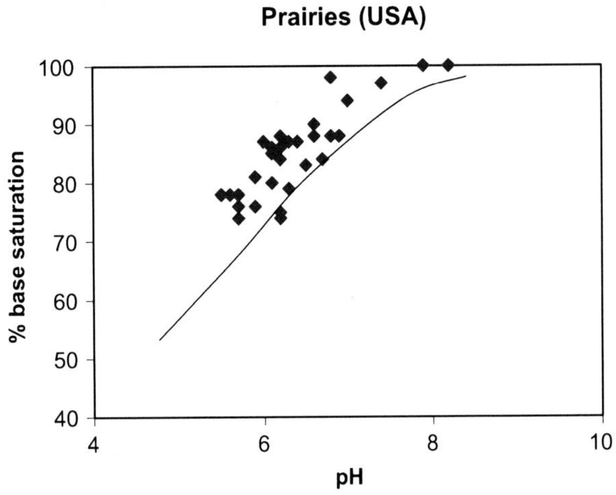

Fig. 2.11 Plot of data given by Ruhe (1984) for base saturation (% of base saturation sites occupied by cations) and pH of prairie soils

zone the chemistry is dominated by the plant and biologic activity which largely determines the pH of infiltrating solutions and their Eh. This of course affects the surface properties and retention characteristics of the soil materials. Sposito (1989) shows the relations between pH and anion–cation abundance fixed on soil materials with mineral and organic components where Na and Cl are good examples (Sposito 1989, p. 138). Low pH favors anion uptake and high pH favors retention of cations as would be expected. Oxoanions such as phosphate are retained at low pH while in certain cases complex materials such as borates can be found to be strongly absorbed at high pH, near 8. The surface sites can be selective concerning the cations adsorbed, just as is the case for absorbed ions fixed within silicate clay minerals such as smectite. Data given by Sposito (1989) and Alloway (1995) indicate general trends of selectivity for transition metal ions on soils rich in oxides (oxisols) and manganese oxides which favor the heavier metals in the series, Cu compared to Ni or Fe. Models for selectivity can be found in Sparks (1998). However, the complexity and variability of soil materials, which concern the minerals present such as the clay minerals and oxide minerals as well as the organic materials, and their proportions are such that models are rather difficult to apply for a given situation in a natural setting.

Overall one may speculate as to what are the relative proportions of interlayered clay (absorbed) ion in soils and what are the relative proportions of adsorbed (charge variable site attractions) minerals in soils. Then it is necessary to assume that the organic materials will have surface charge variable sites similar to those on clays and oxides. If we assume that the pH effect on cations is rather irrelevant for interlayer (fixed charge) site absorption but certainly matters for surface (charge variable) attraction, one can look at the data given by Ruhe (1984) for base saturation (major elements such as Na, Ca, Mg, K) on prairie soils in the great plains of the United States to estimate the response of soil materials to changes in pH. Figure 2.11 shows that there is a strong relationship between pH and base saturation (the presence of cations Na, Ca, Mg, K) of ions fixed on clays in natural prairie type soils.

This pH dependence of base saturation exhibits a distinct slope with less saturation at lower pH values. If one extrapolates to pH values of < 4, more than 50 % of bases are lost from the potential saturation sites. This means that the charge variable sites, those affected by pH, are relatively important as uptake sites for these soils especially as the soils become more acid. This being the case, the materials responsible for charge variable attractions in prairie soils which are largely rich in organic material are then probably to a large extent organic materials. Hence one can deduce that organic materials have a significant effect on cation retention in soils. This is an argument proposed by Sposito (1989). However, organic material evolves in soil zones, becoming more and more carbon rich and less chemically reactive as the material is more mature. Biologic effects tend to result in a more polymerized form of organic matter, producing various forms of humic material. Also the type of organic matter present adds to the complexity and capacity to attract cations to surfaces. Thus organic material evolves as a function of time and its surface chemistry varies with time also.

In order to assess the relative importance of organic matter as an ion absorber compared to clays and oxides in soils one can plot the clay content and content of organic matter as a function of measured exchange capacity, CEC, the number of sites available for fixing cations to soil material, is measured. To this end it is useful to observe the relationship of clays to organic matter in different types of soils: new soils and evolved soils. In some conifer forest soils reported by Collignon (2011) one sees a reciprocal relationship between organic matter and clay content. Here the organic matter is concentrated at the surface of the soils. In young prairie soils, relatively recent Alpine prairie soils (Mariotti 1982; Righi et al. 1999), there is a positive relationship between organic material (OM, or organic carbon, OC or carbon weight percent) and clay contents (Fig. 2.12). The same is true for older soils in dry prairies in Utah (USA) and Iran (Graham and Southard 1983). This contrasts with the negative correlation between organic material content and clays in the conifer forests. Thus we see that both organic matter and clay content increase in prairie soils going upward in the profiles.

The measured CEC as a function of organic matter (Fig. 2.13) shows that in the conifer forest soils one finds a good correlation between organic content and cation exchange capacity but not for clay content (Fig. 2.13). In conifer forest soils apparently most of the cation exchange is due to organic matter. In the Alpine meadow and prairie soils one finds a poor correlation between organic content and exchange capacity (Fig. 2.13). In the semiarid prairie soils (Mollison, USA and Iran) there is no systematic correlation. It seems then that the type of organic matter, conifer forest, and young Alpine soils and prairies will play a role in its ability to fix cations. If we consider the relationship between clay content and CEC there is a negative correlation in the conifer forest example, a poor correlation for mountain meadow and a good correlation for alpine prairie soils as well as for other prairie soils. It seems that the type of organic matter, young in acid conifer forests,

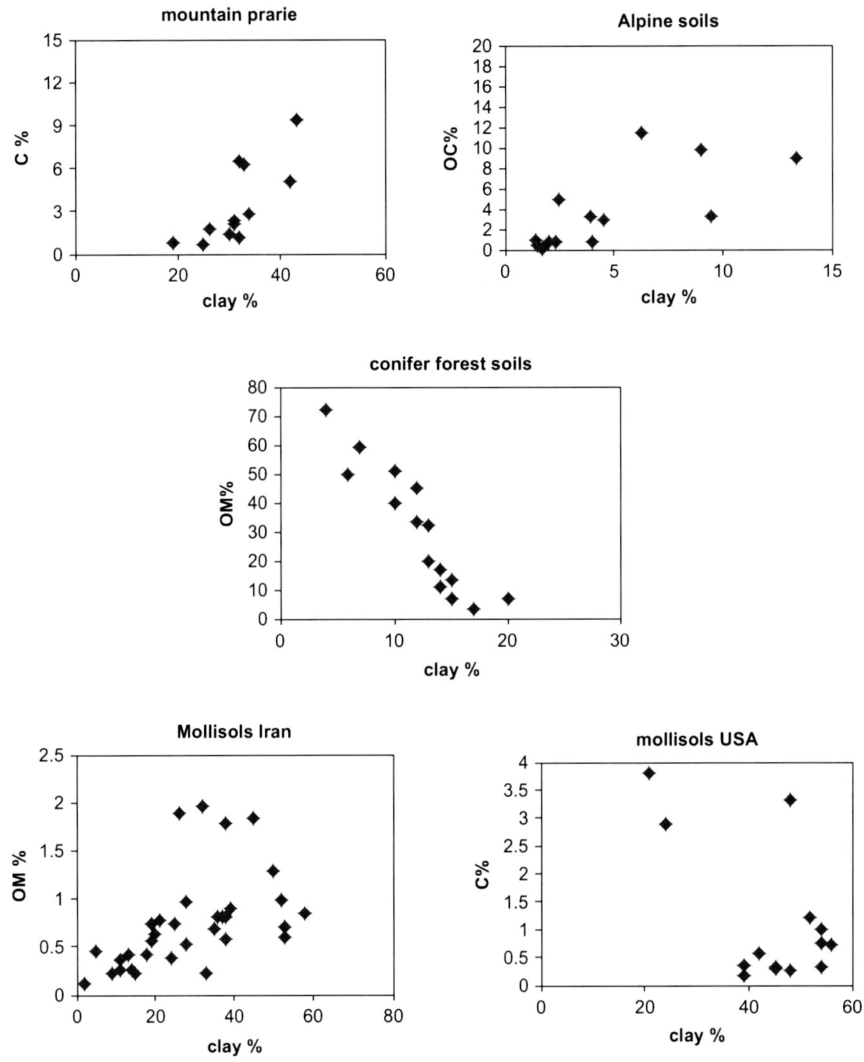

Fig. 2.12 Plots of clay % and organic material (*OM* organic material, *OC* organic carbon or *C* carbon %) from various sources (Mariotti 1982; Righi et al. 1999) concerning mountain prairie soils: conifer forest soils (Collignon 2011) and semiarid prairie soils (Graham and Southward 1983; Mahjoory 1975)

dominates in the cation exchange whereas its impact decreases as the soils become older, Alpine prairie compared to Mollisols.

If the above observations can be generalized, it would appear that the organic matter in young and acidic soils has a significant role in attracting cations to its

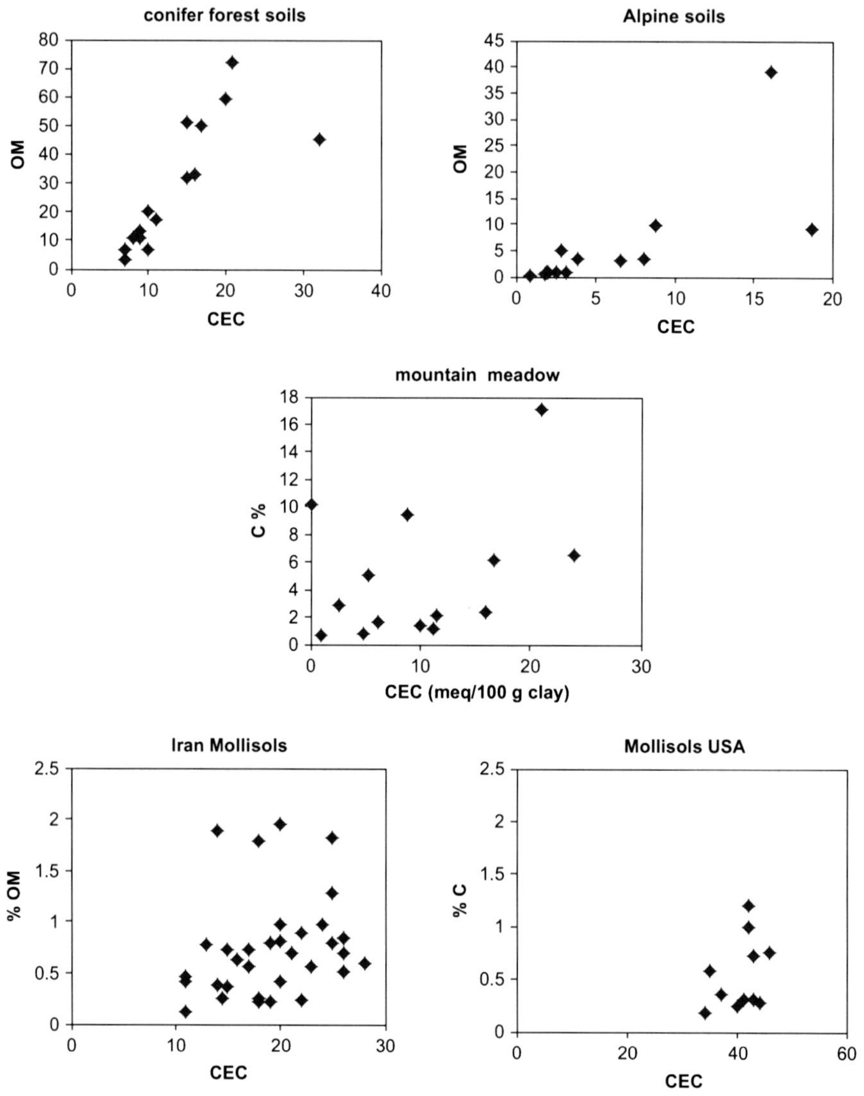

Fig. 2.13 Relations of organic matter and organic carbon and cation exchange capacity (CEC milli equivalents charge/100 g) for mountain Alpine soils (Righi et al. 1999; Mariotti 1982), conifer forest soils (Collignon 2011), and semiarid prairie soils (Graham and Southward 1983; Mahjoory 1975)

surface whereas as the soils become more basic, prairie-type, the clays have a greater impact on cation attraction (Fig. 2.14).

Thus we can summarize by saying that organic matter can in fact be an important contributor to cation retention in soil materials, but this seems to occur in young materials such as forest soils or other sites of high organic content of moderately

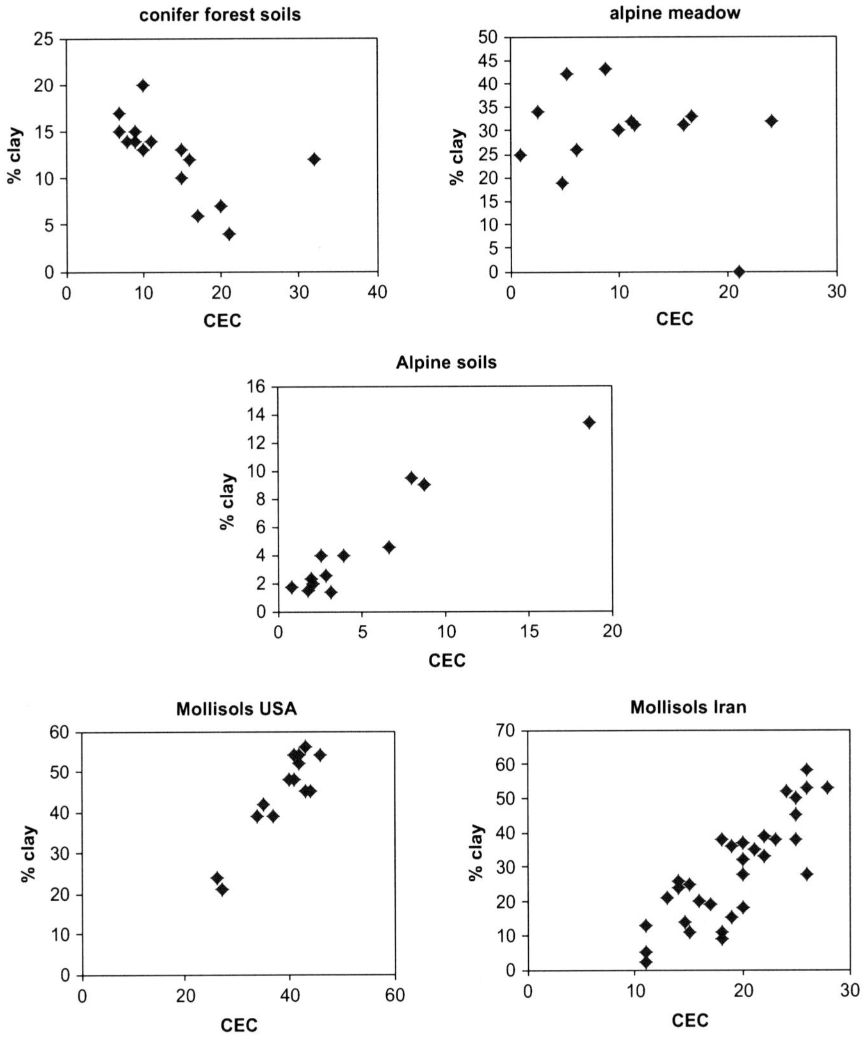

Fig. 2.14 Relations of cation exchange capacity (milli equivalents charge/100 g) and percent clay content of conifer soils (Collignon 2011), mountain Alpine soils (Righi et al. 1999; Mariotti 1982), and semiarid prairie soils (Graham and Southward 1983)

transformed material. In a large range of prairie soils, alpine areas to semiarid steppe prairies, it appears that organic matter is less important than clay mineral content (silicates and oxides) in terms of cation uptake. In all cases a large portion of the exchange sites are pH sensitive and hence fixed by variable charge interactions. The difference between interlayer cation exchange in clay minerals which is not affected by pH and the variable surface charge is very important concerning the movement and site speciation of minor elements in the surface environment.

2.5.7 *Surface Precipitation of New Phases*

Precipitation reactions occur from solutions that are oversaturated with respect to a mineral phase. Solubility constants for precipitation in bulk solution are tabulated in many textbooks. Precipitation of minerals subsequently sequesters many trace metal ions (Ni, Cu, Co, Zn, U, etc.) by adsorption, coprecipitation, or diffusion into the mineral lattice. Many of the hazardous metals migrate to aquifers because of human activity. With seasonal cycles in pH and redox conditions, the (hydr)oxide and carbonate host minerals dissolve, releasing trace metals into aquifer systems and threatening ecosystems. A detailed understanding of the processes of precipitation and dissolution is essential for a quantitative prediction of contaminant fate and transport.

Over the last years field experiments and laboratory studies on clay minerals indicate the importance of the retention of trace elements by phyllosilicates, either by sorption on the surface or by incorporation in a clay structure (Ross 1946; Manceau et al. 2000, 2004; Scheinost et al. 2002; Lee et al. 2004; Isaure et al. 2005; Panfili et al. 2005). Polarized EXAFS has been used to reveal the reaction mechanisms [see Schlegel and Manceau (2006) and references therein]. There is a difference in the reaction kinetics between di- and trioctahedral clay minerals. Zn sorbed at a moderate pH of 6.5 on hectorite led to the formation of mononuclear complexes at the edges of the clay mineral (Schlegel et al. 2001a). Zn phyllosilcate can nucleate from these surface complexes and grow epitaxially in only 5 days at near-neutral pH (7.3) and dissolved silicon concentrations relevant to geochemical systems (Schlegel et al. 2001b). Dähn et al. (2002a, 2003) studied the interaction of Ni with montmorillonite and found that Ni was incorporated into the structure of a neo-formed phyllosillicate after 2 weeks at pH 8 and after 1 year at pH 7.2. In the trioctahedral hectorite all octahedral sites are filled with Mg, so that the divalent Zn can easily be incorporated (Decarreau 1981, 1985). In the dioctahedral smectite only two-thirds of the octahedral sites are occupied. Grauby (1993) and Grauby et al. (1993) found a limited solid solution between Mg and Al phyllosilcates, suggesting that the divalent cations have low affinity for dioctahedral frameworks. This indicates that the difference in reaction kinetics can be related to the differences in crystal structure between di- and trioctahedral clays.

There is also a difference in the reaction kinetics between di- and trioctahedral clay minerals. Zn sorbed at a moderate pH of 6.5 on hectorite leads to the formation of mononuclear complexes at the edges of the clay mineral (Schlegel et al. 2001a). Zn phyllosilcate can nucleate from these surface complexes and grow epitaxially in only 5 days at near-neutral pH (7.3) and dissolved silicon concentrations relevant to geochemical systems (Schlegel et al. 2001b). Epitaxy refers to the method of depositing a monocrystalline film on a monocrystalline substrate. The deposited film is denoted as epitaxial film or epitaxial layer. The term epitaxy comes from the Greek roots epi, meaning "above", and taxis, meaning "in ordered manner." So, "epitaxial" can be translated "to arrange upon."

Dähn et al. (2002a, 2003) reacted Ni with montomorillonite. Ni was incorporated into the structure of a neoformed phyllosilcate after 2 weeks at pH 8 and after 1 year at pH 7.2. This might be explained by the smaller degree of supersaturaion at pH 7.2 compared to pH 8. Differences in the nucleation kinetics, also in favor of the trioctahedral clay hectorite, were observed by Schlegel and Manceau (2006). They assume that this difference may be caused by the structural similarities between the Zn-Kerolite and the trioctahedral clay. In the trioctahedral hectorite all octahedral sites are filled with Mg, so that the divalent Zn can easily be incorporated (Decarreau 1981, 1985). In the dioctahedral smectite only two-thirds of the octahedral sites are occupied. Grauby (1993) and Grauby et al. (1993) found a limited solid solution between Mg and Al phyllosilcates, suggesting that the divalent cations have low affinity for dioctahedral frameworks. This indicates that the difference in reaction kinetics can be related to the crystal structure between di- and trioctahedral clays.

2.6 Summary

Ions in aqueous solution are the key to surface geochemistry. The initial stages of interactions between water and rocks are processes, which dissolve minerals, crystallize new minerals, and leave some elements in solution. These elements will move in or with water at the surface of the earth and this will redistribute elements according to their tendency to form new solid phases or to remain in solution. Ions in solution are cationic or anionic units or oxoanions in which oxygen is combined with a cation to form an anionic unit. Oxoanions are often the result of changes in oxidation states of a given element where the more oxidized state attracts oxygen ions and the ensemble is in electronic disequilibrium with an overall negative charge.

In aqueous solution the water molecules directly attached to the metal ion are said to belong to the first coordination sphere. A second and perhaps third sphere of water molecules is formed around this complex of cation and inner-sphere water molecules. Less is known about anions and oxoanions. The attraction of water molecules to cations is important with respect to the intensity of their attraction to a charged surface such as a silicate or oxide mineral. Such attraction also occurs to organic matter found at the surface of soil zones. The relative intensity of attraction and retention of inner-sphere water molecules determines relative abundances in waters or on solid phases. There is a continuous competition for uptake sites in or on solids between ionic species in solution.

Also there is competition between ions in solution of different sorts for the sites on or in solids via absorption or adsorption (in or on the solid phases). Some ions are attracted more than others and the proportion taken up by a phase depends upon their relative attraction as well as their relative amounts in solution. Conditional selectivity relations vary as a function of solution concentration of dissolved ions, temperature, or the specific type and composition of the sorbing minerals.

On surface sites (variable charge sites) there is competition for adsorption sites from either protons or hydroxyl ions, depending upon the pH of the solutions. Cations are less favored on positively charged sites at low pH and oxoanions are less favored at high pH on negatively charged sites on adsorbers. The pH effect is extremely important in surface environments concerning soils and sediments containing organic materials with strong biologic activities, which can control pH. It is also important to remember that the charge variable surface adsorbers are not only mineral in character but also organic matter can attract and fix ions from aqueous surface solutions.

The mobility of elements is furthermore affected by the oxidation state of the elemental ion concerned. Higher oxidation state leads to more strongly charged ions, which can become bound to oxygen atoms producing a negatively charged ion. The effect of Eh is then potentially of great importance for elements that can have different oxidation states in their ionic forms. Oxidation state is controlled by redox conditions that in general are varied by biologic activity in soils or sediments.

Given the different chemical bonding relations comparing absorbed ions (outer sphere cation–water bonding) which are indirectly attracted to inner structural or surface mineral sites by electrostatic forces between the hydrated cation and the absorber surface and adsorption (inner sphere bonding with water) where more direct chemical bonding between cation and ions in the mineral structure occur, one can expect that the chemical stability of the two types of chemical relations of hydrated cation and solid material will be quite different. Absorbed ions (outer sphere cation–water molecules complexes) will be easily exchanged for other ions in solution. Outer-sphere absorption will lead to mobile cation relations in aqueous solution. On the other hand adsorption, chemical bonding between cation and atoms in the mineral structure of the adsorber, will lead to much more stable relations and the result will be a lower tendency for ion exchange (Maher et al. 2012; Tinnacher et al. 2011). Therefore the tendency to find solid materials with hydrated ions on their surfaces in chemical equilibrium with the ambient solutions will be higher for outer-sphere absorbed ions than for those of inner-sphere complexes fixed on mineral surfaces.

2.6.1 Controlling Factors

Most oxides and hydroxides exhibit amphoteric behavior. The net charge on the molecule is affected by the pH of their surrounding environment and can become more positively or negatively charged due to the gain or loss of protons (H^+).

From the above one can conclude that the fate of dissolved species in solution generated by mineral dissolution and phase changes under surface chemical conditions is of great importance in the movement of elements from solution to solids and the shift of solid composition as solutions migrate at the surface. The problem is one of chemistry, generated by the dissolution of minerals of high temperature origin, and the subsequent variations of these parameters (especially Eh and pH) by microbial activity. Retention and release of elements are of significant environmental importance (to surface geochemists) in that these actions determine where elements will be found at the surface and under which chemical conditions.

Redox conditions are the determining factors of ionic configuration (cation, oxoanion) for ions in solution. The more oxidized an ion, the higher the probability to form an oxoanion and change from a positively charged ion into a negatively charged ion. This causes a reversal of attractions between ions and solids in the alteration and soil environments.

The pH of the altering and soil solutions determines the competition between ions in solution and charged sites on soil materials. The association of ions with solids or their retention in solution is largely a function of solution pH.

Eh and pH are key factors to ion migration and movement in the alteration process and transport processes at the surface of the earth.

2.7 Useful References

Brown G, Calas G (2012) Mineral-aqueous solution interfaces and their impact on the environment. Geochemical Perspectives 1 (4 and 5):742

Dzombak DA, Morel MM (1990) Surface complexation modeling. Hydrous ferric oxide. Wiley, New York

Lützenkirchen J (ed) (2006) Surface complexation modeling. Wiley, New York

Stumm W (1992) Chemistry of the solid–water interface. Wiley, New York, p 427

Tan M (1998) Principles of soil chemistry. Dekker. p 513

Sposito G (1989) The Chemistry of soils. Oxford Univ. Press, New York, p 277

Sposito G (1994) Chemical equilibrium and kinetics in soils. Oxford, New York, p 269

Salomons W, Förstner U, Mader (eds) (1995) Heavy metals: problems and solutions. Springer, Berlin, p 412

Glossary

Actinides Series of chemically similar metallic elements with atomic numbers ranging from 89 (actinium) to 103 (lawrencium). All of these elements are radioactive.

Anion A negatively charged ion (NO_3^-, PO_4^{2-}, SO_4^{2-}, etc.)

Alkalinity The capacity of water for neutralizing an acid solution.

Amphoteric Reacting chemically as either an acid or a base.

Amorphous material Noncrystalline solids.

Bentonite A clay usually formed by the weathering of volcanic ash, and which is largely composed of montmorillonite-type clay minerals. It has great capacity to absorb water and swell accordingly.

Calcareous Refers to materials, particularly soils, containing significant amounts of calcium carbonate. It also describes rocks composed largely of, or cemented by, calcium carbonate.

Cation A positively charged ion (NH_4^+, K^+, Ca^{2+}, Fe^{2+}, etc.) in the soil that is electrically attracted to the negatively charged sites on soil colloids (clay and humus).

Cation exchange capacity (CEC) The capacity of soil to hold nutrients for plant use. Specifically, CEC is the amount of negative charges available on clay and humus to hold positively charged ions. Expressed as centimoles of charge per kilogram of soil (cmolc/kg).

Coordination sphere The central metal ion plus the attached ligands of a coordination compound.

Extended X-ray Absorption Fine Structure (EXAFS) A technique for investigation of the immediate environment of metal atoms in crystals or solutions, e.g., Fe–S bond distances in pyrite. The X-ray energy is varied and the fine structure of the absorption spectrum is recorded indirectly as fluorescent radiation.

Exudates Soluble sugars, amino acids, and other compounds secreted by roots.

Ferric Containing iron in its +3 oxidation state, Fe(III) (also written Fe^{3+}).

Ferrous Containing iron in its +2 oxidation state, Fe(II) (also written Fe^{2+}).

Fungicide A substance or chemical that kills fungi.

Inner-sphere adsorption complex Surface complex in the formation of which an ion or molecule to a solid surface where waters of hydration are distorted and no water molecules remain interposed between the sorbate and sorbent.

Ion Charged entity resulting from the loss or gain of one or more electrons from an atom or molecule.

Heavy metals Metallic elements with high atomic weights, e.g., mercury, chromium, cadmium, arsenic, and lead.

Humus Humus is a complex substance resulting from the breakdown of plant material in a process called humification. This process occurs naturally in a soil. Humus is extremely important to the fertility of soils in both a physical and chemical sense. It is a highly complex substance, the full nature of which is still not fully understood.

Humic substances A series of relatively high-molecular-weight, yellow to black colored organic substances formed by secondary synthesis reactions in soils. Humic substances are products of biochemical decomposition. They are complex substances, which are resistant to further decomposition. Consequently they tend to accumulate in the soil. Most humic substances are dark and are hence responsible for the dark soil color that is commonly associated with soils of high organic matter content.

Hydrogen bond Intermolecular attraction between a hydrogen atom in a polar bond with an unshared electron pair of an electronegative atom in sufficiently close proximity.

Hydration sphere Shell of water molecules surrounding an ion in solution.

Inner-sphere solution complexes These are solution complexes that closely associate with the charged mineral surface (chemisorption), often forming specific bonds with the mineral surface.

Layer A combination of sheets in a 1:1 or 2:1 assemblage.

Metal(oxyhydr)oxide Minerals composed of different structural arrangements of metal cations. In soils principally Al(III), Fe(III), and Mn(IV) are in octahedral coordination with oxygen or hydroxide ions. Metal(oxyhydr)oxide are the by-products of weathering.

Mononuclear The simplest types of coordination compounds are those containing a single metal atom or ion (mononuclear compounds) surrounded by monodentate ligands.

Outer-sphere surface complex Surface complex in the formation of which waters of hydration remain between the sorbate and sorbent.

Oxoanion An oxyanion is an anion containing oxygen. Oxoanions are formed by many of the chemical elements. Nitrate (NO_3^-), Nitrite (NO_2^-), sulfite (SO_3^{2-}), and hypochlorite (ClO^-) are all oxyanions.

Redox reactions Any chemical reaction in which the oxidation numbers (oxidation states) of atoms are changed is an oxidation–reduction reaction. Shorthand for reduction–oxidation. Oxidation (loss of electrons, gain of oxygen) involves an increase in oxidation number, while reduction (gain of electrons, loss of oxygen) involves a decrease in oxidation number.

Particle size The diameter, in millimeters, of suspended sediment or bed material. Particle-size classifications are: Clay (< 0.002 mm); Silt (0.002–0.02 mm); Sand (>0.02 mm).

Phyllosilicates This is the name given to silicate minerals having a layer type of atom arrangement. The term derives from the Greek φυλλον (= sheet). The principal phyllosilicates can be classified on the basis of their layer structures and chemical compositions into the following groups: kaolinite-serpentine, pyrophyllite-talc, smectite, vermiculite, mica, brittle mica, and chlorite

Soil The natural dynamic system of unconsolidated mineral and organic material at the earth's surface. It has been developed by physical, chemical, and biological processes including the weathering of rock and the decay of vegetation. Soils are the natural medium for the growth of land plants. Soil comprises organized profiles of layers more or less parallel to the earth's surface and formed by the interaction of parent material, climate, organisms, and topography over generally long periods of time. Soils differ markedly from its parent material in morphology, properties, and characteristics.

Soil Fertility Soil fertility is defined by the Soil Science Society of America as "the status of a soil with respect to the amount and availability to plants of elements necessary for plant growth" (Soil Science Society of America 1973).

Soil Organic Matter Soil organic matter is the fraction of the soil that consists of plant or animal tissue in various stages of breakdown (decomposition).

(Soil) pH The pH of soil indicates the strength of acidity or alkalinity of the soil solution which affects the soil constituents, plant roots, and soil microorganisms.

Soil is neutral when pH is 7, it is acid when pH is <7, and it is alkaline when >7. The pH scale is logarithmic, so a difference of a unit is a tenfold difference in acidity or alkalinity (e.g., pH 5 is ten times more acid than pH 6).

Sorption General term for the retention of a solute in contact with a solue without implication to a retention mechanism. This term includes adsorption, absorption, precipitation, and surface precipitation.

Solute A dissolved substance.

Surface precipitation Three-dimensional growth of a species on a surface. This mechanism differs from adsorption in that the retained species directly interact with each other on the surface and can even have the solid structure grow away from the original substrate.

Topsoil Topsoil is the surface layer of soil containing partly decomposed organic debris, and which is usually high in nutrients, containing many seeds, and rich in fungal mycorrhizae. Topsoil is usually of dark color due to the "organic matter" present.

Toxicity Refers to a harmful effect on a plant (or animal) from the alteration of an environmental factor.

Transition elements A (loosely defined) group of 38 elements with specific chemical properties. Examples of transition metals include Iron (Fe), Zinc (Zn), Nickel (Ni), Copper (Cu), Silver (Ag), Manganese (Mn), etc. The name transition comes from their position in the periodic table (groups 3–12). These elements are very hard with high melting points and high electrical conductivity and characterized in most cases by variable oxidation states and magnetic properties

Weathering The breakdown of rocks and minerals at the Earth's surface by the action of physical and chemical processes generated by their contact with water.

Chapter 3
Weathering: The Initial Transition to Surface Materials and the Beginning of Surface Geochemistry

Geology has had a hard time in establishing itself in the modern world (i.e., post-classic age of the ancient Greeks). With the advent of Christianity came the spread in Europe of ideas and dictums of the Middle Eastern theology based upon late Bronze Age ideas and dogma. The dogma of the Christian Old Testament holds that the earth was created at one time and only one event after its creation (the great flood) altered its surface and aspect. The ideas of erosion of rocks and deposition of altered materials and others were essentially outlawed by the ecclesiastic followers of these texts. The surface of the earth was immutable.

Then came the Renaissance and the revival of older understanding of natural phenomena. It was apparent to anyone who wished to look at the events that volcanic activity would change the aspect and masses of material at the surface of the earth without the help of the deluge. Scientists of the Italian awakening and their disciples determined that layers of material contained old forms of life that no longer existed, but they were found to a succession and hence a time series. The science of sedimentology was in the making. Layers of sediments soon to become rocks were piled one upon the other in a temporal sequence and pushed up to make mountains. All well enough, but what happens after the erection of mountains? They cannot continue to accumulate forever. Renaissance scholars came upon the ideas of the dissolution of rocks and their eventual precipitation by looking at caves and caverns in carbonate materials where features of dissolution and precipitation could be clearly seen. The understanding of dissolution and precipitation mechanism fascinated them to say the least. This led to the vogue of representing Gods and mythical creatures inside of caverns with stalactites and stalagmites present showing the process of destruction and re-generation. The gardens of the Piti palace of Florence show this fascination. The representation of caverns with depositional features became a standard decorative theme during this period. The ideas of dissolution and precipitation of dissolved matter are outlined by Bernard Palissy in his text destined to enlighten the sixteenth century royal court of France where he gave courses in natural science (Pailssy 1563). Thus the basic concepts of surface geochemistry were at the basis of the new sciences explored by the modern world.

A. Bauer and B.D. Velde, *Geochemistry at the Earth's Surface*,
DOI 10.1007/978-3-642-31359-2_3, © Springer-Verlag Berlin Heidelberg 2014

The process of dissolution and eventual deposition of dissolved materials is the foundation of surface geochemical thought.

The primary configuration of the materials at the surface of the earth is one of contact of solids with water and air. In the air one finds oxygen and carbon dioxide, which are dissolved in small quantities in the water present. The incorporation of these gases and their interaction with water molecules develops aqueous solutions that have active oxygen present and a slightly acidic solution due to the dissolved carbon dioxide which produces the acidic state. The importance of a slightly acidic solution is the activity of hydrogen ions, H^+. Water is the medium in which solid–solid transformations take place, either by dissolution and recrystallization or by exchange of ions. The transformation process needs water to be accomplished at a reasonable rate. The greater the amount of water present the greater the reaction rate in that the solutions are further from equilibrium with the solids and movement of ions and electrons is accelerated by the lack of equilibrium or the importance of the reaction potential. Most stable surface minerals contain water or hydrogen ions. This is not usually the case for high temperature minerals. Minerals from low grade metamorphic rocks or diagenetically altered sediments do usually contain hydrogen ions or water molecules but not as much as those stable in the alteration zone. The oxygen in the water tends to combine with metal ions forming ions of higher oxidation state. Iron is a notable example. When iron is oxidized, it usually leaves its former mineral structure to form an oxide or more likely a hydroxide. The result of surface alteration and re-equilibration with surface chemical conditions is in general to segregate mineral elements into hydrous silicates, called clay minerals, to form transition metal (Fe and Mn) oxyhydroxides, and to release a significant amount of ions as hydrated species in the aqueous altering solutions.

In studies of weathering, recent interest has been largely centered on the rate at which materials are affected and as a measure of the amount dissolved into the ambient aqueous solutions [Brantley et al. (2008) for instance]. The assumptions are that alteration is a function of water–mineral dissolution dynamics, where minerals are essentially free entities in contact with water. However the state of natural rocks is more complicated than this as one might suspect. Take for example basalt weathering. It is known that the smooth surface of a lava flow resists alteration, more than adjacent more crystalline rocks (Rasmussen et al. 2010) even though theoretically the basalt should alter faster in that the high glass content would be highly unstable under surface conditions. In fact macrocrystalline rocks such as granites alter more rapidly. One reason is most likely the structure of the rock material that develops under thermal stress. Large crystal grains by their anisotropic thermal expansion properties tend to dislocate at crystal edges from adjacent minerals. In doing so they create a micro-passage which allows water to infiltrate into the rock surface. As the water remains in contact with the silicates for a longer period of time than simple runoff fluids, the interaction of silicate and un-saturated aqueous solution can effect mineral dissolution (see Velde and Meunier (2008), Chap. 2). The key to understanding the initial and microscopic mechanisms of rock alteration is the residence time which the water remains in contact with the rock (Sect. 1.1). The glassy basalt flow surface is more isotropic in its thermal expansion properties and will have a much lower tendency to crack under the stress of daily temperature change. In the same way it has been noted that carbonates are less affected by surface dissolution than silicates in the Seine (France) river basin (Roy et al. 1999) largely because the carbonates are fine grained and hence the differential expansion of the grains will be smaller on a local scale and effect less dislocation at grain surfaces. Despite the inherent solubility of carbonates in slightly acidic atmospheric aqueous solutions, one often sees salient geomorphic features of carbonate rocks in mountain landscapes whereas silicate rocks are more eroded. These facts are difficult to take into account in using mathematical models for the dissolution rates of rocks.

Probably the most reliable measure of surface water–rock interaction is the formation of alterites, i.e., the distance in a profile down to the rock–alterite interface or the rock—C horizon level. Here one finds the total effect down to the initial stages of alteration. However at this interface one encounters a portion of alterite, but most of the material is still in the form of broken rock fragments. Hence the overall alteration process, i.e., total transformation can only be estimated by the material evacuated from the alteration zone by moving water. This integrates the water–rock interface, the continued rock–water interaction in the alterite zone, and the interaction under the influence of the soil plant zone where the chemistry is usually significantly different from the alterite zone. These differences are outlined in White et al. (1995, 2008).

A second aspect emphasized by Jenny (1994) is that climate is a major factor in alteration rate. This incorporates rainfall and temperature which determines the biological interaction factor. Rainfall is important but contact time is very important also. If it rains once a month or five times giving the same amount of precipitation, the effect will not be the same. Rainy climates do not necessarily have a high total rainfall. Snow on a mountain top during the winter does not have the same effect as rainfall down in the plains. These are obvious observations but they are at times overlooked.

3.1 Alteration Processes: Oxidation, Hydration, and Dissolution

Weathering is the breakdown and alteration of rocks and minerals at or near the Earth's surface into products that are more in equilibrium with the conditions found in this environment. The interaction of air and water with rocks is called alteration which is a term used to designate the approach to chemical equilibrium attained at the surface of the earth. The causes of disequilibrium are geologic. The basic action of geology is to change the geographic place of materials at the surface. The major feature noted is mountain building. Here rocks, masses of chemical components forming coherent solids materials, are brought up above sea level to form mountains and other surface features. The material moved to the surface was in a state of chemical and physical equilibrium at conditions of significantly higher temperatures and pressures than those found at the rock–water–air interface. The re-equilibration occurs in several different steps with the formation of intermediate phases. Two mechanisms of chemical adjustment are of highest importance: oxidation and hydration of the high temperature rock minerals. The high temperature minerals are most often less hydrated than surface phases and they contain less water or hydrogen ions than those stable at the surface. The metal elements, those of variable oxidation state, are in a more or less reduced state. Iron, for example, is in the metallic state in the core of the earth, in the divalent state of oxidation in most minerals in the crust of the earth, and it becomes trivalent, as the rock minerals which contain it are found at the surface. For the most part iron is in the trivalent state in surface minerals, those not affected by organic action. Hydration of high temperature minerals means that there is an introduction of hydrogen ions into the structure of surface phases, and occasionally one finds water molecules present. The change in oxidation state of metallic ions and the introduction of hydrogen into new minerals are the key factors to understanding mineral change. A third modification is equally important and this is the dissolution of minerals into the aqueous phase and their transport to the sea. This interaction is one of incongruent mineral change in an aqueous environment.

Thus the three means of attaining chemical–mineral equilibrium in an aqueous environment in contact with the air are oxidation, hydration, and dissolution.

3.1.1 Air and Water: Interaction of the Atmosphere and Aqueous Solutions

The major difference between aqueous solutions found within rocks and water at the surface is the influence of the atmosphere. In surface waters there is an effect of gaseous dissolution of oxygen and carbon dioxide. Oxygen dissolves in small quantities but is present as an agent of oxidation when water and rock interact. Carbon dioxide enters the aqueous media to create an anion and cation

$$CO_2(aq) + H_2O(l) \leftrightarrow H_2CO_3(aq)$$

$$H_2CO_3 \leftrightarrow H^+ + HCO_3^-$$

The presence of the hydrogen cation is one of the major factors in surface geochemical change. This ion is extremely active due to its small size; it can diffuse easily into chemical structures composed of larger ions, such as silicates. We will consider these two major factors in surface geochemistry.

3.1.2 Oxidation

The process of addition and combination of oxygen to minerals is relatively easy to understand, taking iron as an example. In most high temperature minerals, iron is in the divalent state. In the water solutions oxygen is in the elemental state, O_2. Oxidation means basically that oxygen is combined with the iron within the mineral structure or in the formation of a new phase. We can outline such reactions considering the high temperature silicate mineral iron olivine. Fe_2SiO_4, where iron is covalently linked to silicon and oxygen. Reaction with oxygen gives, schematically

$$\underset{\text{Fayalite}}{Fe_2SiO_4} + 1/2O_2 + 2H_2O \rightarrow \underset{\text{Hematite}}{Fe_2O_3} + \underset{\text{(in solution)}}{H_4SiO_4}$$

The Fe_2O_3 produced by these reactions is relatively insoluble, and precipitates to form hematite. The iron, which has released an electron to become oxidized, transfers the electrons to oxygen to reduce it to a higher electronic density ion. The oxidation causes a change in ionic radii which facilitates bond breakage (Fe^{2+} 77 pm, Fe^{3+} 63 pm, for example). As iron ions and oxygen ions are unstable together, they combine to form a new compound where the electrons form a balanced compound. The compound is iron oxide, a mineral of low solubility in aqueous solution and hence one finds it as a new mineral, hematite. However the silica in the mineral is relatively soluble in aqueous solution especially in the noncrystalline form, and it tends to combine with water to form silica ions.

Such a reaction is relatively common in the sense that a portion of the high temperature rock minerals are transformed into new solid mineral phases and another part of the mineral is dissolved into the aqueous solution. Similarly, a clinopyroxene upon oxidation gives Fe oxides, and Ca is released in solution according to the reaction:

$$\underset{\text{Clinopyroxene}}{2CaFeSi_2O_6} + 1/2O_2 + 10H_2O + 4CO_2$$

$$\rightarrow \underset{\text{Hematite}}{Fe_2O_3} + 4H_4SiO_4 + 2Ca^{+2} + 4HCO_3^-$$

3.1.3 Hydrolysis

Another chemical action is that of hydration, where water is added to the crystal structure of a mineral, usually creating a new mineral, often called a hydrate. Hydrogen ions from the slightly acidic aqueous rain water solution often exchange for cations in the high temperature minerals. A clear example of the interaction of slightly acidic solutions can be seen in the analyses of the composition of a thirteenth century stained glass window from Angers France (data of the authors). In Fig. 1.19 analysis points from the altered surface to the unchanged glass composition are plotted showing the change of relative elemental density.

One can compare the relative losses of the more soluble or more ionically bonded ions which indicate a succession based upon ionic charge and ionic charge density. The monovalent ions are displaced by hydrogen cations most easily followed by divalent ions. The smaller ions are more easily removed by diffusion, Na compared to K and Mg compared to Ca.

One can consider the alteration of feldspar. Here ionic exchange with elements in the solid phase replaces mineral elements with hydrogen forming what are called hydrous phases, usually classified as clay minerals. For example, one can consider the case of Plagioclase interacting with water to form the clay mineral Kaolinite

$$\underset{\text{K-feldspar}}{2\,KAlSi_3O_8} + H_2O + 2\,H^+ \rightarrow \underset{\text{kaolinite}}{Al_2Si_2O_5(OH)_4} + 2\,K^+ + \underset{\text{aqueous silica}}{4\,SiO_2}$$

Hydrogen cations are consumed by the reaction and note that both the K^+ ions and the SiO_2 are soluble and can thus be carried in the aqueous solution, leaving behind the clay mineral kaolinite ($Al_2Si_2O_5(OH)_4$).

3.1.4 Hydration

Hydration is the absorption of water into the mineral structure. A good example of hydration is the absorption of water by anhydrite, resulting in the formation of gypsum

$$CaSO_4 + 2\,H_2O \leftrightarrow CaSO_4{\cdot}2H_2O$$

Hydration expands volume and also results in rock deformation. Hydration mainly affects the surfaces of the rock rather than changing the mineral structure throughout. Chemical combination of water molecules with a mineral leads to a change in structure. Hydration is the process of combination of H_2O with a mineral that is either anhydrous or less hydrous than the final product. The more hydrated a mineral is, the easier it dissolves in water, which in turn facilitates weathering of rocks.

Note the differences between congruent and incongruent weathering processes. Whereas the dissolution is mostly congruent, the hydrolysis and oxidation examples represent an incongruent process.

3.1.5 Biological Weathering

This form of weathering is caused by the activities of living organisms—for example, the growth of roots or the burrowing of animals. Biological weathering involves the disintegration of rocks and minerals due to the chemical and/or physical agents of an organism. Roots of threes are probably the most significant agents of biological weathering, as they are capable of breaking apart rocks by growing into cracks and joints. Plants also give off organic acids that help to break down rocks chemically. The types of organisms that can cause biological weathering range from bacteria to plants to humans.

Important processes with physical weathering character are for example the simple fracture of particles because of animal burrowing or by the pressure of growing roots. Another possibility is the breaking of particles, by the consumption of soils particles by animals. Biological activity can cause the movement and mixing of materials. This movement can introduce the materials to different weathering processes found at distinct locations in the soil profile.

Chemical processes like dissolution can be enhanced by the carbon dioxide produced by respiration. Respiration from plant roots releases carbon dioxide. If the carbon dioxide mixes with water carbonic acid is formed which lowers soil pH. Cation exchange reactions by which plants absorb nutrients from the soil can also cause pH changes. The absorption processes often involves the exchange of basic cations for hydrogen ions. Another important process in the dissolution of minerals in soils is chelation. Organic substances, known as chelates, produced by organisms have a catalytic effect on mineral dissolution. Organisms can influence the moisture regime in soils and therefore enhance weathering. Shade from aerial leaves and stems, the presence of roots masses, and humus all act to increase the availability of water in the soil profile.

3.1.6 Rocks and Alterite Compositions

What then are the results of alteration of rocks as seen by the chemistry of the materials left behind as solids which are called alterites?

The process of attaining chemical and phase equilibration of rock (high temperature minerals) and surface chemical constraints occurs at the rock–water interface. Rain water is oxygenated and slightly acidified by carbon dioxide which is dissolved in it. This rainwater is the agent and the medium of transformation of rock to alterite (equilibrium mineral assemblage at the surface). If a rock component is totally unstable, it will dissolve initially in the aqueous solution. This means that the movement of the aqueous solution causes the rock to be transported elsewhere as dissolved material. This erosion has been the subject of many studies in the past (White et al. 2008; Degens 1965; Dethier 1986, for example). Different methods have been devised to measure the rate of dissolution and displacement of dissolved rock material from mountain to sea levels. These studies integrate the actions of the alteration of rock and alterite material by contact with infiltrating rainwater throughout the alteration zone. The rock–water contact with the alteration of rock and altered materials (alterite material) reflect also the interactions in the soil zone where plants determine the solution chemistry. The methods used are measurement of stream water chemistry and a consequent deduction of the loss of material in solution compared to the bedrock of the area of drainage. This is of course a very good measure of the overall result of weathering by dissolution; but it integrates several stages and contexts of chemical interaction between water and rock material, which ignores these new phases produced and the action of biological agents. In such considerations it is interesting to look at the materials that alter at the first stages of alteration, the saprock zone of an alteration profile and compare them to the rock composition. In this instance the influence of relative mineral equilibria is minimal in that the greatest part of rainwater which reaches the rock–water interface has had a rapid movement through the large passageways of the profile to reach the rock with a minimal amount of dissolved material present. The water is highly un-saturated with rock elements. This aqueous solution has a maximum capacity to dissolve the minerals in a rock without coming to equilibrium with new mineral phases stable at the surface.

Another approach used to assess alteration is the comparative analysis of the alterite material in order to assess the stage of extent of alteration. This is called the Chemical Index of Alteration [see Scott and Pain (2008) for example] which assumes that feldspar alteration is the key to understanding the change in mineral alteration. Calcium feldspars often produce kaolinite directly upon alteration (see Velde and Meunier 2008), but other more silica-rich phases are also produced which are not directly related to feldspar transformation. Uses and drawbacks concerning this method of assessing alteration are given by Scott and Pain (2008).

Our initial approach is to assess the changes in chemical components, their relative proportions, in the rock and saprock of several alteration profiles in order to assess the earliest stages of water–rock interaction, where the difference in relative mineral stability between high temperature rock minerals and low temperature aqueous solutions is the greatest. The elemental distributions between rock and saprock are considered in two categories, major elements and minor elements, i.e. those forming the major phases present and those accommodated in these phases or on them.

3.2 Weathering (Water–Rock Interaction)

3.2.1 Initial Stages of Weathering: Major Elements

Geologists and geochemists use a terminology which concerns the relative abundance of different elements in a given rock sample. In a very general way this classification identifies the elements, which compose the largest part of the mineral phases present. They would be the elements which, by their relative abundance, will provoke the presence of the different specific minerals found in a rock. This is essentially a concept based upon rocks which can readjust the phases present to the conditions of temperature and pressure to which they are subjected. New conditions produce new mineral phases, which attempt to come to physical and chemical equilibrium by shifting elements from one phase to another and creating new ones. Major elements are those whose abundance is important in the formation of the minerals present. Minor elements are those which are included in major element phase in small but variable quantities, and trace elements are present in very small quantities in different phases. In general a major element is one whose presence in oxide weight percent (wt%) is above 0.5 wt% while a minor element will be present in quantities of 0.5–0.1 wt%. Trace elements are present in lower quantities, usually in the range of hundreds to tens of parts per million (ppm). However it is obvious that such definitions are dependent upon the sample observed. In igneous rocks, it is estimated that phosphorous is present in the range of 0.3 wt% oxide, or in shales 0.17 wt% (Mason 1958, p. 151), but in sedimentary phosphate deposits, it will be present as the major oxide along with calcium. Thus major and minor elements are generally expressive of the sample at hand. However, trace element usually describes elements in concentrations of <0.1 wt%.

Thus major element usually signifies that it is present in quantities high enough, when combined with others, to form a specific phase or several in the sample. Minor element usually means that an element is incorporated in small amounts into a solid phase where abundant elements provoke presence of the solid material. Minor elements are usually present in fractions of a percent of the solids. The term trace element usually means abundances in very small quantities [parts per million (ppm) or parts per trillion (ppt)] where the element is in a phase in small quantities or perhaps just adsorbed onto the surface of a mineral and is expressed as atom weight percent.

We will consider the elements designated as being of major abundance in silicate rocks to be Na, Mg, al, Si, K, Ca and Fe, Mn most of them form silicate minerals retaining their relative abundance in the earliest stages of weathering. These elements are not responsible for the formation of all of the minerals present but for about 99 % of them. The rocks considered by geologists to be of magmatic origin represent materials with highly unstable minerals formed at high temperatures. The comparisons made are between the initial rock and the rock in its earliest stages of weathering, and therefore the formation of new minerals stable at the surface derived from highly unstable phases crystallized from molten materials. Table 3.1 gives the initial data for 15 examples of the early stages of rock weathering. An investigation of relative change in elemental content, compared to the total weight percent of the elements present, is given. This expression is the weight percent of the element investigated in the early alterite compared to that in the rock. In this manner, a value of more than one indicates a relative gain in the element upon initial alteration, while a value below one indicates a relative loss of the element.

A first comparison can be made for the two major silicate mineral forming cation elements Al and Si which are the most covalently bonded to oxygen in an oxyanion structure and hence considered to be the most chemically stable elements forming the dominant phases at the surface. Figure 3.1a shows that the general trend is a loss of silica and a gain of alumina. This is to be expected according to the observation of pedologists such as Pedro (1966) who indicate that the major overall trend of alteration is silica loss and accumulation of alumina. This is evident in the extreme

Table 3.1 Element ratio data: relative concentrations of elements in bedrock and alterite zones (C horizon) for various rock types

Alterite/rock element abundance										
Source	Rock type	Climate	Na	Mg	Al	Si	Fe	K	Ca	Mn
Fontanaud	Ultrabasic	Temperate	0.5	0.74	0.96	1.4	1.19	1	0.5	2
Iledefonse	Gabbro	Temperate	1.15	0.87	1.09	1.01	0.95	0.67	1.01	0.5
Proust	Amphibolite	Temperate	0.59	1.25	0.87	1.01	1.34	1.21	0.56	1
Proust	Glaucophane schist	Temperate	1.33	1.23	1.08	0.93	1.13	0.67	0.72	0.5
Meunier	Pagerie granite	Temperate	0.5	1.3	1.09	0.96	1.19	0.91	0.61	
Meunier	Rayerie granite	Temperate	0.7	1.3	1.2	0.92	1.7	1.4	0.63	
Loughnan	Basalt	Tropical	0.46	0.86	1.04	0.95	1.06	0.6	0.91	
Oh and Richter	Granite	Tropical-temperate	0.29	0.83	1.69	0.9	2.18	1.17	0.05	1.5
Oh and Richter	Diabase	Tropical-temperate	0.7	0.79	1.16	1.06	1	0.67	0.64	1.17
Oh and Richter	Schist	Tropical-temperate	0.88	1.07	1.02	0.98	1.08	1.01	1.06	1.07
Valeton	Basalt	Tropical	0.03	0.07	2.04	0.56	2.22	0.04	0.01	1.3
Leumbe	Trachyte	Tropical	0.78	0.6	1.51	0.76	1.5	0.43	0.5	0.78
Egli	Granite	Alpine	0.98	0.91	0.98	0.97	0.91	0.98	1.02	1
Chesworth	Basalt	Temperate	0.58	0.92	1.12	0.95	1.06	0.49	1.08	1.04
Navarette	Andesite	Tropical	0.33	0.6	1.34	0.9	1.16	0.55	0.34	1

Most examples are from temperate climate regions. Data from Fontenaud (1982), Ildefonse (1978), Proust (1976, 1985), Meunier (1980), Loughnan (1969), Oh and Richter (2005), Valeton (1972), Leumbe et al. (2005), Egli et al. (2001), Chesworth et al. (1981), Navarrete et al. (2008)

cases of alteration, tropical soils, for example, which form oxyhydrated minerals of aluminum. It is also evident from the data set that this trend exists for materials in the initial stages of alteration in various climates and for various rock types. Thus the trend of silica loss and alumina gain is a fundamental part of alteration geochemistry at the rock–water interface.

The complete hydrolysis of the silicates causes oxides and hydroxides to form. The geochemistry of iron at the surface is determined by a change in oxidation state (oxidation) and hydration. Hydration is also the rule for aluminum dissolved in solutions. In these cases the iron leaves the initial silicate minerals of the rock and when oxidized by interaction with surface chemistry forces forms a new oxyhydroxy phase. Iron essentially leaves the silicate and sulphide minerals in most instances. This is seen in the red–orange color common in most altered rocks. In the initial stages of alteration alumina leaves the anhydrous minerals, forming hydrous ones usually considered with silica (see Velde and Meunier 2008). If one considers the relations of iron and alumina in the early stages of alteration, Fig 3.1b, it is clear that the increase in alumina observed in the relations with silica is accompanied by an increase iron. This fundamental trend effects an increase of trivalent elements in the altered material in the early stages of weathering, Al and Fe, which is quite important.

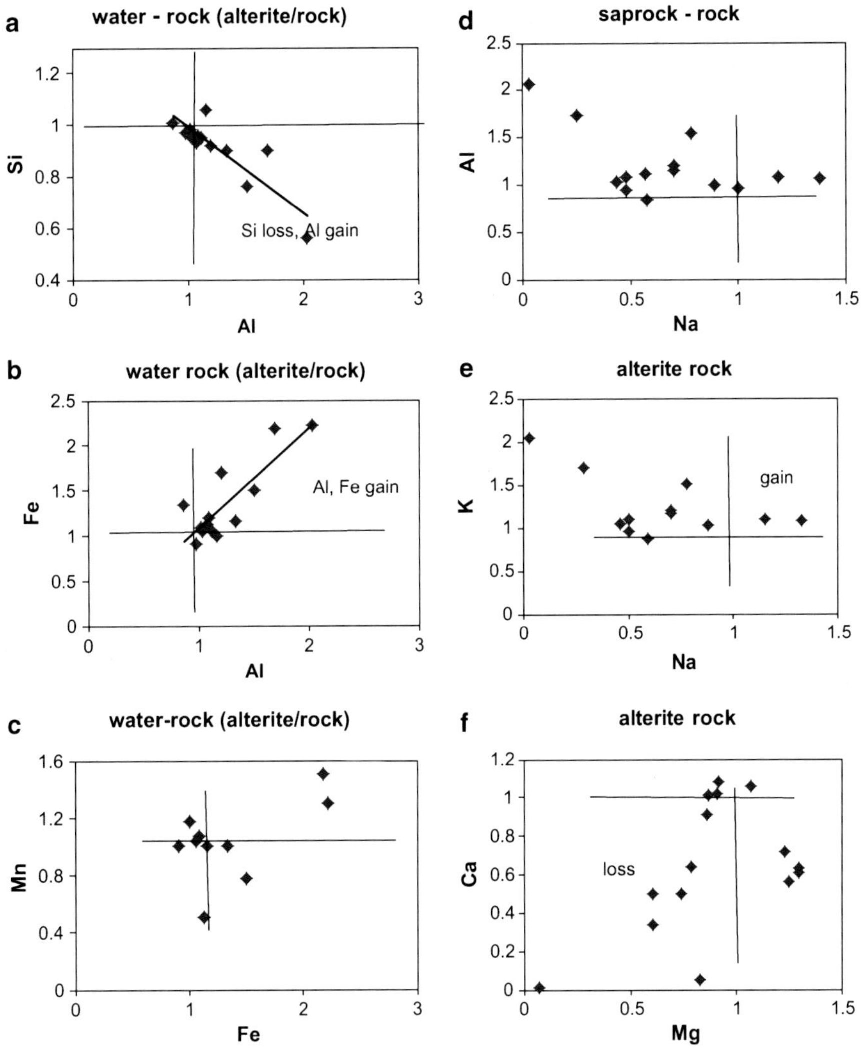

Fig. 3.1 Representations of relative changes in elemental concentration between bedrock and alterite zones (alterite/rock ratios). Values greater than one show a relative increase in the element during the initial phases of alteration or transformation of rock to materials stable at the surface (**a**) Si/Al, (**b**) Fe/Al, (**c**) Mn/Fe, (**d**) Al/Na, (**e**) K/Na, and (**f**) Ca/Mg

The major elements most susceptible to oxidation state change in rock-surface interaction (Mn and Fe) show a general increase in their relative content although manganese is not clearly affected in the same way as iron. In some instances iron content increases while that of manganese decreases compared to the source rock. Hence they seem to be decoupled at times in their chemical behavior. Thus iron and manganese oxides and oxyhydroxides are not necessarily associated in the alteration process (Fig. 3.1c).

In general it appears that as alteration progresses with the accumulation of more alumina there is a concomitant loss of sodium. This is not unexpected in that the sodium is not favored in clay mineral substitutions, whereas potassium and calcium are. There is a significant loss of sodium without accumulation of alumina, indicating that sodium is lost before any significant changes in relative alumina content occur (Fig. 3.1d). However some cases show relative increase in sodium. Sodium is in general not favored at surface interfaces and thus it is largely lost to altering fluids, and it then finds its way into the ocean, which is reflected in the fact that the sea is relatively heavily concentrated in sodium ions. The alkali elements Na and K show a loss of Na and a relative increase in K (Fig. 3.1e). As potassium is seen to be increased in many alterites, it is logical to note that the sea has a relatively low potassium content but a high sodium content.

The other alkali and alkaline earth elements K, Ca, and Mg show enrichment or loss in different cases of alteration. This seems to vary with rock type and alteration intensity (climate). Magnesium and calcium show similar trends of loss but magnesium seems to be relatively retained more than calcium (Fig. 3.1f). Minerals such as calcite or gypsum are typical for soils in climates where evaporation is more important than precipitation.

Hence the strong trend of silica loss and alumina plus iron gain seems to be the most clear trend in the initial stages of weathering for the rock types considered coming from different climates of alteration. The major trends found in all samples reflect the instability of high temperature silicate minerals which are transformed into low temperature silicate minerals and oxides This engenders an increase in alumina, the second most abundant cation in the silicate minerals. A second effect is the oxidation of iron, from the divalent to the trivalent state. In this instance iron leaves the silicates it was found in under the high temperature regime and forms an new, independent phases, as oxides or hydroxides.

3.2.2 Silicate Mineral Transformations: The Origins of Alteration

Silicate minerals common in high temperature rocks such as basalts and granites can be used to explain the alterite geochemical trends for major elements. Several common minerals can be used as examples: alkali feldspars for granites and olivine, pyroxenes for basalts, and plagioclase feldspars for both rock types. The transformations are schematically given as follows:

Alkali feldspar	
2(K, Na)$AlSi_3O_8$ to kaolinite $Al_2Si_2O_{10}(OH)_4$	Loss of SiO_2 and Na, K
2(K, Na)$AlSi_3O_8$ to smectite $K_{0.3}Al_{2.3}Si_{3.7}O_{10}(OH)2nH_2O$	Loss of SiO_2 and Na, K
Olivine	
2(Mg, Fe)$_2SiO_4$ to oxide Fe_2O_3	Loss of SiO_2 and MgO
Diopside	
2.7 $CaMgSi_2O_6$ to smectite $Ca_{0.3}Mg_{2.7}Si_4O_{10}(OH)_2$	Loss of SiO_2 and Ca
Orthopyroxene	
6.6 $MgFeSiO_6$ to smectite $Mg_{0.3}$ $Mg_3Si_4O_{10}(OH)_2$ + oxide 1.6 Fe_2O_3	Loss of SiO_2
Minerals in both rock types	
Plagioclase	
$CaAl_2Si_2O_8$ to kaolinite $Al_2Si_2O_{10}(OH)_4$	No change in silica, loss Ca

In the above very schematic description of mineral alteration, to new surface minerals, the alkali feldspars common to granitic rocks release very much silica; the olivine and pyroxenes (enstatite and diopside) common to basalts release silica when iron is oxidized to form a non silicate mineral. Plagioclase alters commonly to kaolinite, a low silica clay mineral, and maintains its silica content in forming the new phase. In all cases where alkali or alkaline earth elements are present, these are largely lost also. These reactions can be used to explain the change in rock composition as it changes to an alterite.

3.2.3 Rock Alteration: Gain and Loss of Major Elements

The initial alteration is almost on a mineral grain by grain scale within the rock itself. When a sufficient amount of grains are altered, the rock loses its structure and mechanical competence to become an alterite. Several terms are applied in pedology (saprock, saprolite) defining more precisely the stage of alteration. However the processes of alteration are the same, mineral grains become new minerals under the influence of chemically unsaturated rain water which infiltrates the alterite material. Given that rocks can be or are frequently grouped according to a general definition of chemical composition, usually based upon overall silica content, one could expect that the chemistry of an altering rock would influence the alteration products by the minerals it produces.

It is well known that acidic rocks are silica and alumina-rich and contain less calcium and more potassium than basalts, whose composition is more basic, i.e., contain more iron and magnesium. This being the case one can pose the question of their relative geochemical behavior under conditions of surface alteration. Below we compare the alteration data in the initial stages of alteration and in the alteration profile. Considering basic rocks, volcanic and metamorphic, it appears that these materials lose both potassium and calcium in most of the examples given. Figure 3.2 compares the changes in K and Al content for the two rock types, basic and acid. If one considers K and Al, elements generally identified with granitic, acidic, rocks, there is a tendency for both rock types to concentrate Al but potassium can be either lost or gained in the rock types. Here one sees that the

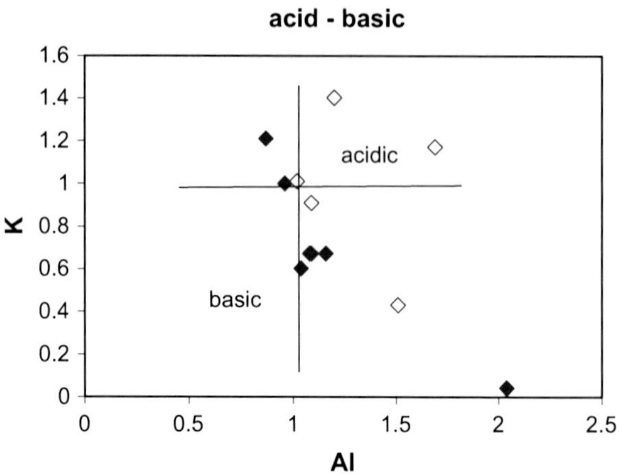

Fig. 3.2 K–Al alteration ratios (rock/alterite concentrations) for the two major compositional types of eruptive rocks

overall chemistry of a rock does not necessarily distinguish it in its alteration chemistry. The variability from one sample in a rock type group (loss or gain in an element) suggests that the system of alteration is complex.

3.2.4 *Rock Types and Element Loss or Gain in the Alterite Material*

The initial stages, where only several percent of the rock is affected by water–rock interaction, show some similarities but also differences in elemental behavior. If one takes a closer look at further steps in the alteration process, the alterite zone, where the altered material, clays, and oxides form several percent to tens of percent of the sample, it is possible to follow the alteration trends by rock type more closely. Again we consider granite and basalt materials as representing the end points of common rock type compositions.

3.2.5 *Granite Alterite*

A comparison is made with the data above, for a variety of rock types under different climates, to more specific cases. Egli et al. (2001) present chemical data for altering granites from an alpine area having several thousands of years exposure to surface chemical conditions. The altered material contains only <5 % new alteration minerals. The granites show the reciprocal relationship of Al and Si as found in the general case, and the correlative increase in Fe and Al found in the general case. However the relations of alkali ion content are better shown in these data for the Alpine granite. Here the increase in aluminum is accompanied by an increase of potassium. Given that there is a positive correlation between potassium

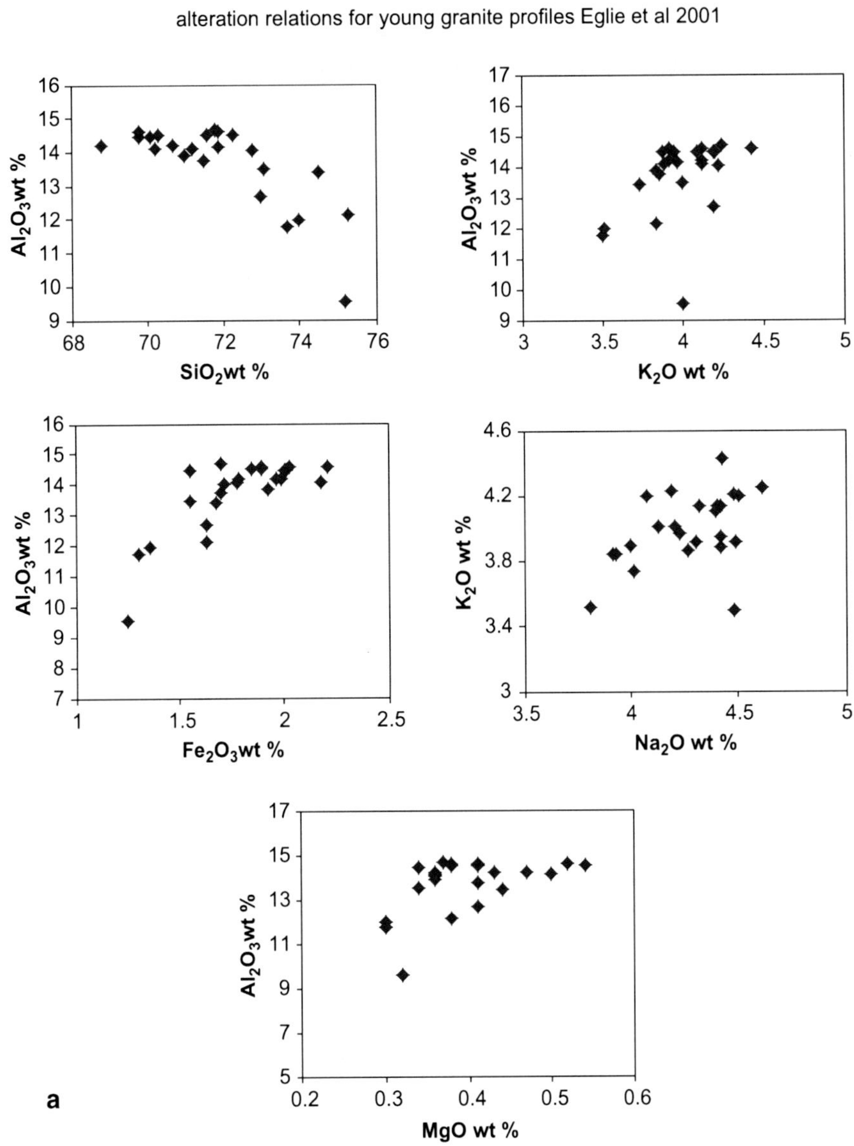

Fig. 3.3 (continued)

and sodium, it appears that in the initial stages of alpine weathering, forming only several percent of new alteration minerals, the alkali ions are not lost to the solution rapidly but kept in the newly formed phases. Relations between Mg and Al indicate the same process. The case of calcium is less well defined. Hence there seems to be an initial capture of alkali ions and alkaline earth ions in the case of granite weathering (Fig. 3.3a).

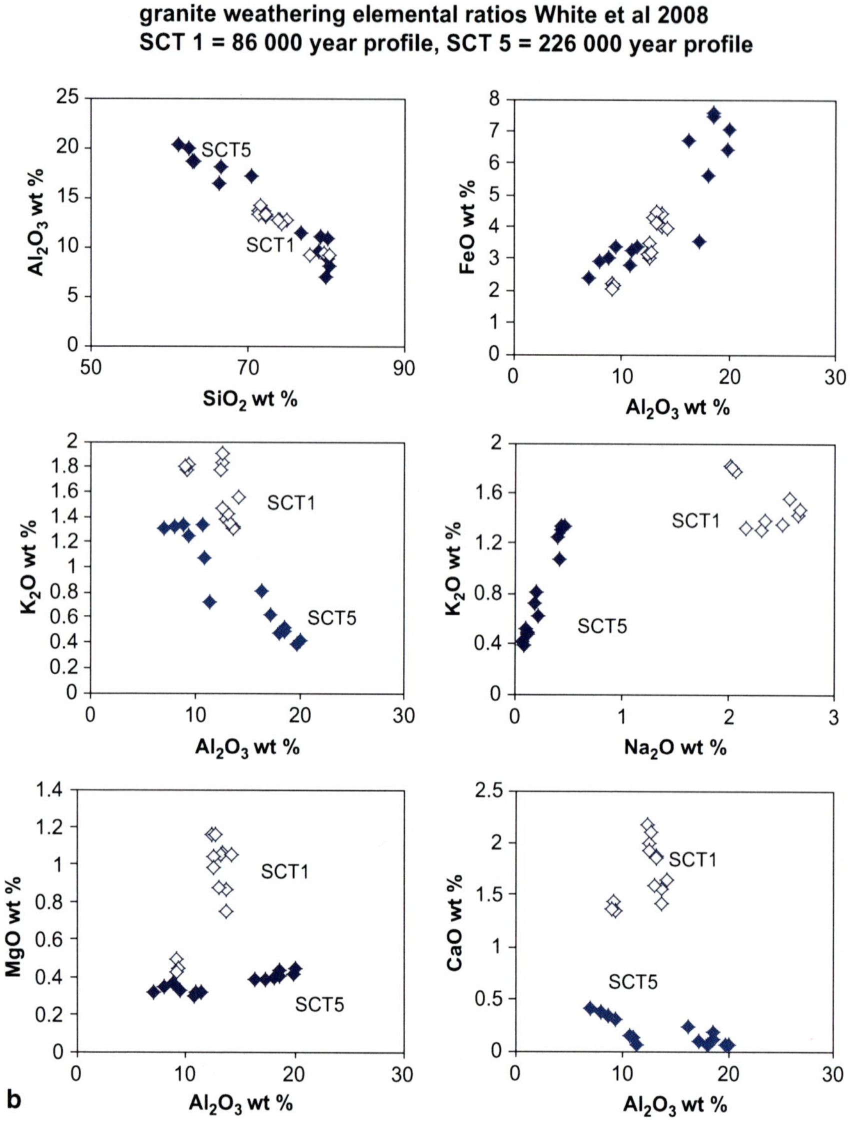

Fig. 3.3 (continued)

The observations of White et al. (2008) on a series of granite based soils found as marine terraces on the central California coast under semi-arid climatic conditions give insight into alteration processes under a different climate, Mediterranean compared to Alpine. The terraces form a time sequence with the youngest having an age of 86,000 and the oldest 226,000 years instead of thousands of years the case

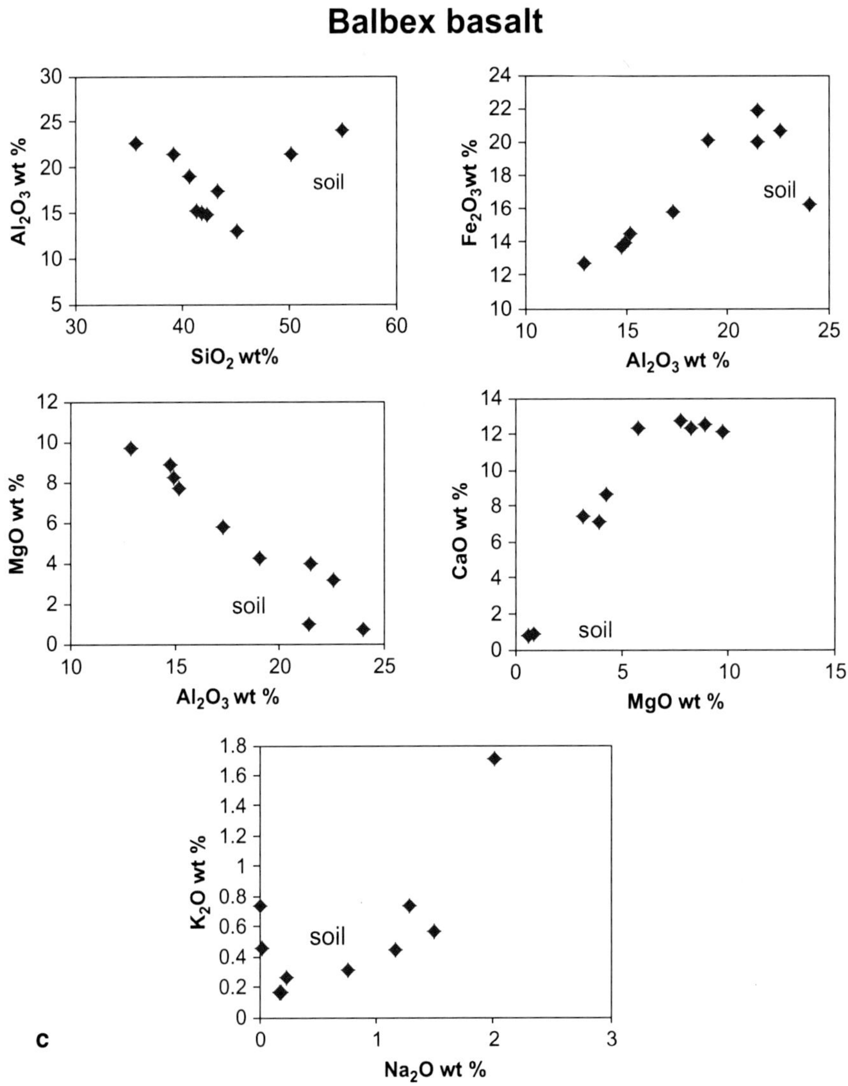

Fig. 3.3 Element ratios in alteration sequences. Comparison of relative element concentrations in granite alteration sequences (alterite materials) for young high mountain alterites (**a**) (Egli et al. 2001) several thousand years of alteration. (**b**) Terrace alteration in a Mediterranean climate over tens to hundreds of thousands of years (White et al. 2000) SCT1 = 86,000 years and SCT5 = 227,000 years. (**c**) Elemental ratios for altered basalt under temperate climates (hundreds of thousands of years of alteration (Chesworth et al. 1981) from the Massif Central France

of Alpine alteration. Figure 3.3b indicates the co-relative variation in major element content comparing samples in profiles of 3 (SCT1) to 6 m (SCT5) depth of different age (SCT1 = 86,000 years and SCT5 = 227,000 years). The trends of Si and Al are the same as in the different initial stages of alteration, with silica loss and alumina gain. They superimpose for the young and older profiles. Iron and silica are related in a positive fashion, as in the initial stages of alteration for the general series. In both cases the trends are extended in the older profile to higher concentrations of Al and Fe due to prolonged alteration. K, Na, Ca, and Mg show different trends depending upon the age of the profile. There is a decided differential loss of some elements comparing young and older profile concentrations. In the younger profile (86,000 years) K and Al are negatively related. Potassium is lost progressively as alteration proceeds and as more aluminous alterite minerals are formed. This effect is also suggested in the very young Alpine alteration sequence of Egli et al. (2001). The abundance of Na shows no clear correlation to K abundance in the younger profile, but there is a correlation in the older profile after most of the Na has been lost to alteration. Gradual loss of K is followed by further loss of Na. This two-step process is more clearly expressed in the relations of Mg and Ca. In both there is a trend of strong loss as Al increases in the younger profile but a very great loss of Ca and Mg is seen in the older profile which does not follow alumina content.

In summary, concerning the alteration of granitic rocks, it appears that the initial phases of alteration and the subsequent stages (hundreds to hundreds of thousands of years) shows an increase in alumina with a loss of silica. Relative iron content is increased along with alumina. Alkalis behave in different manners depending upon the species. It appears that both Na and K are held in the alterite in the initial stages of alteration to be lost gradually in the thousand year range and very strongly reduced in longer time scales. However, potassium is lost more gradually than sodium. This is most likely due to the fact that K is preferred in clays of medium to high charge (see Chap. 2). Mg and Ca are lost initially as alumina increases, and are almost absent in the older alterite materials.

These observations suggest that the clay minerals, i.e., minerals formed under surface water–rock chemical conditions, change in their composition with respect to the elements Na, K, Mg, and Ca as a function of time and alteration intensity indicating that the early formed minerals can eventually become unstable under prolonged surface chemical conditions.

3.2.5.1 Basalt Alterite

The chemical data for weathered basalt in the alterite zone under moderate climate conditions in Central France (Chesworth et al. 1981) indicate that the trends found in granite are similar to those found for basalt. Silica is lost and alumina and iron increase. Magnesium decreases as aluminum increases. Loss of alkalis and alkaline earth ions is correlative. However in the case of basalts, potassium is lost much more rapidly than sodium, by approximately a factor of two. This is significantly different from the trend for the granites of California where K and Na are lost at different rates (Fig 3.3b).

It appears that the overall trends of loss of elements from basalt upon weathering with the creation of minerals at equilibrium with surface chemical conditions is quite similar in overall nature to those of granite alteration products.

One can expect then that the waters draining contacts of surface alterite and rock, of various compositions, acid to basic, will contain significant amounts of Si, Ca, and Mg with varying amounts of Na and less K. Silica, the main structuring element in silicates, will be present in the initial altering solutions and alkali plus alkaline earth elements follow as silicate minerals are dissolved and reacted to form new minerals in equilibrium with surface chemistry. This schema is valid for acidic rocks, it seems, but less regular in the case of basic rocks. This is especially true for magnesium which can enter into clay minerals when it is sufficiently abundant and alumina much less abundant. A very interesting observation is that the trend of chemical weathering is similar under different climatic conditions which begin with the initial stages of alteration to be finally accomplished in older alterites or those formed under severe climatic conditions of alteration (wet tropical).

3.2.6 Weathering Profiles and the Soil Zone

The above summary indicates major trends in movement of the major element abundances in the water–rock alteration process. However observation of chemical data for alteration profiles from the rock–alterite interface up to the soil zone (alterite–plant interaction zone) indicates a more complicated elemental movement. This is seen in the data for basalt in Fig. 3.3c where Si, Fe, and K seem to change in relative abundance in the soils, increasing above the end of the alterite trend values. This suggests that the soil zone does not follow the overall alteration trends. This is surprising in that the soil zone is where rain water, highly unsaturated with respect to soluble elements from rock alteration, is in initial contact with alterite material.

If one looks at the elemental content at different levels in a soil–alterite–rock sequence, it is very frequent that the abundance of certain elements does not express loss in the upper parts of a profile where the contact of alterite with unsaturated rain water is the most intense. It is logical to expect that highest amount of weathering in the upper part of a profile. However, certain elements seem to be accumulated, or gain in relative abundance, in the upper portion of an alteration profile. This is particularly the case under conditions of temperate climate alteration. A case in point is illustrated in Fig. 3.4a given by Loughnan (1969) with data for an alteration profile developed on basalt in New South Wales NZ.

Rather notable reversals in elemental abundance are observed for silica, iron, and potassium. Aluminum remains roughly constant in the sequence. The strongest effects are seen in the upper most levels of the profiles, the soil zone. The effect of elemental uplift by plants has been outlined in Velde and Barré (2010) where observations of the type shown in Fig. 3.3 are discussed. In the figure one striking observation is that there is a loss of iron at the surface. This is probably due to the biological activity where insoluble ferric iron can be reduced to the ferrous state

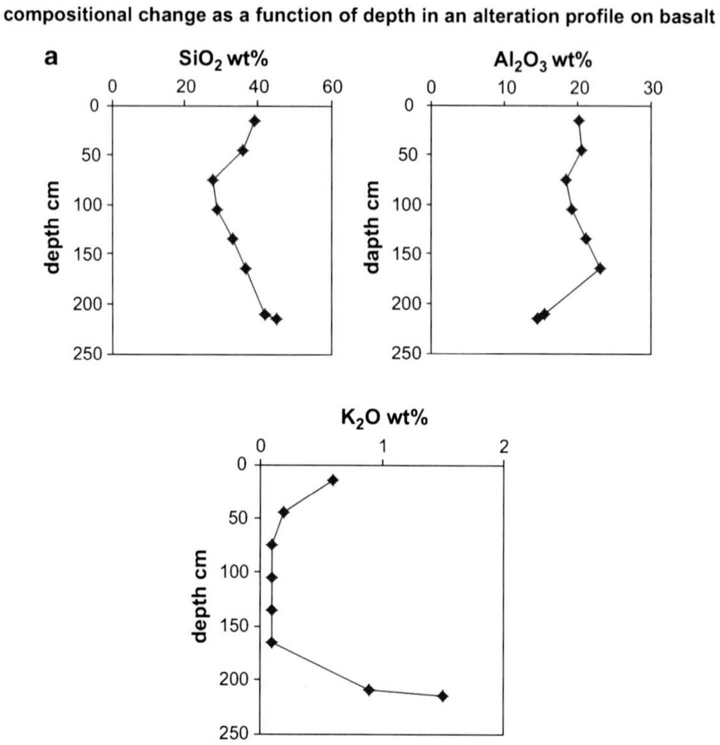

Fig. 3.4 (continued)

during the bacterial transformation of plant litter and root exudates, and this ferrous iron is relatively soluble in aqueous solution. It is then carried downward by aqueous solutions to be deposited in the B horizon or evacuated by pore water to the water table and moved by stream transport further from the alteration area.

The silica content of the soils at the surface is higher than the zone beneath them illustrating the uplift effect of silica via the formation of phytolites. This trend reverses the silica loss trend in the initial stages of water–rock alteration. The same is true for potassium which is seen to increase markedly at the surface due to plant deposition. In the instance of basalt alteration, three of the basic trends in alteration (loss and relative gain of elements) are reversed through the action of plants moving these elements to the surface through root transport and the action of transforming biomass at the surface.

Data from White et al. (2008) for granite alteration under Mediterranean climate conditions (Fig. 3.4b) in Coastal California for two profiles of 86,000 and 226,000 years can be compared. It is clear that the same trend reversals for alumina and potassium are present in the upper, bio-influenced portion of the profile but accentuated in the older profile. The very strong concentration for iron in mid-profile

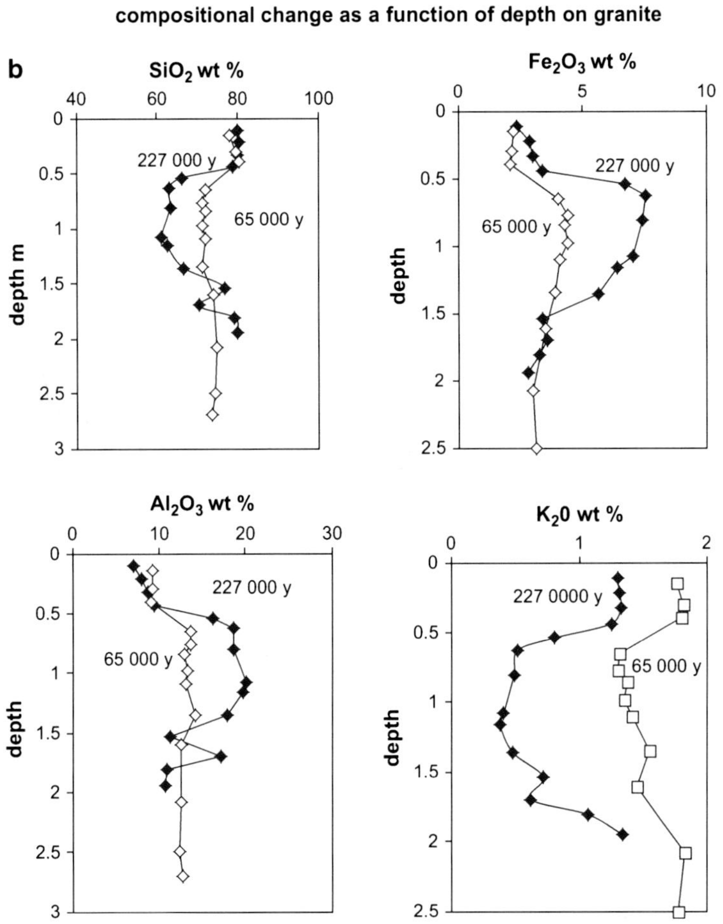

Fig. 3.4 (**a**) Element concentrations for various depths in an alteration profile on basalt in New Zealand under semi-tropical climatic conditions (data from Loughnan 1969). (**b**) Granite alteration compositional profiles on granitic terrace material (data from White et al. 2008)

(over 400 %) suggests an accumulation of iron due to other causes, indicated by Schultz et al. (2010) which is probably biologically driven.

3.2.7 Alterite Chemical Trends

It is striking to find that the major trends of major element loss are followed independently of the rock type, basic basalt and acidic granite, where the starting compositions of the rocks are quite different and the minerals present are quite

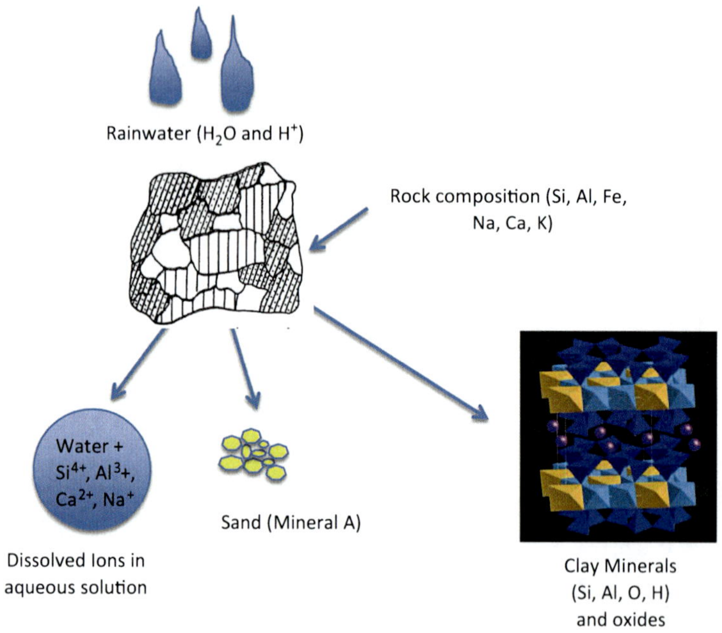

Fig. 3.5 Alteration schema of the results of water–rock interaction. Slightly acidic rainwater interacts with the minerals in a rock (A, B, C) which produce new minerals composed essentially of Al and Si plus hydrogen (clay minerals), and Fe in the form of oxides. Some minerals and mineral grains are little affected and become a more coarse—gained fraction of the alterite (sands) with the dissolved materials taken away in the altering fluids

different also. Plant generated chemical uplift of certain elements modifies their presence throughout the profile (K and Si) while the effect of reduction of oxidized iron is important to its retention in both basalt temperate climate alteration and more dry climate on granite. Therefore it appears that major element chemistry is determined by the same chemical constraints, although the phases produced in the altered rocks through interaction of aqueous solutions and oxidation with the different rock types produce different clay minerals (see Velde and Meunier 2008, Chap. 4). The silicate minerals are transformed to less siliceous phases, with a subsequent increase in alumina in the silicate minerals formed. Equally, the alkali and alkaline earth content of the newly formed minerals is lower than that in the initial rocks. As alteration continues only the most chemically stable high temperature minerals remain, quartz and minerals such as zircon and ilmenite for the most part high temperature oxides. The relations are indicated in Fig. 3.5.

Overall one finds the following chemical relations for most high temperature rocks magmatic and metamorphic, as they alter to form minerals stable at the surface.

Rock minerals	Alterite minerals		Solutions
Silicates	*Clay minerals*	*Oxyhydroxide minerals*	
Na, Mg, Al, Si, K, Ca, Fe, Mn	Al, Si, K (Na, Ca, Mg)	Al, Fe, Mn	Na, Mg, K, Ca

These transfers in the first stages of rock and mineral change are modified by plant action where some of the minerals are concentrated, relatively, at the surface in the soil zone. This action results in the stabilization of clays which are more silica-rich than clays formed under rigorous alteration conditions such as kaolinite (Al–Si silicate) or gibbsite (Al hydoxide) and which can retain potassium and ammonium ions at the clay surface.

The chemical change is accompanied by a physical change. The new materials are less dense than the high temperature minerals. The clays have a density of about 2.5, similar or below most high temperature silicates and oxides, but since they are of small grain size and little physical effort has been exerted upon them, they remain little compacted in the alterite zone and the aggregate is of low density, having a significant amount of water associated with them on the clay surfaces. The oxides tend to be very stable and are often intimately associated with the silicate minerals, as grain coatings or other types of attachment to the more resistant mineral grains such as quartz.

3.2.8 End Member Alterite Products: Laterites and Bauxites

As seen in the chemical trends for initial stages of alteration of different rock types, alumina and iron increase together for the largest portion of the alterite sequences under moderate climatic conditions. From these observations one can then predict that the end point of alteration will be a combination of iron and alumina oxyhydroxides. However this is not entirely true. The most severe cases of alteration produce either iron oxide deposits (iron crusts) or high alumina concentrations (bauxites). Both types of concentrations occur in tropical areas with intense rainfall and hence strong dissolution. The stages approaching the concentration of either Fe or Al are normally called laterite alterations where iron content evident in the alterite is a strong red–brown color indicating the presence of iron oxides. Under these conditions of alteration the Al–Fe concentrations increase but at a certain point they diverge; on the one hand iron is concentrated and Al lost while in the other Al increases at the expense of Fe. The case of laterite (Fe) formation is well documented by Tardy (1993) and McFarlane (1976) and that of bauxites (Al) summarized in Valeton (1972). In the case of laterites, the iron crust found at the surface in old deposits in western sub-Saharan Africa is seen to be strongly affected by the action of specific biomes. Forest growth appears to be an agent of the destruction of this iron crust (summarized by Tardy 1993, p. 211). This suggests that the accumulation of iron in a laterite crust is subject to specific chemical conditions engendered by specific plant regimes. As seen in Fig. 3.4b above, the

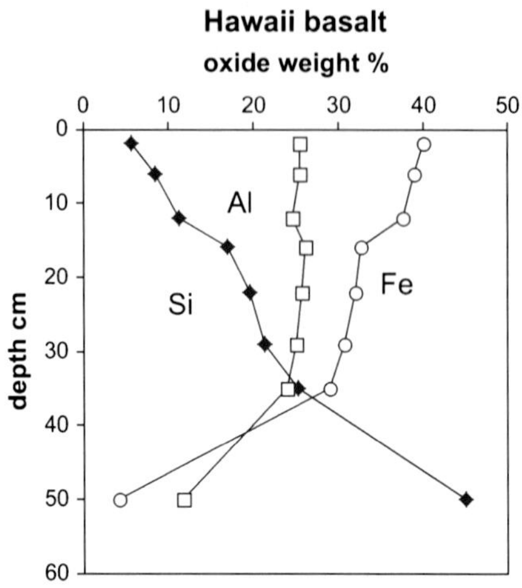

Fig. 3.6 Hawaii basalt profile. Relative change in concentrations of Si, Al, and Fe in alteration profiles for basalts in Hawaii (Valeton 1972)

concentration of iron in a soil profile can be strongly reduced under semi-arid grassland conditions in the soil zone under moderate alteration conditions. However basalt soil profiles on Hawaii, under grass vegetation and more intense alteration, show different chemical relations in the upper part of the profile. In this case of intense weathering, under prolonged tropical seasonal variations (savannah type climate with periods of high rainfall), there is a continued increase in iron at the surface (Fig. 3.6).

Data for two bauxite deposits reported by Valeton (1972) show the evolution of iron and alumina content for the latter stages of alteration toward bauxite, expressed by the aluminum hydroxide mineral gibbsite. Silica continues to decrease and alumina reaches a roughly constant value relatively deep in the profile. The relations of silica and iron are of course strongly anti-correlated. Here the chemical direction is toward the formation of a laterite soil, with eventually a high iron content to the exclusion of all other major and minor elements except alumina.

The comparative abundances of the elements Fe and Al in some alteration profiles are shown in Fig. 3.7. The upper diagrams show concentration trends for basic rocks (those dominated by Fe oxide content over alkali and alkaline earth elements) under temperate and dry tropical climates. Here there is a strong increase in both Fe and Al where oxide weight percent reaches above 40 %. In other instances there is an increase in Fe_2O_3 and a decrease in Al_2O_3 under humid tropical climate conditions. These profiles produce iron-rich layers at the surface of the alteration sequence. In other instances one finds that the profiles lose iron toward the top and alumina content increases. These sequences would produce bauxite mineralogies with less iron present. The climates in these instances are humid. From the descriptions given by the authors of the papers from which the data were taken, it is not clear just what type of vegetation is present or was present during the periods of alteration development. The problem often encountered in the case of iron accumulations is that the alteration is old and the climate has changed since its period of accumulation. This is evident in that the iron crusts at the surface often are being destroyed by new vegetation, especially forest biomes.

If we take the example of iron content in the profile of basalt alteration under conifer forest conditions and granite under scrub land cover (Fig. 3.4a and b), there seems to be a distinct loss of iron from the surface zone which would be logical to ascribe to biological activity. Tentatively we propose that the increase in alumina at the expense of iron in the upper layers of a soil profile would be due to the plant regime present at the surface. This is of course a combination of climate, rainfall intensity, and frequency which are the major driving forces of alteration but the stability of iron compared to aluminum oxide could well depend upon the oxidation state of the iron whereas the alumina is unaffected by plant regime due to its stable oxidation state of +3.

The clay mineralogy in sequences of more and more intense alteration is one where smectites dominate the alterite initially. Smectite is the most silica-rich surface clay mineral with an Al/Si ratio of <1. Successively one finds kaolinite and illite, Al/Si $= 1$, and eventually gibbsite, the silica free clay mineral. The proportions of these clays change with the stage of alteration, becoming more silica poor toward the top of the profile or with weathering intensity. In cases where iron is less, present probably due to biological action, the clay assemblage is essentially gibbsite and the material is called bauxite. When iron is stable it is strongly concentrated and the alterite is called laterite. In either case, the stage of ultimate alteration is due to the passage of large quantities of rain water through the alteration profile. This occurs in the humid tropics of course, and all the more so in situations of prolonged alteration of long periods of time due to a lack of tectonic movement of the basement rocks. As it turns out these two situations occur together in the central parts of Africa and South America.

The examples of alteration in profiles of the Haute Volta (Africa) based on basic rocks (gabbro and meta gabbro) show different trends in Fe and Al accumulation depending upon their position in the topography of the area (Pion 1979). Two profiles show either alumina accumulation or accumulation of iron depending upon their position in the topography under tropical dry climatic conditions (Fig. 3.7b). In these two profiles the clay mineralogy does not seem to be very different, being composed mainly of kaolinite with some smectites present. The profile which is less altered (Tankiédougou 6) shows the normal parallel increase in Fe and Al, while the second profile shows the reverse trend. Here it is apparent that the specific topographic site and most likely biome can determine the evolution toward laterite or gibbsite accumulation.

The end term of water–rock alteration is then one of low silica content and essentially a hydrated trivalent ion-oxide assemblage as a final stage. The loss of silica, alkalies, and alkaline earths is almost complete as is the loss of elements of lesser abundance, which were present in the original rocks either acidic or basic. The overall chemical consequences of water–rock alteration appear to be rather independent of the composition of the starting material. The concentration of either iron or alumina in the alterite seems to be dependent upon geographic conditions, drainage, and more likely upon the plant regime prevalent at the surface.

Valeton (1972) describes numerous occurrences of bauxite deposits. In those of recent origin, it seems apparent that there is an association of laterite (iron and

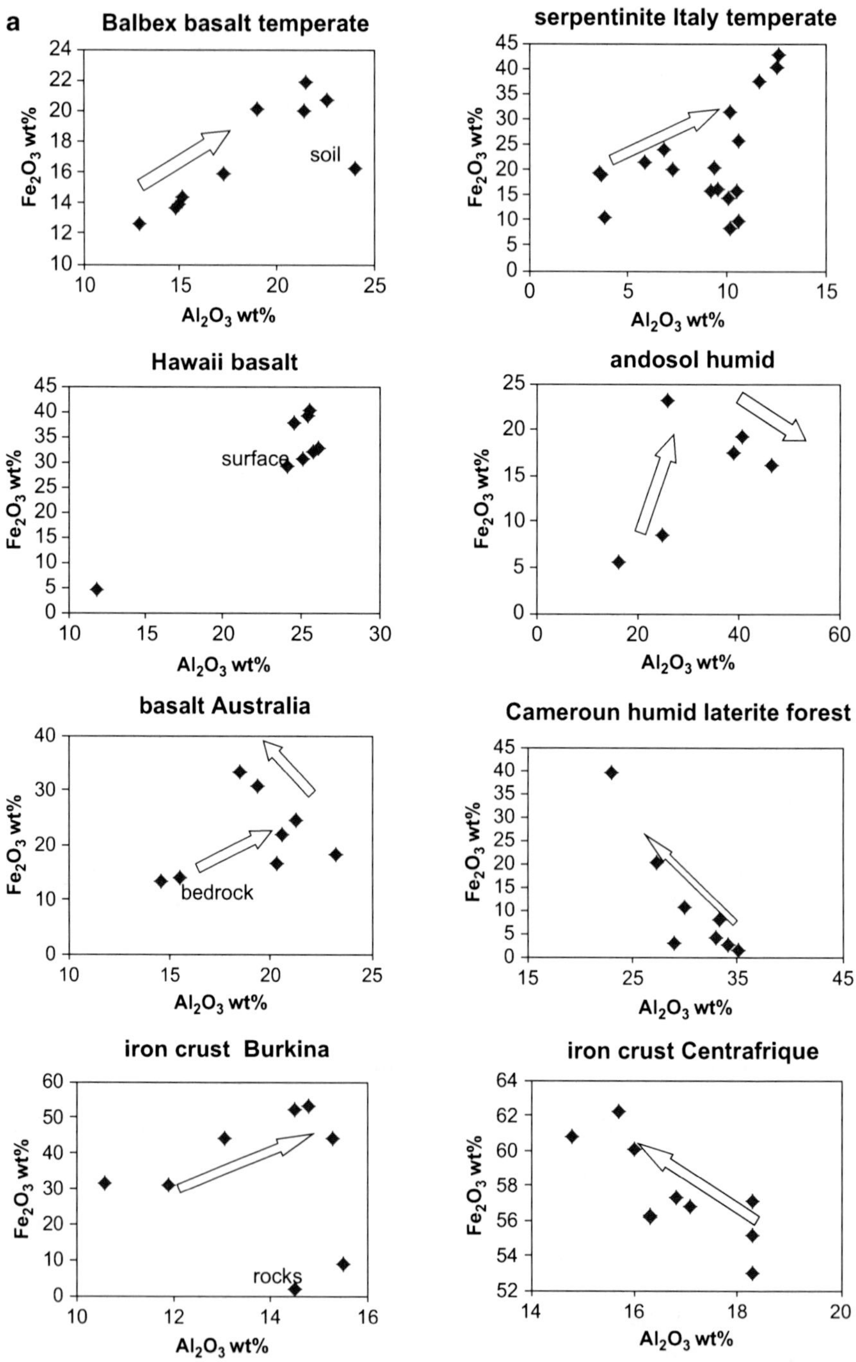

Fig. 3.7 (continued)

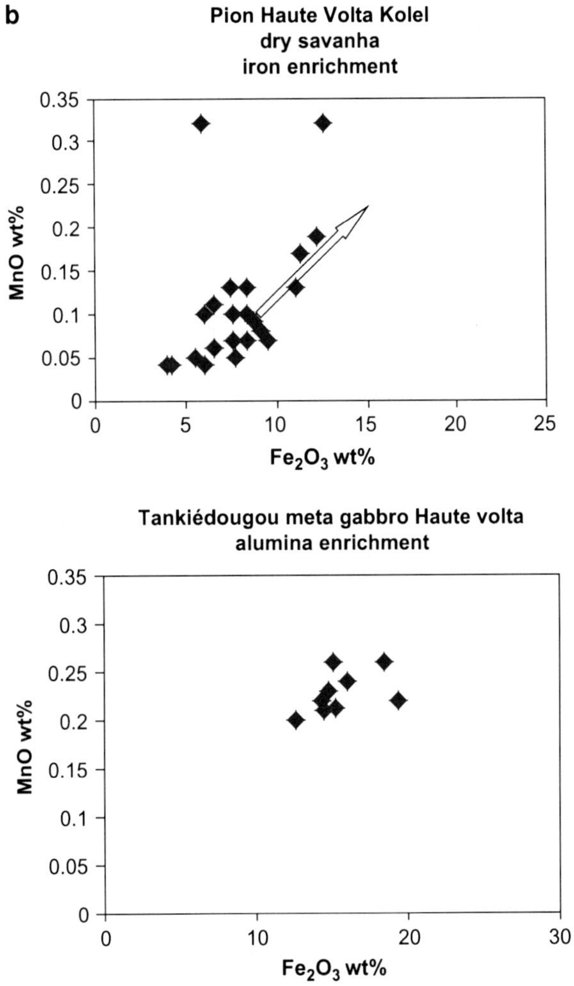

Fig. 3.7 (**a**) Fe–Al relations for different stages of alteration under different climates. Basalt and serpentine (iron-rich rocks) from temperate climates (Chesworth et al. 1981; Bonifacio et al. 1997; Valeton 1972). Alteration under more humid climates (Loughnan 1969; Leumbe et al. 2005) and ultimate stages of alteration from laterites and iron oxide-rich ferricrete or iron crust materials (Temgoua 2002; Beauvais and Mazaltarim 1988). The *arrows* indicate compositional trends with alteration intensity. (**b**) Gabbro and meta gabbro alteration profile data from Haute Volta savanna climate (Pion 1979)

alumina accumulation) and at times iron crusts (iron accumulation) with bauxites (alumina accumulation). The spatial relations of the iron and alumina accumulations in a landscape seem to demand upon slope (drainage) and position in the landscape where one might suspect that different types of vegetation occur in response to micro-climatic variations. Overall the iron accumulation seems to occur at the surface whereas the alumina accumulation is present at depth but accentuated at the surface.

It is not all that clear as to when the correlative increase of iron–alumina in the alteration sequence begins to select either iron or alumina giving in the end either laterite (iron oxyhydroxide) or bauxite (alumina hydroxide). In temperate climate forest (conifer) soils there is a strong decrease in iron content near the surface. In savannah soils where iron oxide has accumulated, new installation of forest biomes tends to break up the iron oxide crusts. These two observations may indicate that the concentration of either iron or alumina could be due to specific plant regimes under climates of intense weathering (wet or dry tropical). Not all intensely weathered rocks concentrate one or the other of the oxides, which indicates that the tendency to form mono-element deposits in the soil zone could be the effects of specific chemical conditions engendered by plants. Perhaps the most important observation is the concentration of Al and Fe at the expense of Si and then the other chemical components of rocks begins in the early stages of weathering and continues on throughout the various stages. This tendency seems to be independent of the type of clay mineral formed, smectites or kaolinite, though the first formed minerals are usually smectites which are more silica-rich than kaolinites which follow in the order of alteration intensity and eventually gibbsite (aluminum hydroxide) is present. The clays then seem to follow in a general manner the loss of silica and the continued enrichment of iron and alumina.

It is not clear whether iron and manganese always are concentrated in the extreme terms of alteration. In the data for temperate climate alteration, Mn increases in relative abundance as alteration begins. However in more advanced alteration materials, such as fericrete or laterites, Mn is not well correlated with iron indicating that there is no simultaneous concentration of both elements. One can assume that iron and manganese do not follow the same chemical trends in weathering throughout the cycle of rock alteration. In the soils of temperate climates, Fe and Mn tend to be found in hydroxide form. However, both elements can form divalent, di-tri valent or trivalent oxides. In the laterite and more developed iron concentration, alterites iron can be found as an oxide (Fe^{3+}). The pH and Eh values of valence change are different for the two elements with Fe tending to be more easily oxidized (i.e., lower pH and eH values). However both Fe and Mn can form oxides and hydroxides of different valencies and given that the structures are similar one can expect strong associations of different minor and trace elements with these two elements of major abundance which appear to occur together in altertie formation (see Degens1965).

Krauskopf (1967) summarizes calculated stability data for ions, oxides, and hydroxides (pp. 249–251) which gives an idea of the types of materials one can expect under different conditions of alteration at the surface. In general pH will range from 4 to 9 and Eh (redox potential) from about −0.2 to 1.0. In Fig. 3.8 these ranges of conditions are represented showing the conditions of change from dissolved ferrous iron (divalent) to ferric iron oxide (trivalent) and those of dissolved manganese (divalent) to tetravalent oxide. In general values of low Eh and pH favor ionic forms of metallic elements, low Eh and high pH favor hydroxide forms, and high Eh and higher pH favor oxides. The transition in stability of ionic and solid forms of different transition metals show similar slopes of reaction but the

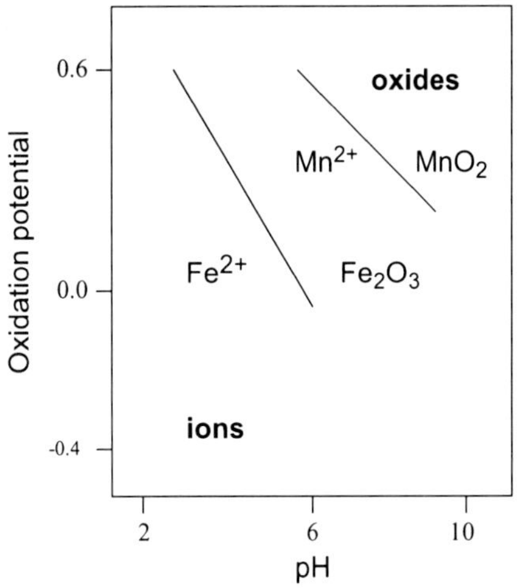

Fig. 3.8 General relationship of chemical states of the metal elements Fe and Mn as a function of oxidation and pH. High pH and oxidation potential lead to oxide phases; lower oxidation potentials lead to dissolved ions in solution or hydroxyl ion complexes. Based on Krauskopf (1967, p. 250)

sequence as a function of pH–Eh depends upon the characteristics of each element. It is sufficient to consider the major elements concerning the formation of oxides in that the minor and trace elements will generally not be sufficiently abundant to form a solid phase due to their low concentration and inability to saturate the aqueous solution.

However, the minor and trace elements will most likely be associated with the major elements Fe and Mn in the oxide or hydroxy-oxide solid phase form. Therefore it is important to realize that Mn forms oxides at significantly different conditions from those of iron oxide formation. For a large part of basic conditions iron oxide will be present while Mn will remain in the soluble ionic form. However $Mn(OH)_2$ can be present and associated with iron minerals.

3.3 Rock Weathering: Minor Elements

The major elements in the composition of a rock form the very largest part of the mineral phases present. Minor elements of <0.5 wt% are not often present in a sufficient concentration to form a separate mineral phase. To a large extent minor and especially trace elements are found in substitution in the structures of the phases formed by major elements. However this often repeated maxim is not always true and probably more often invalid concerning elements that are of minor abundance but not suited to substitution in the major element phase formed at the surface. Often minor and especially trace elements are found in mineral phases that have not reacted to the changes in chemistry of the surface and they are found in the "sand" fraction of un-reacted minerals. Such minerals as zircon, a zirconium silicate, sphene a calcium titanium silicate, perovskite a calcium titanate, monazite a rare earth phosphate, among others tend to remain present throughout the alteration process and are found in the more coarse fractions of alterite minerals.

Hence for certain elements there will be a constancy in the amount of minor elements present, which does not reflect transfer to the new alteration minerals. There is also the possibility that some of the minor elements are in chemically inactive phases and another portion in new minerals, with the result that one will see two trends of element association, one due to the initial rock and the other to the new minerals present. These aspects are treated in more detail by McQueen (2008).

However, there are often strong relationships of some minor and trace element concentrations, which change as a function of the extent or intensity of alteration. As we have seen in the observation of major elements, those essential in forming the major mineral phases present in alterites, some elements are lost while others are concentrated. One can consider the effects of weathering on minor and trace element abundance as they are affected when some elements are lost and others concentrated. Our approach is to make cross correlations of elements found in alterites in order to see which minor or trace elements are associated together and with which major alterite mineral phases. Essentially in alterites there is a concentration of Al and Fe where the alumina is initially directly associated with silicate minerals (clay minerals) and iron can be associated with a hydroxyl phase. Manganese can be compared to Fe to see if they are concentrated together as oxides or whether they form in different patterns. Given the high capacity to fix transition metal and other elements by the Mn, Fe oxides such an exercise is important in order to follow the migration paths of these elements from rock to alterite.

Two groups of samples will be considered here, whole "rock" or alterite material containing new minerals and un-reacted phases and clay fractions extracted from the alterite. Unfortunately the clay fraction data is relatively rare. Such material is the best to observe the loss and gain of trace elements in the alterite material.

3.3.1 Major, Minor, and Trace Element Affinities

Certain element groups can be used to follow the pathways of transformation, migration, and fixation of minor elements in alterite material. This is especially true for the ionically bonded cations, those elements of high solubility. These elements are the least strongly bonded to solids and will be the most reactive to changes in their chemical environment.

3.3.1.1 Alkali and Alkaline Earth Elements

The elements most often lost to solution by alteration of high temperature minerals are the alkali and alkaline earth elements Na, K, and Ca which are major elements in many of the phases they occur in. We will consider the relations of minor elements of similar chemical characteristics compared to the abundance of these elements.

Elements of major abundance
Na, K, Ca
Elements of trace abundance associated with the above
Li, Rb, Cs, Sr, Ba

The alkali element potassium and the alkaline earth calcium are frequently found in major phases in soils, especially or almost exclusively in smectites and illites where they are held within the mineral structure but can be exchanged for the elements in solution when the chemical activity of these species is sufficiently great. Hence K and Ca are associated with the alteration process where they enter the new minerals but are nevertheless less abundant than in most of the rocks that are the basis of the alterite. These elements can be accompanied in minor abundance by certain trace elements. There is a limited amount of data available on the relationship between these elements and alkaline and alkaline earth elements of minor abundance in rocks as they occur in alterite materials.

3.3.2 K–Rb

Rubidium is in many cases associated in high temperature minerals and has similar chemical characteristics and fills the necessary roles in minerals as does potassium. In the alteration process it appears that this similar chemical role is continued in that K and Rb are often found to vary together. Figure 3.9 shows the relations on a altered granite (White et al. 2008) and on acidic volcanic soil materials (Martinez Cortez et al. 2003). The data for altered granites having been subjected to alteration for different periods of time is very instructive (data Fig. 3.3b).

It is apparent that as alteration advances, the relations of K to Rb concentration change somewhat. In the older sequence there is a relative increase in Rb content compared to K. However in both sequences of alterite, there is a very well-defined relation between the two elements. It is possible that the change in Rb compared to K is due to a stronger contribution of alterite minerals in the older material where the younger shows some non-altered material in the clay fraction, such as muscovite whereas the older minerals show the effects of new minerals being present.

3.3.3 Ca–Sr

Strontium is similar in chemical properties to calcium, in charge and there is not too great a difference in ionic radius. Figure 3.10 shows Sr–Ca relations for the same series of alterites. The results are similar for the homogeneous volcanic series and the granites. The Apennine soil samples, which are of different rock types, indicate the importance of initial rock composition but more strongly that there is a strong relation between Ca and Sr, due to substitution in the same mineral chemical sites.

It seems clear that Sr is easily substituted for Ca in various surface minerals and this element is closely associated with the element of major abundance, calcium.

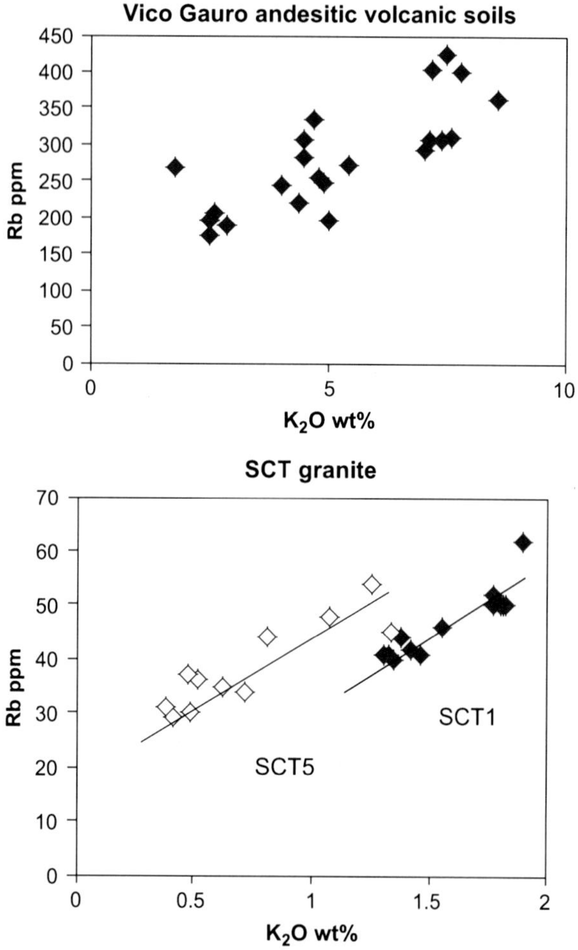

Fig. 3.9 Examples of potassium and rubidium abundances in altered acidic Volcanic material (Martínez Cortizas et al. 2003) and granitic materials (White et al. 1998, 2008). SCT1 = 86,000 years alteration and SCT 5 = 227,000 years alteration)

3.3.4 Ca–Ba

These two elements are divalent but less compatible concerning ionic radium. In general they form different minerals such as carbonates or sulfates, with low solid solution between them. Figure 3.11 shows data for the granites of California reported by White et al. (2008). In the young series of alterites there is a general correlation of relative abundance of the elements but in the older alterites the concentration of Ba is not at all related to that of Ca, changing by a factor of three while Ca changes by 30 %.

This observation seems to be in accord with the occurrence of rather low solubility of Ba and Ca in non-silicate minerals. Thus we can assume that Ba will not be present as a substitution for Ca in surface mineral structures. In the data set

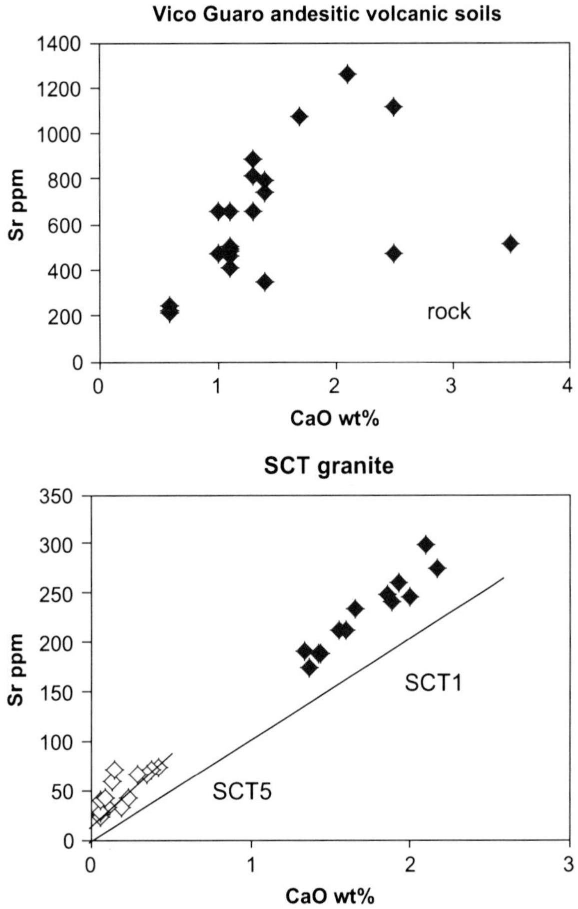

Fig. 3.10 Relations of Ca and Sr in alteration series on acidic volcanic materials (Martínez Cortizas et al. 2003) and granite (White et al. 1998, 2008). SCT1 represents an 86,000 year alteration profile and SCT5 a 227,000 year alteration profile

from the Apennine mountains which are multi-source concerning the initial rock type being altered, there is no clear relation between Ca and Ba. Thus it appears that Ba and Ca are not related geochemically in alterite material although they are found in limited substitutions in such silicate minerals as plagioclase feldspar.

3.3.5 *Li*

This element is present in many geological environments in small quantities but is difficult to associate with substitutions following major elements. Lithium is a light element which has been traditionally difficult to analyze chemically in the presence of other elements in complex systems such as rocks. Clay mineralogists know that it can substitute in the clay structure in vacant sites compensating for charge imbalance created by other substitutions. It can be observed as an exchange ion in or on

Fig. 3.11 Relations of Ba and Ca in alteration series on a granite (White et al. 1998, 2008). SCT1 represents a 86,000 year alteration profile and SCT5 a 227,000 year alteration profile)

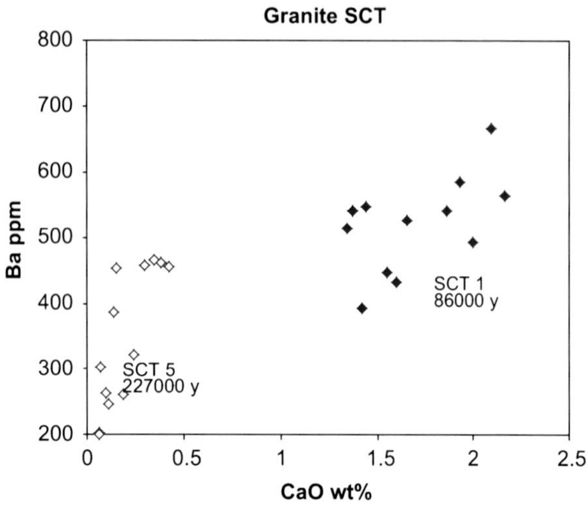

the clays. However we have little information on its presence as a substitution in small amounts for elements of major abundance in alteration minerals.

A very interesting data set given by Mosser (1980) indicates that Li is present in greater concentrations in clays of alterites for a wide variety of eruptive rock types ranging from acid to ultrabasic. The clay minerals can be dominantly kaolinite or smectites with no apparent change in the tendency to increase Li content over the parent rock. Since kaolinite is in principle structurally devoid of alkali or alkaline earth elements, the Li is not likely to be present in substitution for these elements. However it is possible that it is situated in the vacant octahedral site where it should create a charge surplus on the structure. The compensation of this charge surplus by an anion can only be speculative.

Data from the Apennine andic soils of various rock type origin (F. Terribile personal communication) indicates that in the three rock source types present in the 35 soil profiles investigated the relations of Li are most closely related to Al content of the soil and hence probably the clay minerals present (Fig. 3.12). These samples represent the different stages of initial alteration under conditions of temperate climate chemistry. Here whole sample analyses indicate a relation between Li and the alterites dependent upon the type of starting material. In all cases Al increases with Fe and Mn with Fe. It seems that there is a possibility of Li being associated with Al and perhaps Mn although there is much more Al present than Mn in the alterite material.

Lithium seems to be held in the alterite material. There are such minerals as sodium birnessite ($Na_4Mn_{14}O_{27} \times 9H_2O$), which could well fix the lithium in their lattice (B. Lanson, personal communication). Therefore one can expect retention of Li in the alterite material either by association within manganese minerals or aluminous clay minerals.

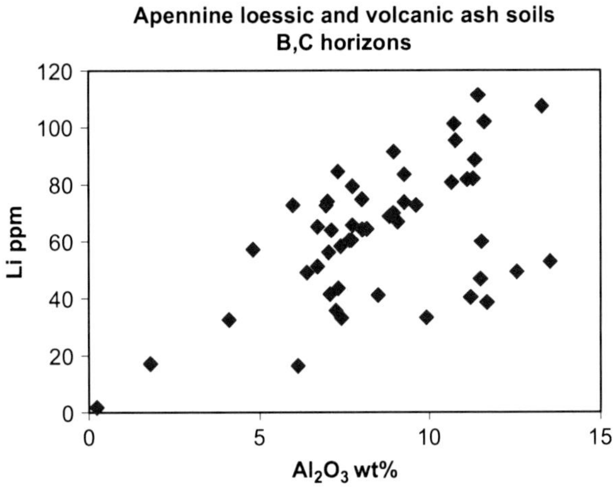

Fig. 3.12 Li–Al relations in B, C horizons of Apennine alterites of volcanic loessic materials [F. Terribile personal communication based on data reported in Mileti et al. (2013)]

3.3.6 Cs

This element in the alkali element sequence is little studied in surveys of alteration (water–rock interaction) sequences. The information one can obtain is from rock-sediment ratios (Mason 1966; Jones 1966) which indicate an increase in the presence of this mineral in the solids preceded by alteration and transported by rivers. This would be in accord with the idea of substitution of Cs in potassium-bearing clays. However the concentration factor of potassium is significantly less than that of Cs or in fact of Rb the other substituting ions in a potassic crystallographic site. The same can be said for Li in fact. Thus the retention of Cs in or on solids seems to a general tendency.

The discussion above is limited to the initiation of alteration and the intermediate stages of transformation of rocks into materials stable at the surface. Looking at the final term of alteration the pattern shows differences compared to the initial steps of transformation. In relations of K and Rb, for example, Temgoua (2002) concerning humid forest laterites the data indicates that the relations of K and Rb are rather random, with no pattern discernible. The same is true for Ca–Sr relations in iron crusts in Burkina Faso (Boeglin and Mazaltarim 1989), yet Ba and Sr show a reasonable interrelation in these samples. This clearly indicates that the loss of silicates has changed the substitution relations between major and minor elements, but there is a persistence of retention of some of the minor elements in the hydroxyl-oxide mineral assemblages. However the data for ferricrete (iron crust) laterite indicates no relation whatsoever. Thus one must consider that the extremes of alteration can well efface any silicate mineralogical relations with minor element abundances, at least in the case of iron accumulations. Valeton (1972) does report values of several hundreds of ppm Sr content for bauxites, which are values comparable to alterites in less evolved materials. Thus Fe oxides do not retain the alkali elements but perhaps Al hydroxide-rich alterites do retain these elements.

Through atomic bomb testing since 1945 and the reactor incident in Chernobyl, radioactive fission products have artificially entered and spread worldwide throughout the atmosphere. Of the remaining fission products in the long term only the isotopes Strontium-90 and Cesium-137 have a significance. Both isotopes behave physiologically like the important elements for organisms calcium and potassium. The interactions of radioactive isotopes of Cs with soils are of concern in environmental studies, due to the high transferability, wide distribution, high solubility, long half-life (half-lifes of Sr-90 = 28 years and Cs-137 = 30 years), and the easily assimilation by living organism. The spatial distribution of radiocesium in the soil depends in particular on its adsorption on soil minerals. The vertical rate of spreading of radiocesium in undisturbed soils is relatively low (with the exception of sandy soils and tropical laterites). The cause of this low depth penetration is the selective sorption of cesium by the crystal lattices of clay minerals (especially smectite and illite). Cations with low hydration energy, such as K^+, NH_4^+, Rb^+, and Cs^+, produce interlayer dehydration and layer collapse and are therefore fixed in interlayer positions. According to concurring results from several authors, at least 80 % of Cs activity remains in the upper 15 cm of soil (ANPA 2000; Kühn 1982; Ritchie and Rudolph 1970; Squire and Middleton 1966; Zibold et al. 1997).

3.3.7 Transition Metal Elements

The geochemical relations of these elements are potentially difficult to follow during alteration in that they can occur in several types of high temperature phases, silicates susceptible to be readily altered, or in less reactive minerals such as oxides or phosphates. Those that occur in sulfides will be rapidly altered and released. Two sets of data are used here, alterites of low intensity of transformation, soils from temperate climates, and materials representing the ultimate stages of alteration : laterites, iron crusts, and bauxites. These latter materials will show the relations of minor elements with oxyhydroxides of Fe or Al. One must keep in mind that some fo the materials in the moderately altered alterites will be un-reacted and some will have changed phases upon alteration. In the iron crust and bauxite alterite one can expect that the former high temperature phases will have reacted largely.

The transition metals are generally associated with Fe and Mn in high temperature minerals (Co, Ni, Cu, Zn, Cr, V) some of which are dominantly divalent ions under conditions of surface alteration (Co, Ni, Zn), while others are normally trivalent, (V, C) and others which are found in both states, (Cu). They will be possible candidates for substitution in iron oxide structures in the trivalent state. They can be related to manganese compounds also (see Chap. 25 in Wedephol 1969).

The very important part of the distribution of these transition metal ions in or on Fe or Mn oxide minerals is that the Fe can change oxidation state while Mn can also be found in different oxidation states as hydroxide or oxide but it is normally found in lower oxidation state than Fe. Hence transition elements associated with Fe or

Mn oxyhydroxides will be redistributed into different mineralogical structures as iron or manganese changes oxidation state. This will inevitably change the distribution of these elements in their mineral substrate and promote either capture or release of these minor elements into the solutions with which they are in contact. In the alterite zone the oxidation potentials attend to be stable, oxidizing in nature, and hence the relations of the transition metal minor elements could be expected to be stable.

3.3.8 Oxides and Associations of Elements

The two most abundant transition metal oxides, forming at times major phases in alterites, are iron and manganese. However Fe is usually much more abundant in surface rock materials than Mn, usually by more than an order of magnitude. The two elements can be found in soils as local concentrations in nodules (Dixon and Weed 1996). These concentrations often contain important quantities of minor elements, especially the transition metals. To a large extent the accumulations are made of oxidized Fe and Mn in varying amounts. This is the expression of the oxidation of Fe and Mn as high temperature rock minerals are transformed into alterite phases. If the Fe and Mn remain in the reduced state, they are taken into solution and moved out of the alterites zone. The iron oxyhydroxide goethite is known to accommodate significant amounts of transition and heavy metals (Co, Ni, Cu, Zn, Cr and Cd, Pb); see Manceau et al. (2000) and Kauer et al. (2009) for example. Some manganese phases can also accommodate significant amounts of transition and heavy metal ions (Lanson et al. 2002). However the accumulation of Mn and Fe as oxide materials is not necessarily simultaneous in the alteration process (alterite zone) and can be strongly influenced by soil chemistry in the upper part of the alteration profiles. These oxyhydroxide minerals can incorporate significant amounts of several elements into their structures as well as selecting them from aqueous solution as adsorbed ions on edge sites.

The incorporated ions will not be available for cation exchange. However elements fixed on active sites of the oxyhydroxide surfaces will be exchangeable, due to temporal variations in solution concentration of exchange ions or especially hydrogen cations (changes in pH). To a certain extent the fixation of transition and heavy metal ions by oxides is more stable than that effected by clay minerals where the absorbed ions remain for the most part exchangeable and liable to removal by changing solution concentrations and pH conditions.

3.3.9 Importance of Oxidation State (Solubility of Oxide)

The oxidation state of transition metal elements is extremely important. In the lower oxidation state, 2+ instead of 3+ or 3+ instead of 4+, the elements will tend to be soluble in aqueous solution. When more oxidized they form insoluble oxides or are incorporated into oxide minerals such as goethite (Fe) or birnessite (Mn). In fact Mn can effect a change in oxidation state of Co, for example, at its surface and thus incorporate this minor element into the manganese structure (Manceau et al. 2000). The reduced ionic state of elements generally favors their solubility in aqueous solutions and as a result a movement out of the immediate environment where the element is found. However drainage waters seem to be less rich in dissolved elements than one would suspect compared to the elemental abundance in the alterite materials. The evacuation from rocks as dissolved materials in ground and stream waters is treated in more detail in Chap. 5. It is apparent that the iron and manganese oxyhydroxide minerals can be a reservoir for minor element metals and heavy metals formed in the alteration processes.

Our initial approach here is to compare the abundance of minor and trace elements in alterite zones with the potential carriers, which form the major phases in the materials, silicates and oxides or oxyhydroxides, Mn, Fe, and eventually Al. In this exercise we hope to demonstrate the relative loss of elements compared to the source rock which is altering and eventually the location of the material carrying the minor and trace elements. The investigation of extreme cases of weathering, laterite (high Fe concentration) and bauxites (high Al concentration), can give an idea of the overall importance of chemical affinity. In Fig. 3.13 several situations in alteration sequences are shown. Initially the first alterite compared to the bedrock (Fig 3.13a) indicates that iron increases but manganese abundance is less systematic. This suggests that Fe and Mn are not always coupled in their chemical reactions and production of secondary minerals. The alterites of a large number of Apennine soils from high altitude indicates a general trend of increase in both Fe and Mn. Similar climatic conditions developing alterites on serpentines (rocks with almost no alkali or alkaline earth content), show simultaneous increase in iron and manganese. By contrast, the acidic granite based soils, developed under Mediterranean climate soils in California, show decrease in Mn compared to Fe in the young series (86,000 years). In the older sequences of soils (227,000 years) based upon terrace materials, the manganese content was below detection limits ($<$100 ppm). However by contrast a strong decrease in Mn content is seen in soils based upon basalts in Hawaii where iron content increases with alteration.

In alterite of extreme weathering, forming laterite iron accumulation and bauxite, contrasted patterns are apparent. The high iron materials show dispersed and non-relational concentration patterns. However in the bauxite (alumina concentrations) the decrease in Fe is accompanied by a decrease of Mn. The quite varied relations of Fe and Mn content in different climatic and alterite state suggest that the general chemical constraints do not affect the dissolution and precipitation of iron and manganese in the same way. Different rock types or climate variables can determine the accumulation or loss of iron and manganese.

If the major element phases associated with transition metal and heavy metal elements do not follow the same chemical paths of accumulation, what can one see in the relations of minor elements associated with Fe or Mn? Two elements generally assumed to be of very similar geochemical properties, Co and Ni can

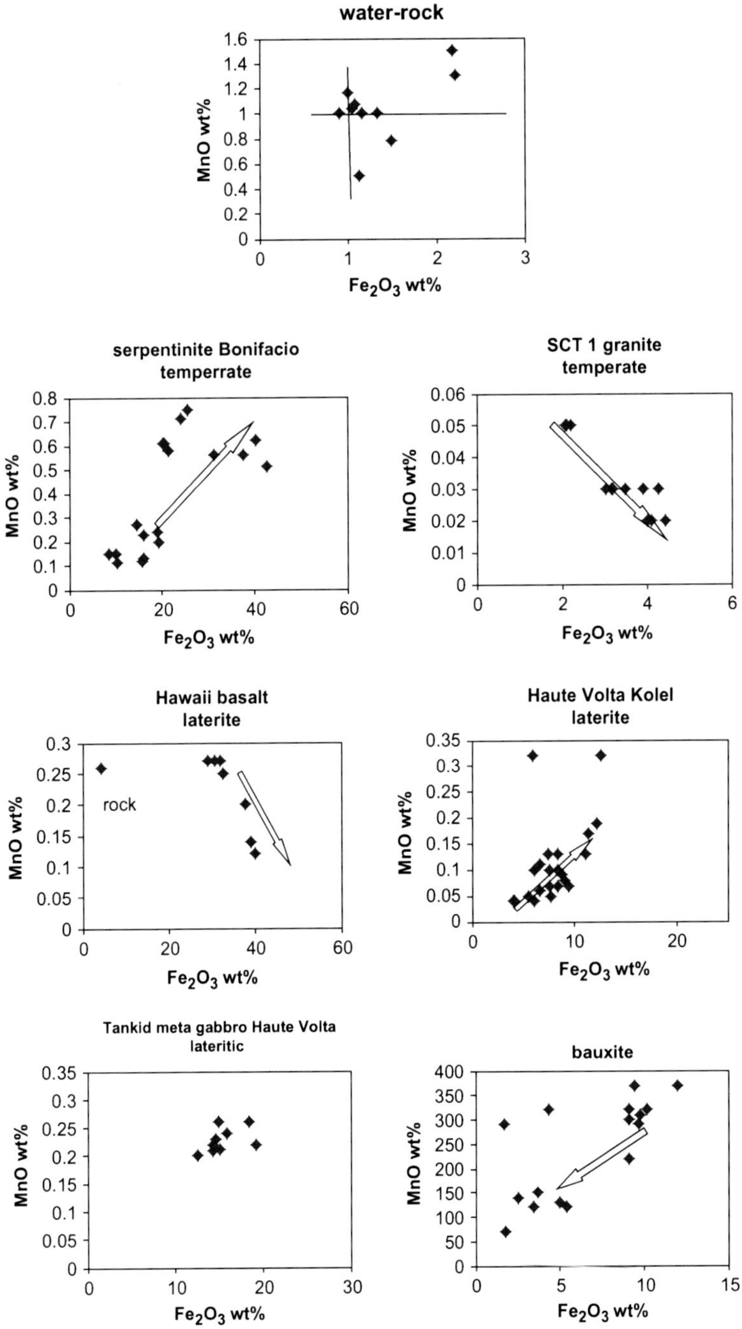

Fig. 3.13 Mn–Fe relations in alteration series for different types of rocks under different alteration intensities. Initial data is for the water/rock interface of initial alteration (data from Table 3.1) and other rock types in an alteration series. *Arrows* indicate increasing intensity of alteration. Basalt and serpentine (iron-rich rocks) from temperate climates (Bonifacio et al. 1997; Valeton 1972) and

be used as an example of geochemical affinities during the alteration process. In Fig. 3.14a there is a positive correlation of Co and Ni in iron crust and bauxite alterites and hence one can assume that the elements are associated throughout the alteration cycle. Most of these elemental concentrations is related to the iron–manganese oxide fractions of the samples.

3.3.10 Co

The relations of Co and Fe or Mn (Fig. 3.14b) indicate positive inter-relations with both Fe and Mn. This is somewhat surprising, but the data seem to substantiate that the higher the amount of Fe or Mn present, the higher the amount of Co. In bauxites Co is related more clearly to Fe content than Mn. In series with high concentration of iron, Co appears to be related more clearly to Mn in some cases as well as in altered serpentinite. This seems to indicate that when iron is lost (bauxites), Co is related to the iron oxides. However in situations of abundant iron oxide Co appears to be related to the Mn present, even though there is little presence (<0.2 %). It seems then that the attraction of Co ions to oxides is dependent upon certain chemical relations that we can only speculate upon at the moment.

3.3.11 Ni

Nickel seems to be associated with both manganese and iron in iron crust alteration materials but slightly less so in bauxites though there is a general relation of correlation for both elements. This is then a similar affinity such as that of Co.

The two transition metal elements, Co and Ni, then can be seen to follow very similar paths of chemical affinity over the whole spectrum of alteration intensity, being associated with Fe and Mn.

3.3.12 Zn

For the data sets available here, Zn does not seem to be related systematically to Fe, Mn or Al in the alterite zone. No clear relationship can be seen in the initial stages of alteration nor in the highly altered iron or aluminum hyroxy-oxide alteration facies.

Fig. 3.13 (continued) those producing laterites Pion (1979). Leumbe et al. (2005) and ultimate stages of alteration from laterites and iron oxide-rich ferricrete materials (Boeglin and Mazaltarim 1989; Beauvais and Mazaltarim 1988)

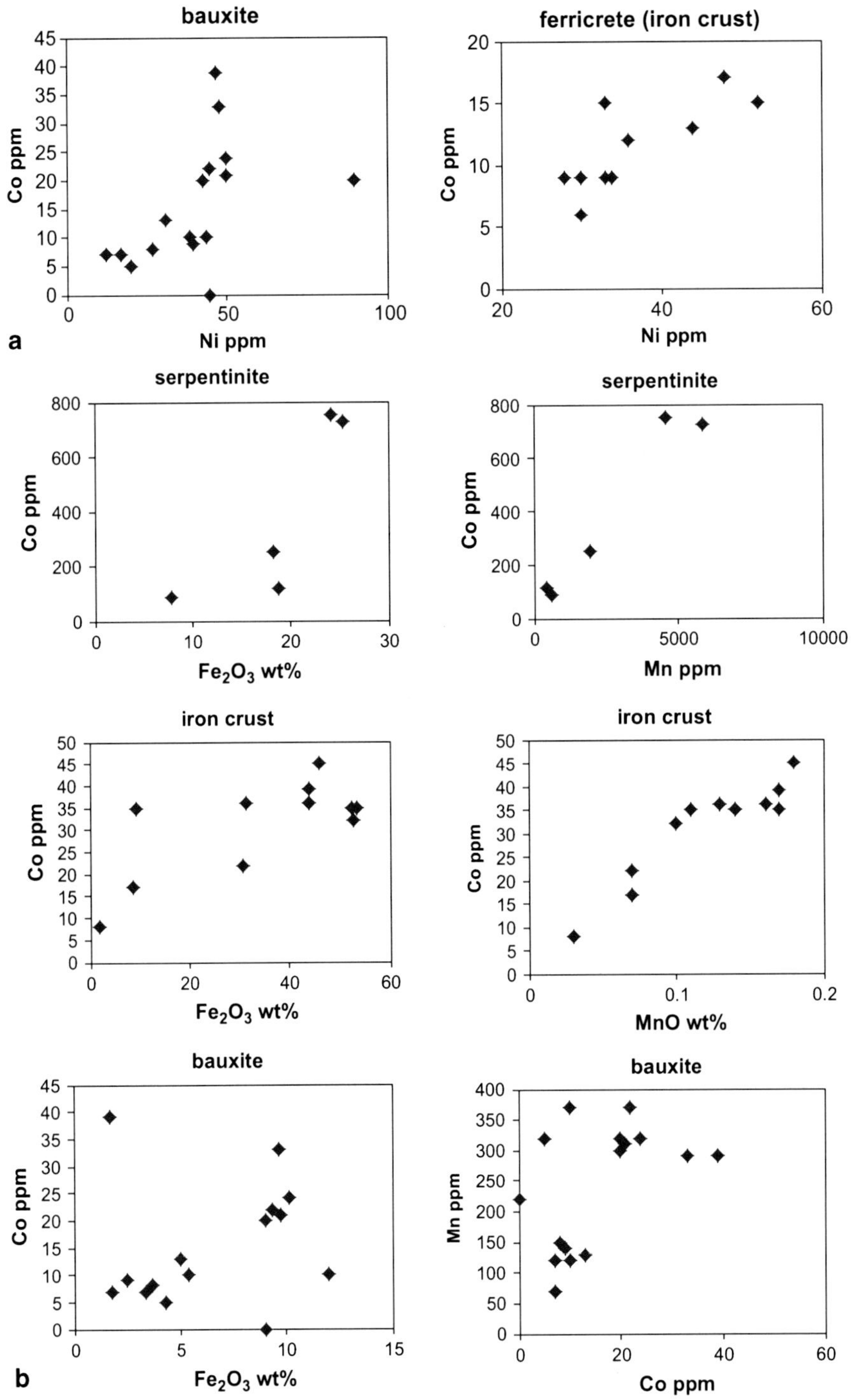

Fig. 3.14 (continued)

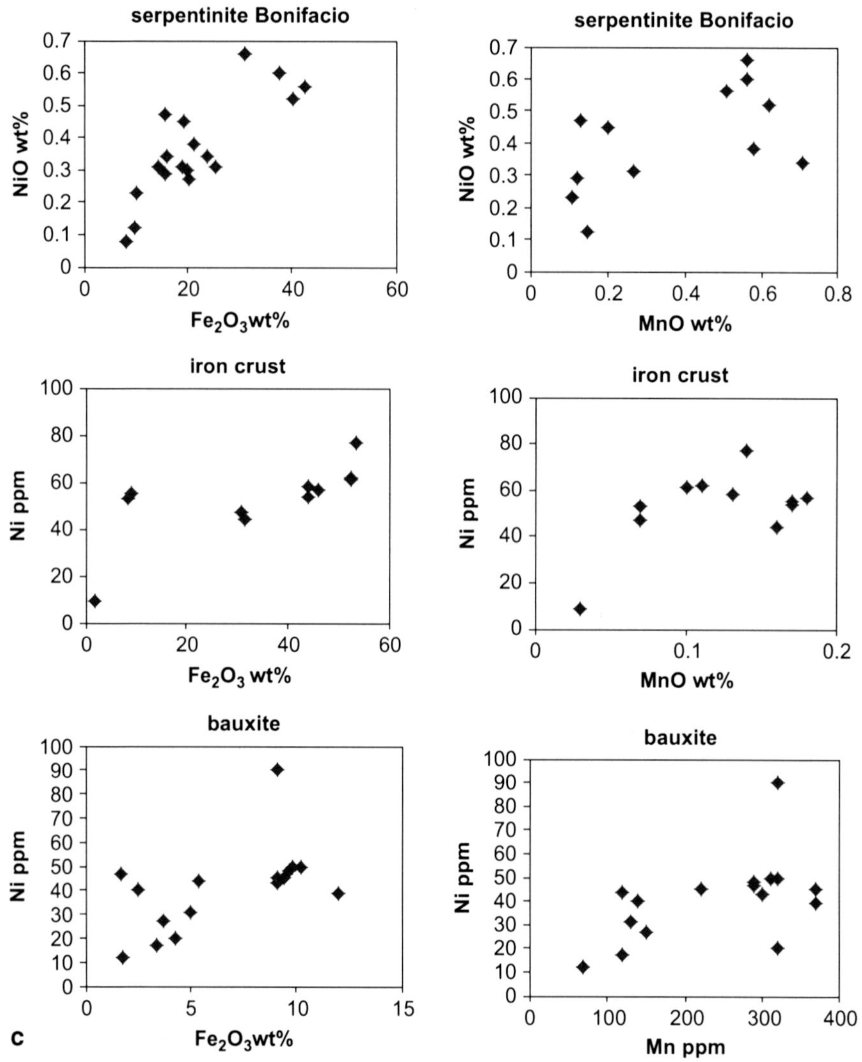

Fig. 3.14 (**a**) Co–Ni concentration relations in alterites producing bauxite or iron crust (ferricrete) concentrations (Valeton 1972; Beauvais and Mazaltarim 1988). The two minor or trace elements are closely related in concentrations. (**b**) Relations of Co and Fe or Mn in different alteration series [Serpenitine temperate climate (Bonifacio et al. 1997; Caillaud et al. 2009) iron crust concentrations, Boeglin and Mazaltarim 1989] and bauxite (Valeton 1972). (**c**) Relations of Ni concentrations and Fe and Mn relations in alterites producing bauxite or iron crust (ferricrete) concentrations (Valeton 1972; Beauvais and Mazaltarim 1988)

Zn is most likely associated with sulfur in high temperature rocks, which upon surface alteration would release the Zn into solution. It is not evident how much of the Zn is removed in aqueous solution from the alterite, but given the lack of associations with other elements, it is not possible to localize its chemical affiliations

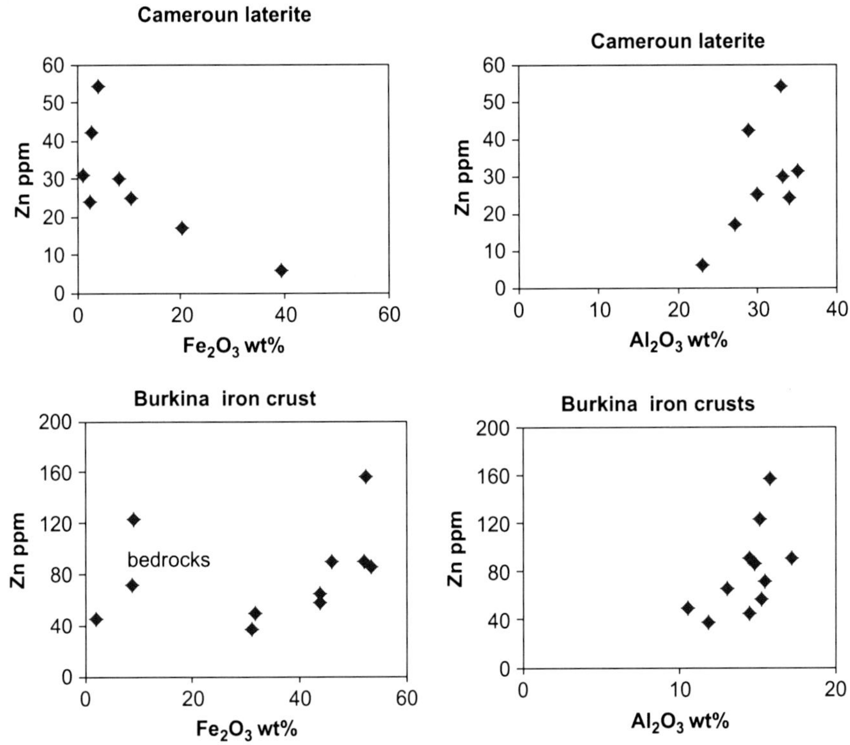

Fig. 3.15 Relations of Zn and Fe, or Al concentrations in laterites and iron crusts (Pion 1979; Boeglin and Mazaltarim 1989)

in the alterite material. In the initial phases of alteration (Apennine soil alterite series) there is a general trend to higher lead and zinc values, probably reflecting the initial mineral associations of high temperature minerals. In Fig. 3.15 some relations are shown for Al and Fe related to Zn in different alteration intensities: in laterite and iron crust types. Manceau et al. (2004) give detailed information indicating that Zn is present in two sites in siliceous smectite. In the samples investigated Zn was found to be present in the 2:1 structure itself and as an exchange cation sites. It is clearly associated with the silicate minerals.

In looking at the data in Fig. 3.15, it is clear that the accumulation relations of the minor elements compared to major elements (Al and Fe) do not show linear relations that cross the origin of the graph. This means that the attraction of the minor elements is low at low iron and aluminum content and much higher at contents of more than 20 % oxide weight percent. This indicates that at a certain point the Al and Fe in the alterite has become an independent phase that has a stronger attraction for these minor elements. Thus one can expect that the minor element relations compared to those of major abundance that eventually produce an independent phase (hydroxide of oxide) will form nonlinear relations when plotted on against the other. This is an important point to keep in mind.

3.3.13 Cu

Cu is a metal which is often associated with sulfur in different minerals in high temperature rocks. These forms should not be prevalent in the stages of alterite formation, sulfur tending to be oxidized into an oxyanion form which is highly soluble, and would not be expected to remain present in the solid alterite phases. In alterites in some instances it appears that Cu is associated with Fe and at the same time there is a correlation with Mn. In these instances, the relations between Mn and Fe accumulation are not clear in the host alterite leaving the question of chemical affinities open to doubt. Nevertheless Cu does seem to be associated with iron oxyhydroxide materials.

3.3.14 V and Cr

In the above short resumé it appears that the divalent metallic ions can often be associated with either Mn (predominantly divalent in its soil oxide form) or Fe (predominantly trivalent in its oxide form in alterites. Co, Zn, Ni, and Cu can be considered to be predominantly divalent in the ionic form under surface alteration conditions. One can consider the elements vanadium and chrome, generally considered to be trivalent under surface conditions

These elements in their ionic forms in alterites are generally considered to be trivalent. However vanadium, a widely spread element in nature, may occur in four different redox states, from divalent to pentavalent. It appears from cross-plotted data that V is more likely to be associated with Fe than Mn. Since iron tends to be trivalent in alteritre oxides, the association of V and Fe seems to be quite logical. The absorption of vanadium on the hydroxyl-oxide of iron, goethite, is well documented (Peackock and Sherman 2004). However, no clear trends are found in bauxite elemental relations for V, Cr, and Fe (Figs. 3.16 and 3.17).

The fate of chromium in soils has been extensively studied (Fendorf 1995). Chromium is derived from both anthropogenic and natural sources, and is mainly present in two stable oxidation states (III) or (VI). Cr(VI) is toxic to both plants and animals, being a strong oxidizing agent, corrosive, and a potential carcinogen. In contrast, the trivalent species is not toxic to plants and is necessary in animal nutrition. Chrome follows the same pattern as vanadium, being more often associated with iron than manganese in alterites. The higher concentration of iron in laterites shows this tendency rather clearly. In bauxites there seems to be no clear association with either Mn or Fe. It appears then that one can identify V and Cr with Fe accumulations in alterites.

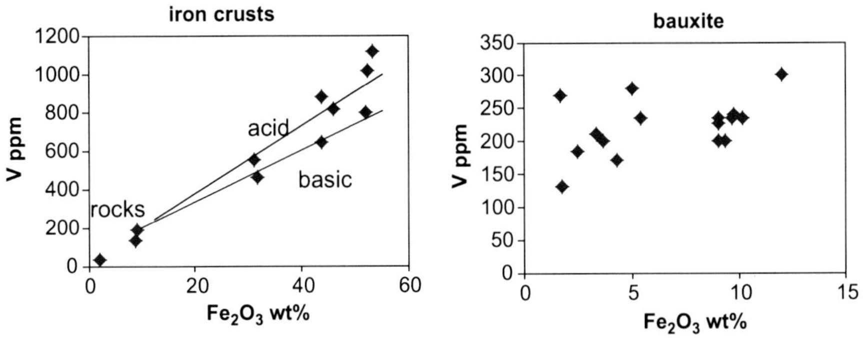

Fig. 3.16 Vanadium relations relative to Fe concentrations (Boeglin and Mazaltarim 1989 and bauxites Valeton 1972)

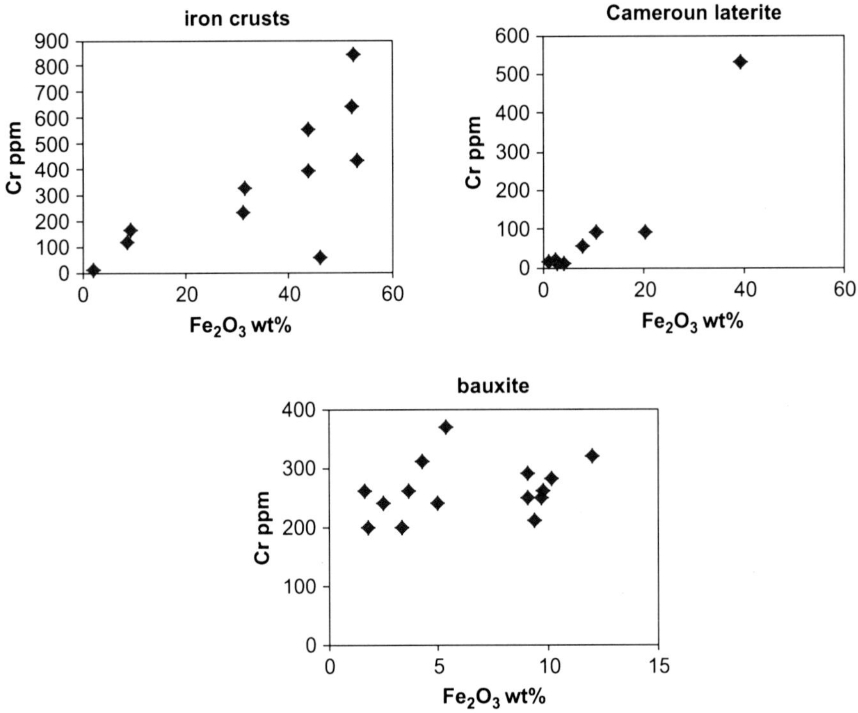

Fig. 3.17 Chromium concentrations compared to Fe concentrations in iron crusts, laterites, and bauxites (Boeglin and Mazaltarim 1989; Pion 1979; Valeton 1972)

The minor element relations of the trivalent transition metal ions V and Cr seem to bear out general logical conclusions that trivalent trace elements will be found associated with trivalent (Fe, Mn) elements in oxide phases in alterites.

3.3.15 Some Heavy Trace Elements Bi, Cd, Sb, Sn, Pb, As, Hg

These elements are often considered in studies of surface (anthropomorphic) pollution, but are not often present in the tables of analyzed elements in studies of alteration and alterites. They act for the most part as cations in surface environments. The following discussion is based essentially upon the data set for Apennine Italian soils where many elements were analyzed at different depth in the alteration profiles.

Two elements which are rather difficult to define geochemically are lead and mercury. The problem stems from their chemical characteristics to a certain extent but above all is due to their prevalent presence at the surface due to airborne or more direct human activity. It is rather characteristic to see lead concentrations increase toward the surface of alterite sequences as well as it is for mercury. The data from the Appenine alterites at depth (B and C horizons) shows a reasonable correlation between Pb and Al content but none for Pb and Fe. Most likely the lead is associated with the clay minerals, which is reasonable in that the divalent lead ion in a hydrated association can easily substitute in the smectite interlayer position in the place of such cations as Ca or Na. Mercury is another problem in that alterite data is scarce and in cases where it has been analyzed in a soil profile it is generally founding strong concentrations at the surface (soil A horizon) and can be suspected of secondary deposition through human activity either airborne or agricultural practice (pesticides, etc.).

Overall it appears that for the samples considered here there is a stronger likelihood of associations of heavy metal ions in alterites with the clay minerals than with oxides. These samples represent the less mature alteration sequences, where concentration of Fe and Al are limited and far from the laterite or bauxite range.

The initial objective is to determine which of the alterite products, iron or alumina-rich, can be associated with a given heavy metal element in the alterite material. Cross plots of these elements with Al or Fe give very similar results, vague positive correlations but roughly the same for either major element. This is not a clear relationship for the major residual elements, which represent either silicate minerals (Al) or oxides (Fe) in the initial stages of alterite formation. Since the heavy elements considered can form oxides, sulfides, or phosphates at high temperatures, it is difficult to identify their interrelations when the rock has not totally adjusted to surface conditions. Some of the heavy trace elements can be found in residual materials, though most likely the sulfide phases where they could have been found would have been affected by oxidation to produce sulfate oxyanion materials, which are highly soluble. One exception in the group cited is arsenic which can itself form an oxyanion complex. If one compares the relative abundances of these elements to lead in young soils formed on loessic volcanic and sedimentary materials, one finds roughly positive relations for all (Fig. 3.18).

Since all of these elements can form sulfides at high temperatures, one can suspect that their surface geochemistry is related to the oxidation of the sulfide

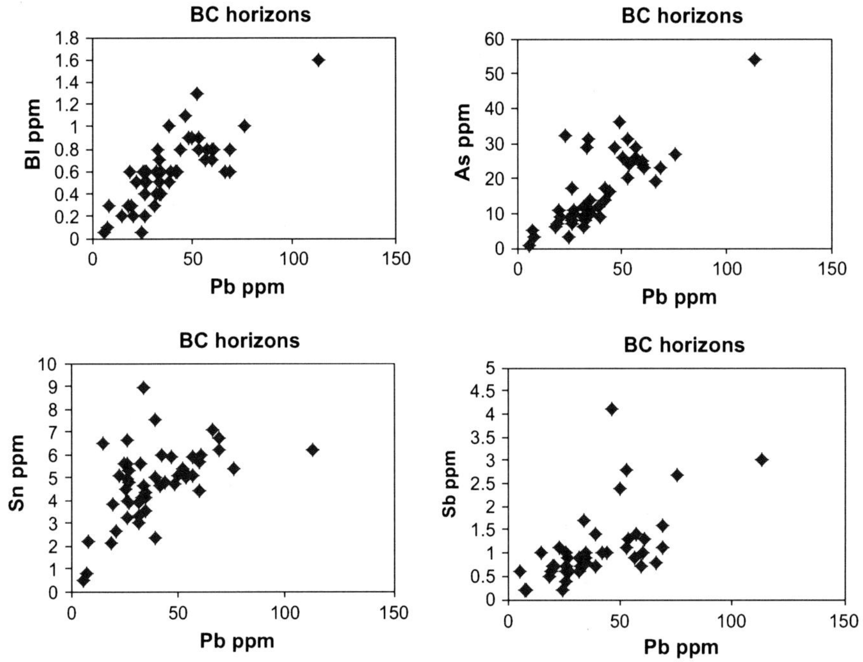

Fig. 3.18 Heavy element abundances in young alteration materials (B, C horizons) on loessic volcanic ash and sedimentary rock materials (data from F. Terrible Univ Fredercio II, Napoli reported in Mileti et al. (2013). Here Bi and As, Bi, Sn, and Sb are clearly related to lead content. Little segregation has occurred between solid materials and altering fluids as the initial materials were transformed into clay minerals, i.e., very little difference in elemental associations are seen between recently deposited material (A horizon) and the alterite material in the B, C horizons

upon alteration and subsequent transportation and initial adsorption on the alterite phases in the form of a sulfate oxyanion complex. Iron can be reasonably attributed to associating Bi, As, Sn with the hydroxyl-oxide. The heavy metal minor elements Cd and Sb cannot be systematically attributed to an attraction to either Al or Fe; however parts of the samples show an apparent association with Al (clays). Correlations with manganese are difficult to interpret due to the low concentration of this element in the samples. There is not a good correlation of Fe and Mn in the samples. Gao and Mucci (2001) found that the phosphate and arsenate forms of oxyanions are attracted to goethite at low pH but much less so at higher values (>7). Thus the observed relationship of As with Al is reasonable.

Overall there is no clear evidence of the geochemical associations in alterites for the heavy elements. However one can find data in the literature for their occurrence in soils, fluvial transportation, and sediments. In these instances both oxide and clay materials are mixed together and can be potential carriers of these elements, be they of alteration origin or be added into the surface materials by human interaction.

3.3.16 Elements in Refractory Phases (Very Low Solubility and High Chemical Stability)

Certain elements are known to form minerals at high temperatures that are specifically difficult to destabilize. These materials frequently end up in beach sands, along with metastable quartz (almost pure SiO_2). However certain elements are associated in several high temperature minerals at the same time. One case is Ti and Fe, which can form ilmenite, the oxide of the two elements combined which is very stable at the surface. However both Ti and Fe can be found in silicate minerals which become unstable under surface conditions. The iron leads specifically to mineral instability when it is oxidized at the surface. Both elements are most often associated in the same high temperature silicate minerals.

In Fig. 3.19 one can follow the relative instability of the silicates and oxides under conditions of increasing alteration intensity (water flow and oxidation). The relations of Fe and Ti are very well correlated in alterites from moderate climates (Fig. 3.19a, b, c) but appear to become more incoherent in more evolved alterites up to laterites and bauxite. This could well reflect the regions of instability of the titano-ferrous oxide or the initial formation of an Fe–Ti hydroxy-oxide alterite phase. In the strong alteration series, it seems that Ti concentration reaches a limit of several percent or less, suggesting an eventual loss to the altering solutions. The relations appear to be valid for acidic (granite) as well as basic rocks (basalt) and other intermediate composition materials (for all volcanic rock types).

Another group or type of element of minor abundance is what is designated as the **r**are **e**arth **e**lements (REE). The rare earth elements (REEs) are a set of seventeen chemical elements in the periodic table, specifically the fifteen lanthanides plus scandium and yttrium. Despite their name, rare earth elements are relatively plentiful in the Earth's crust. However, because of their geochemical properties, rare earth elements are typically dispersed and not often found in concentrated and economically exploitable forms. Today REEs are of great economic interest. The REEs such as lanthanum and neodymium are used to make strong magnets, which help to drive the motors in everything from laptops to electric cars and washing machines. These elements are usually of trivalent ionic state in the surface environment and have ionic radii similar to transition metals and elements commonly found in silicates. The substitution of rare earth elements in silicates found in rocks has been studied in detail (McKay 1989). However the amounts found in high temperature minerals are usually very small. Significant concentrations can be found in some phosphates and carbonates following such elements as yttrium (Mariano 1989). Since phosphates are highly insoluble, the most common form of rare earth-bearing mineral in the surface environment is in phosphate form. Given the low concentration of these elements in silicates, the weathering process, and destabilization of the host mineral, it is difficult to follow them using chemical analyses of the alterite materials (McKay 1989). One could expect changes in relative concentration of some REEs during the transfer to adsorption on Mn and Fe phases according to the laboratory studies of Ohta and

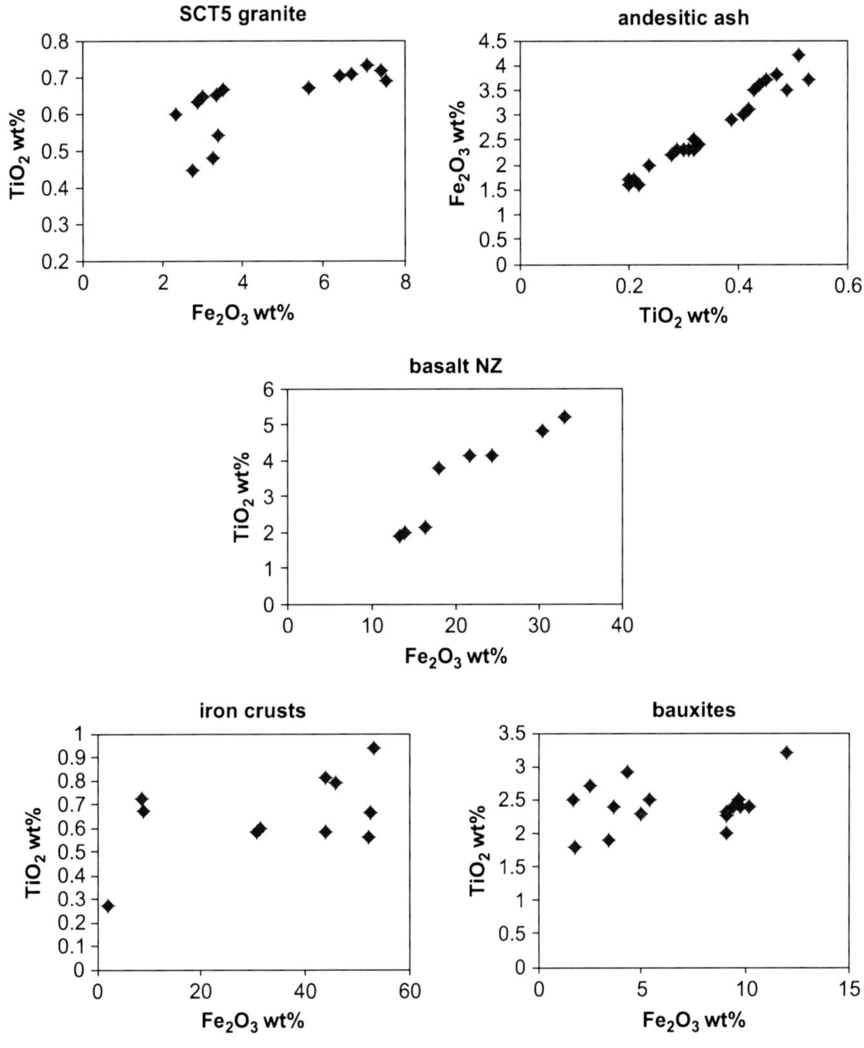

Fig. 3.19 Ti–Fe relations in granite, altered volcanic ash basalt, and highly altered materials (iron crusts and bauxites). Data from White et al. (2008), Martínez Cortizas et al. (2003), Loughnan (1969), Pion (1979), and Valeton (1972)

Kawabe (2001). However most often analyses of alterites reflect the relative abundances of the different elements in the host rock material due to incomplete destabilization of the initial host minerals (McLennan 1989). However under strong oxidizing conditions certain elements, notably Ce is assumed to be able to change valence from 3+ to 4+. This should change its distribution pattern in host phases or independent minerals if they could be formed. Most authors assume that little rare earth material is lost during the mineral transformation of alteration. Singh and

Rajamani (2001) do not find significant change in REE distribution trends in sediments compared to source rocks. Given that there are several phases present in rocks which can contain rare earth elements, their release to the alterite material will depend largely on the relative stability of these minerals in which they are held. Differential dissolution of host material will give very different spectra of rare earth elements since each host mineral usually has a different spectrum of rare earth minerals in different concentration levels.

Rare earth elements are for the most part to be considered as of low solubility. This means that they form oxides or other phases, especially oxyanion types such as phosphates (Byrne et al. 1991), which are of very low solubility at surface conditions. Studies of the abundance of the different rare earth elements in soils and alteration sequences give contrasting results. It appears that the contrasted results are due to the availability of the complexing anions in the altering materials and the intensity of alteration [see Braun et al. (1993), Nesbit and Markovics (1997), Koppi et al. (1996) for example]. To a certain extent the alterite and soil zones can be considered to retain REE in place by the insolubility of the phases that contain them. This is especially true for highly altered material, before the final laterite stage of weathering where much of the REE-bearing material is finally dissolved (Braun et al. 1993). In some lateritic alterites nevertheless most of the REE material remains in the solid materials, even though much can be due to dissolution and re-precipitation mechanisms which form insoluble minerals such as phosphates within the alteration system (Braun et al. 1998).

One can take the soil alteration of loess and wind-born volcanic ash deposit soils from the Apennines [data supplied by Terribilé, Univ Frederico II, Napoli and Mileti et al. (2013)] as an example of the initial stages of alteration of unstable material. Looking at the bulk data for the Apennine soils formed from volcanic ash input, some relations seem very clear. For example Ce and La, the first and second rare earth elements, appear to be closely related in abundance in alterites. Ce and Hf, light and heavy rare earth elements, appear to have a constant relation also. Thus one can expect little fractionation to have occurred for these elements during the transformation of silicate and other minerals during the alteration process unlike that known in high temperature materials (see Lipin and McKay 1989). One can consider the relative abundance and elemental relations of rare earth elements compared to those of zirconium, an element which forms a very chemically inert mineral in rocks and soils. Ce and zirconium are closely related in the Apennine alterites as well as in the highly altered cuirasse (iron crust) laterite materials (Fig. 3.20). The relationship between Ce and Ti is less clear suggesting that the REE containing minerals resist weathering more than titanium-bearing minerals. Curiously the ratio of Ce to Zr is nearly the same in both geological examples. The absolute values vary little from soil to B alterite horizon in these materials where mineral transformation is more important than mineral dissolution. Laterites are of course the reverse example where dissolution is very important for the silicate material. One can imagine then that the rare earth elements in the Apennine soils are in an essentially insoluble form, and that the amount in high temperature silicate minerals altered to make clays and oxides was minimal.

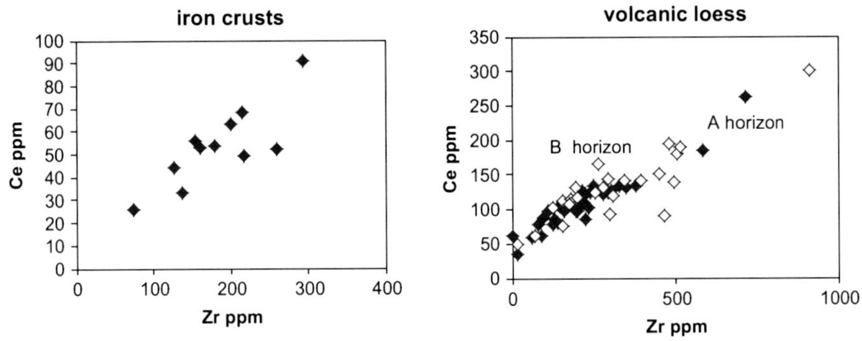

Fig. 3.20 Cerium content related to zirconium in iron crust materials (Mileti et al. 2013) and from volcanic loess soils in the Apennines (data from Terribilé, Napoli and Meliti et al., in press). A *closed symbol* = soil horizon of loess accumulation and B *open symbol* = the alterite horizon)

It appears, from a limited database, that the rare earth element minerals resist alteration more than ilmenite, the Ti–Fe oxide, reputedly chemically inert also. This reenforces the conclusions made by McLennan (1989) concerning the overall effect of weathering on the distribution of rare earth elements in alterites. However a detailed study by Aubert et al. (2001) indicate that weathering of granite under temperate climate forest conditions leads to a depletion of heavy rare earth elements in the alterite. The relative loss is found in the dissolved load of streams in the area studied and is reflected in the fine (clay) material in the upper portions of the profile. This study shows a gradient in rare earth abundance and relative concentration up the alteration profile. There is a smaller change in the lower part of the alterite sequence. This could reflect a change in chemical conditions engendered by the prevalent plant regime in the soil zone. The authors indicate that the change in REE relative element abundance is most likely due to the differential dissolution of the different phases containing the rare earth elements which have selectively included light or heavy rare earth elements. Braun et al. (1998) come to similar conclusions when observing REE in laterites of East Cameroon. Relative loss of heavy and light rare earths is controlled by the dissolution of different minerals and the formation of insoluble phosphates during the alteration process. They suggest that the intensity of alteration can create certain element anomalies such as for Ce. Marker and de Oliviera (1990) indicate that the formation of Fe–Mn oxides during weathering can capture REE during weathering as well as the selective fixation of oxidized Ce by soil vermiculite.

The importance of the relative stability of rare earth host minerals in the weathering of a multicomponent rock and selective release of elements in the series is well summarized in Taylor and Eggleton (2001, p, 139).

Thus the chemistry of alterite conditions is probably not affecting the elements themselves but the stability of the minerals in which they reside in the altering rocks can be a key factor to their presence in the alterite material.

In these samples, rare earth elements found in the residual refractory minerals, oxides and phosphates, are found in the mineral grains of the silt and sand size fraction of transported materials. If they are moved they will be un-reactive during the transport process and should be found in the sand fractions concentrated along the shore lines of large bodies of water. They will then be separated from the clay minerals and associated oxide phases. The concentration of REE in the B alteration horizon is due to the low solubility of the REE containing minerals.

3.3.17 Summary of Minor Element Relations

The geochemistry of minor elements in the alteration process is dictated by the abundance of host phases which can accommodate the elements. The host phases can contain major elements with which the minor elements can exchange within crystallographic sites. Another association is adsorption of the elements of minor abundance onto the host phases either as ions or oxyanions. In both cases the minor or trace elements are substituting for elements of major abundance. The transformation of high temperature minerals, essentially silicates with minor amounts of oxides, phosphates, and sulfides into new silicates (clays) and oxides (Fe and Al) changes the relative abundance of elements in or associated with the solids in going from high temperature rock to surface alterite material. Alteration intensity, from temperate climate to tropical climatic conditions, increases the alteration intensity and gradually eliminates the silicate materials to form either iron or alumina-rich materials. These changes in basic chemistry dictate the presence of minor elements in the more evolved solid alterite materials.

Loss of silicate materials indicates that alkali and alkaline earth elements will be released to altering aqueous solutions. The abundance plots of the elements of minor abundance compared to major elements show the chemical affinities of alkali and alkaline earth elements which is largely due to the similarity in ionic charge, +1 or +2, and their low covalent bonding character. The elements of major abundance such as Na, K, and Ca are present when the silicate clay minerals are sufficiently rich in siliceous clays such as smectites, illites, and vermiculites. When the clays are less siliceous, kaolinite, the alkali–alkali earth content decreases markedly. Therefore these elements and the associated minor element are presenting the early stages of alterite formation but lost as silica from the assemblages.

Transition metal elements of minor abundance are for the most part associated with iron and manganese accumulations. However as alteration intensity increases it appears that Fe and Mn can be retained or released depending upon redox conditions. Iron seems to be the most stable alteration product along with alumina. However the accumulation of one and the exclusion of the other is the final stage of alteration (iron crusts or bauxite). In general transition metal ions are associated with Fe and Mn as alteration increases the iron content of the alterite. The association of Mn and Fe appears to be broken in high intensity concentrations of Fe in iron crusts, where Mn seems to be lost to weathering.

It is important to keep in mind that the attraction of minor elements to oxides and hydroxides will be most important when the oxides are expressed as independent phases. A plot of Cd against Fe for example will not be linear at low iron concentrations because the iron is not expressed in a form that attracts the Cd to its surface. The phases present dictate the geochemical relations in an alterite material.

If we consider that the initial stages of alteration produce smectite clays, containing significant amounts of silica, which can attract metal and alkali ions into the structure as exchange ions as well as fixing them on surface sites, one would expect that the retention of minor elements will be more important in the initial stages of alteration, under temperate climate condition. This is most likely true for alkali and alkaline earth ions but transition metals and heavy metals do not seem to be associated with clays, except perhaps for lead. Overall there is a loss of most elements, major and minor with increasing alteration save Al and Fe, the end points, separately of surface alteration processes.

3.4 Following the Elements

Divalent transition elements (Co, Ni, Zn) tend to follow iron or manganese concentrations. In the data plots there is a tendency to have a constant ratio of host element (major element hydroxyl-oxide) to minor element in the alterites. The oxidized transition metal elements Cr (+2 to +6) and V (+2 to +5) tend to be present in strong accumulations of iron oxides. The formation of predominantly aluminous oxyhydroxides in bauxite breaks these relationships where Mn and Fe are lost from the alterite.

The chalcophyl (elements forming sulfide phases at high temperatures: Pb, Bi, Cd, Sn, Sb) can form oxides, sulfates or chlorides under appropriate chemical conditions. The possibilities of chemical anionic association make them susceptible to changes in associations in the alterites. In fact this multiple association underlines the tendency to remain in cationic form in aqueous solutions where they are mobile and apt to move from one substrate to another.

The elements with a strong tendency to form oxides at surface conditions (rare earth, Ti and Zr) are protected from dissolution and loss in that they are often found in alterites in the high temperature mineral form in the sand fraction of the materials.

A general schema for alteration affinities and the presence of minor elements can be made following the importance of the major alteration phases such as clay minerals, iron–manganese oxides, and aluminum and iron hydroxide. The alterite minerals are accompanied to a greater or lesser extent depending upon the intensity of alteration by the refractory oxide minerals from the initial rock facies. The initial stages of alteration are such that one sees little identifiable segregation and element association based upon the formation of new minerals stable at the earth's surface. However under temperate climate conditions of alteration and the formation of smectite type minerals in the alterite, one can follow changes in the affinities of elements for the new phases. The last stages of alteration and concentration of iron or aluminum eliminate many of the minor elements, as well as major elements or course, and one finds loss of many from the surface material.

During the alteration process there are two forms of transport which can explain loss of specific elements from the alterite. The most evident is dissolution in the altering aqueous fluids and movement out of the alterite zone into the ground water and stream water cycles. A second possibility is movement in the fine grained clay-rich materials where alterite phases can be transported in moving fluids due to their very fine particle size by colloidal transport (colloids are normally in the range of 1 nm and 1 μm). This material also is taken by the fluids to the water table, into streams and further. Thus both dissolved and colloidal material can carry elements out of the alterite. The processes are different. One possibility is in ionic form in solution and the other at surface of solids, or within them. Whatever the chemical reasons the results are the same, loss of material from the surface of an altering rock.

Eventually alterite material will be transported by water erosion mechanisms where the mature material in the soils zone is combined with the alterite material showing different stages of alteration maturity (loss of initial mineralogy). There are nevertheless some minerals that persist even after strong alteration occurs, such as quartz- and zircon-bearing minerals and Ti–Fe oxides among others. These minerals, in the fine sand fraction, can carry significant (relatively) amount of minor elements such as rare earth elements. They do not reflect mineral change nor alteration intensity.

Rates of weathering by water–rock interaction will be variable depending upon the climate. Contact time and volume of water will determine the rate of weathering. Dissolution takes time and is driven by the chemical potential of under saturation with dissolved elements (see articles in Oelkers and Schott 2009). Alteration fronts can be observed in time sequences as presented by Brantley and White (2009) for terrace sediments in central California. One can plot the alteration front as a function of time (see Fig. 3.21) but extrapolation to a rate of alteration assumes constant climatic conditions over the period of observation which is probably not realistic for periods of hundreds of thousands of years. However, the data in the figure indicates a rather regular progression of alteration of the rock–alterite interface for these samples now in a semi-arid climate context. Thus under moderate climatic conditions the alteration front progresses several meters per hundred thousand years of interaction on granitic materials. Numerous factors, such as rock fractures, slope, and drainage can influence the rate of water loss and hence amount of interaction per amount of rainwater entering the system. However it appears that the rate of interaction and alterite formation is reasonably low.

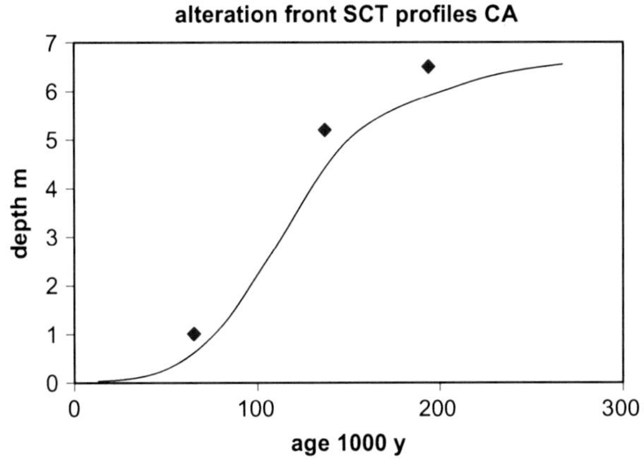

Fig. 3.21 Alteration rate of SCT profiles on granite in semi-arid California. Depth (m) of alteration vs. age of alteration in thousands of years

It should be noted that the data in the figure suggests an "S" shaped curve relationship which is typical of biologically driven reactions (see Velde and Barré 2010, pp. 66–69). Certainly the initial stages of alterite formation is biologically driven, where the low reaction rate slope is visible but in the later stages a more normal chemically driven reaction rate curve is apparent.

The water–rock interaction interface is the beginning of elemental displacement at the surface of the earth and the key to elemental segregation during the process of phase transformation due to dissolution and precipitation of new minerals under surface conditions of aqueous and atmospheric interaction with rocks.

In the end one can attempt to look at the elements by their tendency to be dissolved in aqueous solution or their tendency to remain in the solid state after interaction with air and water in the alteration zone. Numerous compilations have been made to compare the abundance of elements in rocks compared to different forms produced by alteration. Allard (1995) indicates that the relative abundances of elements in ground water shows that few are more plentiful in water than in the rocks. However a few are, which should be remarked. One is Chlorine to be expected in that the ocean is essentially a chlorine anion based solution. A second element of great importance is sulfur. The forms of sulphur–oxygen compounds found at the surface are highly soluble and rarely found as solids at the surface of the earth except in evaporate materials where chlorine is also found. The least soluble or least abundant elements in ground water are those of high valence, tri or quadrivalent, which are present as very insoluble oxide compounds in rocks. Rare earth elements, titanium and zirconium, are examples. Strongly covalent elements such as Al and Si are slightly more soluble, which is to be expected in that they form new minerals from high temperature phase and hence com into chemical equilibrium with aqueous solutions. Monovalent cations such as Na and K are of course highly soluble and well represented in ground waters. Thus the effects of chemical interaction at the surface and the movement of materials in solutions as dissolved matter are dependent upon the chemistry of the elements and the solubility of the new phases that they might form.

3.5 Useful References

Aiken W (2002) Global patterns: climate, vegetation and soils. University Oklahoma Press, Norman, OK, p 435

Black C (1957) Soil-plant relationships. Wiley, New York, NY, p 792

Buckman H, Brady N (1969) The nature and properties of soils. Macmillan, New York, NY, p 651

Drever J (1982) The geochemistry of natural waters. Prentice Hall, Upper Saddle River, NJ, p 436

Foth D (1990) Fundamentals of clay science. Wiley, New York, NY, 360 pp

Hooda P ed (2010) Trace elements in soils. Wiley, New York, p 596

Kabata-Pendias A, Pendias H (1992) Trace elements in soils and plants. CRC Press, Boca Raton, FL, p 364

Loughnan F (1969) Chemical weathering of the silicate minerals. Elsevier, Amsterdam, p 154

McFarlane (1976) Laterite and landscape. Academic, London, p 151

Scott K, Pain C (2008) Regolith science. Springer, Heidelberg, p 461

Sposito G (1994) Chemical equilibrium and kinetics in soils. Oxford University Press, New York, NY, p 269

Tardy Y (1993) Pétrologie des latérites et des sols tropicaux. Masson, Paris, p 457

Valeton I (1972) Bauxites, Developments Soil Science, vol 1. Elsevier, Amsterdam, 226pp

Yaron B, Dror I, Berkowitz B (2012) Soil-subsurface change. Springer, Heidelberg, p 366

Chapter 4
Soils: Retention and Movement of Elements at the Interface

4.1 Background Setting

Soils are developed at the surface zone where atmosphere and rock materials have interacted in the first instances of alteration, specifically with rainwater that is basically very unsaturated with respect to mineral elements. At the interface water moves into the alterite and eventually it moves outward (down hill by gravity) to the water table and into the surface flow of streams and rivers. This movement of water from land surfaces to streams and rivers brings with it the dissolved and some fine-grained material (clays) into the system of material displacement which is called erosion. Erosion does not necessarily mean displacement by mass movement of solids. Dissolved material makes up a significant amount of displacement of material in the geological cycle. The most important aspect of the contact zone is that it is for the most part covered, or at least partially, by plants which form a zone of roots which interact with the alterite material. This is evident in a general way, but in the consideration of most pedologists, soils are the result of the presence of plants. Without plants soils would not exist. In fact plants hold the alteration debris in place against the forces of erosion by water impact. If plants are not present, the alteration material is moved to lower zones and eventually deposited as sediment either in dry wash beds or in river sediments. Accumulation of alteration products at the horizon of altered material is the work of plants, which form and stabilize a mantle of material into which they can reach for mineral sustenance and water resources. The uppermost portion of the alteration profile is frequently divided into the O horizon where organic matter is dominant and just below is the A horizon where silicate and oxide alterite material is dominant in the presence of decomposed organic materials. Eventually the alterite, below the A and O horizons, becomes very thick, tens of meters in tropical zones. Plant roots become a minor portion of the profile at depth but all the same they stabilize the uppermost fine grained material of the alterite mass.

Hence the soil is often considered to be that zone where the remains of plants are present, in the form of soil organic matter, either dominant in amount (O horizon) or

A. Bauer and B.D. Velde, *Geochemistry at the Earth's Surface*,
DOI 10.1007/978-3-642-31359-2_4, © Springer-Verlag Berlin Heidelberg 2014

less important but nonetheless present (A horizon). The soil horizon is essentially one of circulation of water and air. The porosity is maintained by the action of roots, which make cavities in the soil and leave them open as they die and retract. This makes a zone of oxidizing conditions, as most plants need air for their different metabolism functions. Not only water but also air can circulate in the soil (root) zone.

The soil zone is also characterized by a concentration of clay and oxide minerals, which are most often greater than the clay content in the alterite zone of water–rock interaction below it. Clay-sized material (<2 μm) is extremely important for plants in that it forms aggregates when combined with decomposed organic matter which contain capillary water that is a reserve for plants during dry conditions. The clay and oxide minerals present also fix nutrient elements essential for plant growth and development. Clay–oxide mineral content is extremely important for geochemical considerations in that this material is the vehicle of minor element retention or release. As clay-sized materials move fixed minor elements move with them. Clay-sized material (oxide and silicate) content is the key to movement of minor elements at the surface.

Gradually the clay and oxides of the A horizon, which can be continually created from un-reacted rock debris present even in the upper parts of soil profiles, migrate downward into the column below (B horizon) creating a layer of clay-rich material which fixes minor elements adsorbed on them or absorbed in them. Clay migration is due to the movement of rainwater into the soil through passageways kept open by plant root action. Another possibility is the formation of soil aggregates, which contract on drying and leave, temporarily, open passages to new rainfall after dry periods. The process of clay migration is initially relatively rapid and slows its rate as time passes (Fig. 4.1a). At longer periods of soil formation and alteration the B and A horizon join in roughly the same clay content. Typically in alteration profiles of intermediate development, the B horizon has a higher clay content than the A horizon. The porosity of the B horizon decreases due to clay accumulation, when the natural porosity channels become filled with clay.

Plants have root systems of varying density, depth, and distribution. These roots modify the soil structure by creating pores (root passages) and by extending organic soil materials as exudates, which interact with clays creating soil aggregates, which in turn shrink on drying to create pores and passages in the soils. Figure 4.2 indicates some differences in soil porosity due to plant action on soils in samples taken from an experimental plot in Iowa (courtesy of D. Laird, National Soil Tilth Laboratory, Ames, IA, USA). The corn (*zea mais*) planted soils show a rather shallow porosity pattern (Fig. 4.2a) while soybeans show a significantly deeper structure (Fig. 4.2b). The larger, rounded pores are due to earthworm activity. The clays tend to concentrate at the base of the root activity zone forming the B horizon. Different plants have different root structures and will give different porosity with resulting clay distribution patterns.

The work of pedologists has been in part dedicated to the characterization of the soils found at the contact of alterite and plants as a function of climate, which in turn determines the plant types present. The plant population, controlled by climatic

Fig. 4.1 (**a**) Clay content in B horizon of prairie soils (USA) as a function of time (data from Foth 1990). (**b**) Here one sees the accumulation of clays in pore spaces forming successive layers of sedimented clay platelet material which gives a particular optical texture when observed in thin section under the microscope. These accumulations are called cutanes. The microphotograph shown is a site of clay accumulation in a soil where the clay has migrated along fractures and accumulates in recognizable masses where particles are aligned roughly parallel to one another (photo A. Meunier; photograph shows a 7 mm wide zone)

a

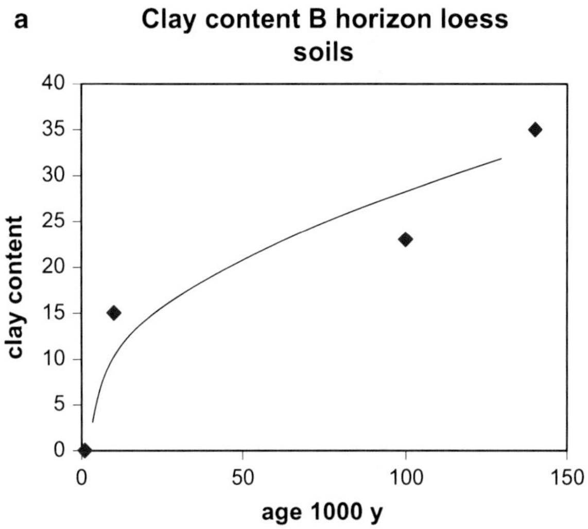

b Cutane clay deposits around cracks in soil

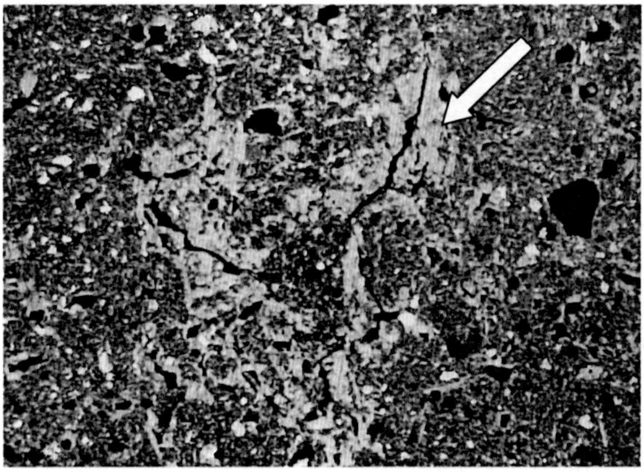

factors, determines many of the chemical and physical characteristics of the soil material in the root zone. The climate, temperature, and humidity determine the conditions of bacterial action, as does the type of plant debris deposited on the surface. Some plant materials are easily degraded by bacterial action while others are more refractory to bacterial degradation. Thus there is a strong interdependence between climatic conditions, plant type, and bacterial action which all condition the basic chemistry of the soil solutions that interact with the mineral debris or alterite material present in the A horizon.

Plants in the soil zone affect the chemistry of the soils and the porosity, which condition the movement of particulate material. Two major chemical variables determined by organic actions are pH and Eh. These chemical variables strongly

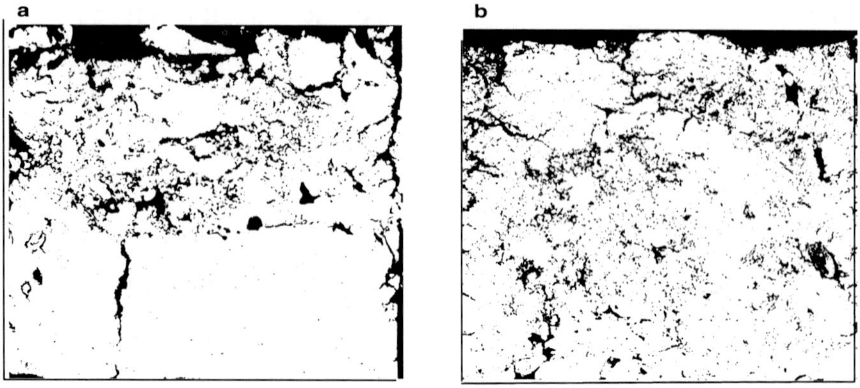

Fig. 4.2 Root porosity structure developed by two plant types. Examples of the effect of the root systems of different crop plants on soil macroporosity (shown in *black*). Structure (**a**) is for a field planted continuously in *zea mais (corn)*, structure (**b**) is for a field planted continuously in soybean. The *zea mais* roots maintain a porosity to a depth of about 10 cm while the soybean roots maintain a structure to near 20 cm. Samples courtesy of D. Laird National Soil Tilth Laboratory, Ames, IA, USA)

affect the stability of mineral phases, by controlling the presence and type of iron present (Eh) and to a certain extent the type of silicate mineral present (pH). The stability of these solids in turn then controls the presence or fixation of elements of minor abundance and can be extremely important to the life cycle of plants and animals.

Movement of material in the alterite profile, below the soil zone, can be considered to be essentially dissolved matter from the water–rock interaction of the soil zone, but some clays are transported to the water table. Clays move downward in the profile from A (soil) to B (subsoil) horizons in pore created by plant root pathways. Some material is moved laterally on slopes through the porous soil structure. This lateral movement is important to consider when dealing with surface geochemistry in that it is an important vector in the movement of alterite materials toward rivers and eventual sedimentation. It includes dissolved elements due to alterite formation (dissolution) and the movement of particulate alterite products, clay-sized particles and colloids down slope to streams and rivers.

4.1.1 Soil Development Types

4.1.1.1 Immature Soils

It is possible to consider the effects of chemical alteration and plant–rock interaction as a function of the age of the soil. If little time has elapsed in order for plants to interact or for rainwater to effect dissolution–recrystallization reactions,

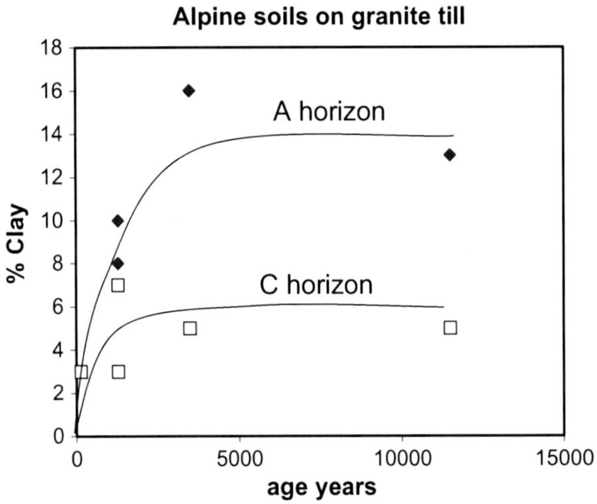

Fig. 4.3 Clay content in alpine soils as a function of the age of the soil in the A horizon and alterite, C horizon (data from Egli et al. 2001). The initial rate of formation is important, but it becomes more stable increasing slowly after 4,000 years or so

such considerations are not appropriate. A bare rock on a mountain top is not a good example of soil formation.

In cold climates little interaction occurs, even for reasonably long periods of time. Egli et al. (2001) show that the clay content of the A horizon of alpine soils, where plant roots have been present, can only change the clay content to several percent of the alterite material after 10,000 years. Other studies of alpine soil alteration indicate higher rates of clay formation (Righi et al. 1997; Egli et al. 2001) of up to 13 % after similar periods of alteration. In all cases there is a significant loss of mineral material to aqueous solutions while relatively little clay-sized material is formed. In Fig. 4.3 the development of clays under alpine conditions is visibly greater in the A horizon (soil) than the C alterite horizon zone. The initial phases of alteration and plant growth show a strong creation of clays, which decreases in rate with time.

In very dry climates little plant activity is possible and soils are almost inexistent in many cases. Erosion, due to infrequent but violent rain storms, sweeps the fragmented rocks (due to physical weathering) from the surface and little accumulation of alterite is observed. Thus situations of extreme climatic conditions (cold or dry) give little evidence of plant and alterite interaction. Materials moved from alteration surfaces of immature alterites contain little clay. Most of the major and minor element change is that of loss to altering fluids when present and active.

Immature alteration stages produce clays at the surface A horizon with less clay found in the lower part of the alteration profile.

4.1.1.2 Temperate Climate Soils

Temperate climate soils are extremely important to human activity in that a very large portion of the population of the world derives its food resources from temperate climate soils. Populations traditionally follow food. Temperate climates form two broad categories of plant cover: forest and prairie. These biomes are present as a function of climate and substrate rock composition. In general, forests, especially conifer forests, are favored by and maintain acid conditions in the soil zone. Acidity is maintained by low activity of bacteria in cold and wet climates. Hence conifer forests are typically found on mountain massifs where acidic magmatic rocks such as granites or metamorphic rocks of similar composition are dominant and the climate does not favor bacterial action in the soil. Prairies tend to be found on sedimentary rock basements or more recent and unconsolidated sediments. These last groups tend to contain carbonate material, and are generally more chemically basic in character. Prairie plants favor basic soil substrates such as carbonates, which they maintain to a certain extent (see Jenny 1979). In temperate climate zones, areas of food production were based upon native plants of the *graminae* types, wheat, rice, and corn. Much of the expansion of agriculture under temperate climate conditions has been devoted to the conversion of former forest areas to those suitable to prairie-type plants (see Velde and Barré 2010, Chap. 5). To a large extent these soils are those found around the areas of important human populations, and hence those which will be susceptible to the transfer of minor elements becoming at times major element concentrations through anthropic actions.

In old alteration profiles of temperate climate origin (hundreds of thousands of years) one finds a migration of clays from the A horizon that have moved downward to what is called the B horizon. The phenomenon is present but of varying importance in soils based upon different rock types. The finer grained the bedrock (sedimentary rock) the lower permeability and relatively less transport is observed. Clay content increases as a function of depth in the upper parts of the profiles for old soils in South-eastern USA as seen in the data of Oh and Richter (2005). The development of a clay-rich horizon is more clearly delimited in soils and alterites developed on coarse-grained rocks such as granite where coarse-grained quartz provides a granular substrate to allow clays to move downward. In the fine-grained shale, much of the material develops into clay-sized materials and movement is slowed by lack of pore and fracture pathways.

The overall result is that alteration is strongly influenced by the actions of plants in soils. Alteration of un-reacted minerals and stabilization by roots, which prevent surface erosion and help to concentrate clay-sized materials near the surface of the alteration zone. Plants then maintain clays at or near the surface where they will be exchangers of elements and other substances. The lower clay content materials of the initial stages of rock–water alteration are typified by element loss while the upper parts of alteration profiles, soils, tend to maintain certain elements in place or bring them to the surface by uplift mechanisms through the action of deep roots (see Velde and Barré 2010).

4.1.1.3 Mature Soils

The end point of alteration is where oxyhydroxides of aluminum, iron, or manganese are present [see Pedro (1966) for example]. These are the mature soils, those which are very old, not having experienced significant surface erosion, or those formed under tropical wet climates where very abundant rainfall has accelerated alteration. In the alteration profiles under warm, wet tropical conditions the organic layer is relatively reduced due to the high activity of bacteria and other agents of organic matter transformation. In tropical regions the alterite zone can reach depths of tens of meters where only fine-grained oxyhydroxide alterite material is present in the clay-sized fraction which represents most of the alterite material.

An example can be seen in the clay content of soils developed on basalt on Hainan Island (China) under subtropical climate conditions (commented on in He et al. 2008) which increases clay content with time at a different rate in the A and B horizon. After longer periods of time under these weathering conditions of strong alteration, the clay content converges with alteration where clay-sized material is present in high amounts throughout the profile (Fig. 4.4). Mature soils formed under conditions of high alteration intensity tend to show less differentiation in the amount of clays present and the different horizons are less well characterized by clay content.

4.1.2 Summary

The different categories of soil development and type depend largely on climate and to a lesser extent on geomorphological parameters. In mountains one tends to find young soils on slopes, where massive erosion occurs with reasonable frequency leaving bare or slightly altered rock to be affected by water–rock interaction. On the top of high mountain chains the low temperatures inhibit alteration processes, chemical and biologic, so that the evolution is slow. Overall climate affects the development of soils: again cold areas have slow rates of evolution and in very dry areas the evolution is even slower. The remaining areas of temperate or tropical climate, those of moderate temperatures and variable but abundant rainfall, show rates of evolution based upon the amount of water–rock contact. The more water flowing into a system the more it will alter the minerals present. Plant activity follows this trend and reinforces it.

One of the major agents of retention of clay-sized alterite material is the presence of plants. These agents of change prefer to keep a relatively high clay content in their direct environment. Root systems function as a means of accumulation of nutrients, such as water and mineral elements, as a source of physical stability holding plants in place and also as an agent of stabilization of clay-sized material.

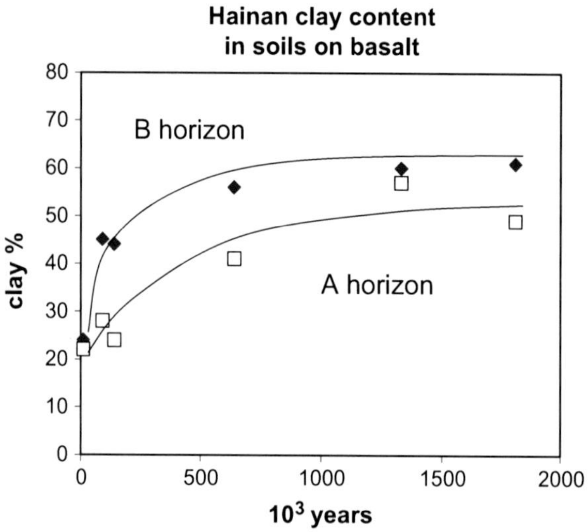

Fig. 4.4 Clay content in soils formed on basalts on Hainan Island, southern China. Clay content of the A and C alterite horizons initially diverges but then converges to the same values after a relatively long period of time of soil formation (1.5×10^6 years) under conditions of sub-tropical alteration (data from He et al. 2008)

4.2 Chemical Uplift by Plants

4.2.1 The Chemical Effects

Plants are known to be important in determining certain parameters, which control mineral transformation and stability in soils. Movement of mineral elements from depth to the soil zone can be called an "uplift effect" (Jobbagy and Jackson 2004) Analyses of pore water in A, B, and C horizons in soils developed on terraces over periods of hundreds of thousands of years (White et al. 2008) indicate that the elements in aqueous solution follow different trends. The trend depends upon the proximity to plant material at the surface. Chemical interactions are different in the two zones. This is shown specifically by ratios of Mg/Ca and Na/K. The authors take these relations to be a strong indicator of the effect of plants on surface chemistry. Weathering rates have been seen to be greatly enhanced by plant activity which can be 10–18 times greater under pine trees than plant free samples in laboratory experiments (Borrman et al. 1998). These observations being the case what chemical parameters do plants control?

4.2.1.1 Potassium and Silica

It has been known scientifically for some time and deduced from practical experience for longer periods (thousands of years) that plants influence the chemistry of the soil zone. This action can be called chemical uplift [see Velde and Barré (2010) for a more detailed discussion]. Observations of alterite and soil chemistry often, but not always, indicate a change in the abundance of certain elements. This is especially true for potassium, calcium, and silicon, elements brought to the surface, which enter the plants and are deposited at the end of the growing cycle of the plant. Different plants concentrate these elements to different extents, but the effect is general for most plant species. Overall the uplift effect creates a buffer zone, which is active against the loss of some elements, especially potassium, and the maintenance of silica-rich species of clay minerals in the soil zone. The end result is to maintain the presence of certain elements as soluble or available cations and to prolong the existence of silica-rich clays, which are strong cation exchangers.

An example of the direct control of plants on the chemistry of the surface A horizon can be seen in the data of Schultz et al. (2010) for terrace soils composed of granitic material of different ages on the coast of central California (USA). Here the climate is of a Mediterranean type, with limited and variable rainfall. The conditions of plant growth are hence variable. Analyses of plant biomass compared to the potassium content of the soils developed upon granitic materials indicate the importance of plants in bringing up material to the surface. In the older terraces, plant growth is reduced, and the potassium content of soils is lessened. A rather strict correlation between biomass and potassium in the soils is observed (Fig. 4.5). These relations indicate that the residence of potassium in soils is dependent upon a continuing supply of potassium from depth brought up by root action. Lower input into the soil zone is directly reflected in the soil alteration layer potassium content. The potassium is not strongly held in the soil clay material and it is eventually extracted by alteration processes when its supply is not renewed.

The elements most notably uplifted by plants are potassium and silica, but others are affected also such as minor elements. The important point to realize is that significant amounts of elements are brought to the surface and stored, at least for some time, in the soil horizon. This material is then not immediately lost to the chemical forces of dissolution at the rock–alterite interface. Not only are clays stored at the surface but also a certain number of major, and also minor elements, will be retained in the zone of plant influence.

White et al. (1998) show that the compositions of pore water in the alteration profiles from Porto Rican tropical forest soils give a strong indication of chemical uplift in that many of the elements normally lost in the initial stages of alteration are found in greater abundance in pore waters in the soil zone (such as Mg, Si, K). Thus this indicates that not only do plants bring certain elements to the surface. These elements are lost in part to the pore water during alteration at the surface (altering solutions) and thus these elements need to be replenished to the growth cycle of the plants.

4.2.1.2 Climate and Chemical Uplift

The uplift effect is sensitive to climate, as plant regimes change. Data from a volcano in Costa Rica (Meijer and Burrman 2003) indicate that the uplift effect is most likely a function of plant activity. In the lower altitudes (higher rainfall and higher temperatures) the increase in concentration of Si and K in the soil zone is

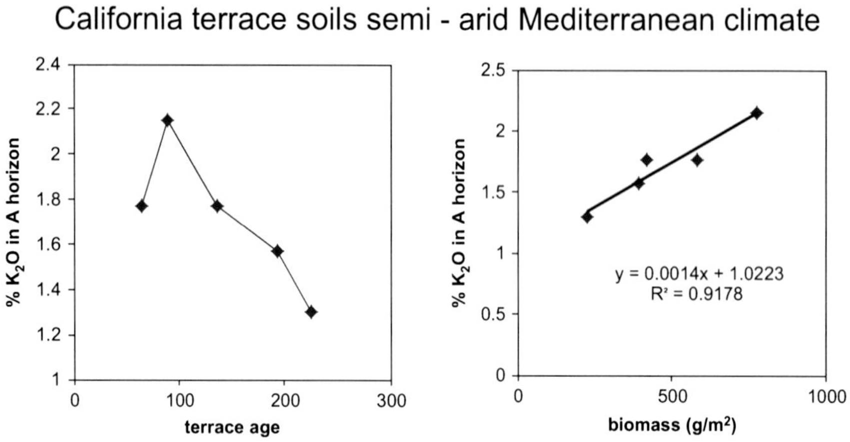

Fig. 4.5 Relationship between biomass and potassium content in the A horizon of soils developed on a series of terraces in central California (Schultz et al. 2010). The initial periods of colonization of the granitic terrace material show an increase in biomass and subsequent decrease marked by potassium content of the soils. Further plant growth appears to exhaust the immediate potassium resources and biomass decreases correspondingly

stronger than at higher altitudes on the same andesitic bedrock materials (Fig. 4.6). Two sets of chemical profile data for basalts developed under temperate climate (central France) and subtropical conditions (New Zealand) are given in Fig. 4.7. Despite the differences in climate and plant regime the presence of silica and potassium initially decreases from the rock composition in the alterite zone, but there is an increase toward the soil zone where plant activity is important. The change in silica and potassium is important in these soils because the bedrocks are specifically poor in these elements, near 40 wt% SiO_2. They neither will contain quartz which, through its metastable presence, typically enriches the surface layers of an alteration profile. Hence in these cases it is clear that the plants reach below the surface root zone with deep roots to absorb Si and K from solutions that are altering the rock mineralogy in the water–rock interaction zone and bring these elements to the surface into the plant biomass where they are recycled to the soil zone upon release from plants and decay of plant materials (leaves, twigs, roots).

Climates are different at different parts of the earth surface. Thus the effect of plants is quite variable. One measure, which can lead to confusion concerning the effect of vegetation, is the yearly average rainfall often used to describe or measure climate. Rainfall does not alone determine the type of vegetation nor the effects of alteration. A steady amount of rain gives different results compared to that coming in short bursts over a year's period of plant growth. Savannah regions in areas of contrasted climate receive as much water as moderate temperate zone plants (northern Morocco and Central Western France for example), but the plant regime is significantly different, one dry grasses and the other deciduous forests. Temperature and rainfall frequency change the plant regime and the alteration intensities.

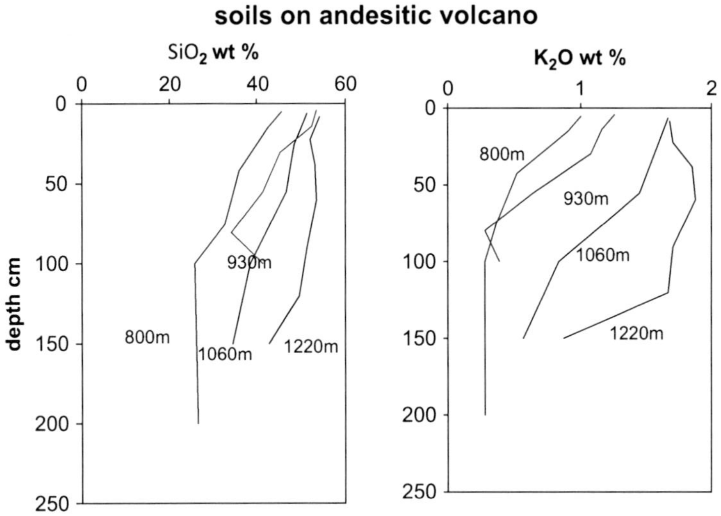

Fig. 4.6 Data from Meijer and Burrman (2003) for the chemical composition of alteration and soil profiles from an andesitic volcanic Porto Rican mountain. The altitudes indicated show different profiles with similar trends where Si and K are increased in the surface horizons. The trends are more accentuated where alteration is highest, at lower altitude and plant activity is higher

Further, different plant species can bring up more or less of different elements, such as silica (Cornelis et al. 2010) as shown in the soils below different species of trees growing under temperate climatic conditions. Uptake of silica can vary by a factor of ten among deciduous and conifer species. Hence the absolute values of chemical uptake are variable according to the parameters of plant growth and plant type, but in general the impact of plants will be apparent in the chemistry of the soil zone.

Thus one can conclude that there is a general trend for enrichment of some major elements in the soil surface layer under the influence of plant growth. This is very frequent but not universally true, when one looks at the data available in the literature. The plant uplift effect is especially important in the temperate climate zones.

4.2.1.3 Time

Overall, alteration reactions will eventually impoverish the alterite–soil system under conditions of abundant rainfall. The basalt series on subtropical Hainan Island China of different ages (described by He et al. 2008) shows an initial increase in clay content whose rate of increase decreases with time arriving at 60 % of the material present in the A horizon at 1.7 million years of alteration time. The potassium content of the A horizon initially increases but gradually decreases as the clays become less siliceous and less capable of fixing alkalis (Fig. 4.8).

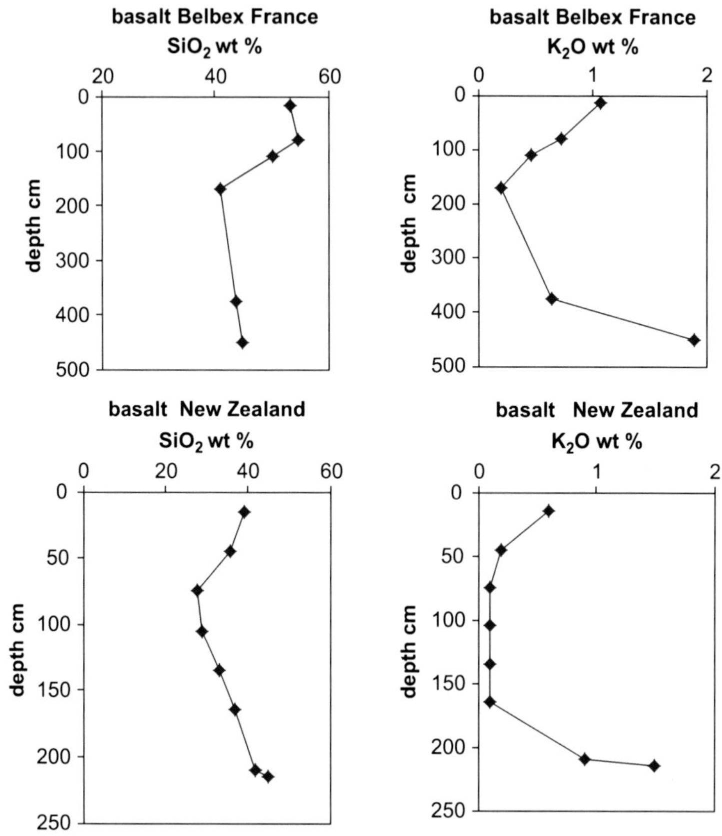

Fig. 4.7 Profiles for alteration zones on basalt from a temperate climate (Belbex, central France, Chesworth et al. 1981) and a basalt altered in subtropical climate in New Zealand (Loughnan 1969). Clear trends for silica and potassium surface enrichment are present. The chemical change is easily seen in that the basalts are silica-poor (without quartz which remains metastably present in soils) and have low potassium contents compared to most eruptive and metamorphic rocks. Below the zone of plant activity it is clear that water–rock alteration produces a potassium-poor material and the potassium in the rocks is largely lost to circulation water in the alteration zones

The plants cannot recover enough potassium from the basalt substrate to replenish that lost by rain water percolation in the soil and alterite. The clay minerals present under conditions of intense or prolonged alteration are not conducive to potassium retention being kaolinite and gibbsite. Silica decreases after an initial increase in the A horizon with the result that the clays become dominantly aluminous hydrates, gibbsite instead of more silica-rich clays capable of fixing potassium.

However in some instances, in the case shown for very old soils from temperate climates, a million years or so (Oh and Richter 2005) the potassium content of soils continues to decrease to the A horizon. Here the amount of uplift material available

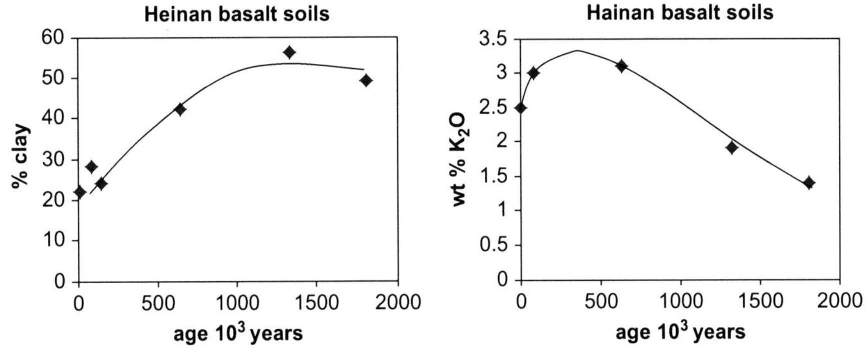

Fig. 4.8 Clay and potassium content in soils formed on basalt on Hainan Island (China) as a function of time. Here the clays increase in the A horizon and initially the potassium content increases also. However the heavy rainfall of the subtropical climate eventually exhausts the alterite resources such that plants cannot continue to enrich the soil zone after some time (data from He et al. 2008)

has been exhausted as in the Hainan case. The clays do not form a substrate that can capture potassium and store it in illitic material, and hence the uplift action is severely limited.

4.2.2 Elements in Soils

4.2.2.1 Fe

An element not often considered to be related to the interaction of alterite and organic plant activity is iron. Schultz et al. (2010) have indicated that iron can be accumulated through the interaction of plants and soil biological action under temperate climate conditions. Loughnan (1969, p. 77) gives data for chemistry of soils on basalt under subtropical climatic conditions. In Fig. 4.9 one sees the increase in Fe in the soil zone (upper 50 cm of the profile) in an alterite–soil sequence on basalt in New Zealand (Loughnan 1969).

Iron content increases in the alterite zone as shown in Chap. 3, but decreases in the soil zone. The same trends are apparent in the data for granite alteration under semiarid grassland conditions (White et al. 2008). Plant action is operative as seen in the increase in silica in this zone, but iron content decreases for soils developed on rocks of contrasting initial chemistry (basalts versus granites). Hence plant uplift of chemical material is operative for some elements, but the plants in this instance decrease iron content, probably by higher acidity and lower Eh in the soil zone. Here it seems that the plant action is one that decreases iron content at the surface while increasing K and Si from depth (see Fig. 4.9). The loss of accumulated iron in the very uppermost horizons is typical of alteration profiles classified as Podsols.

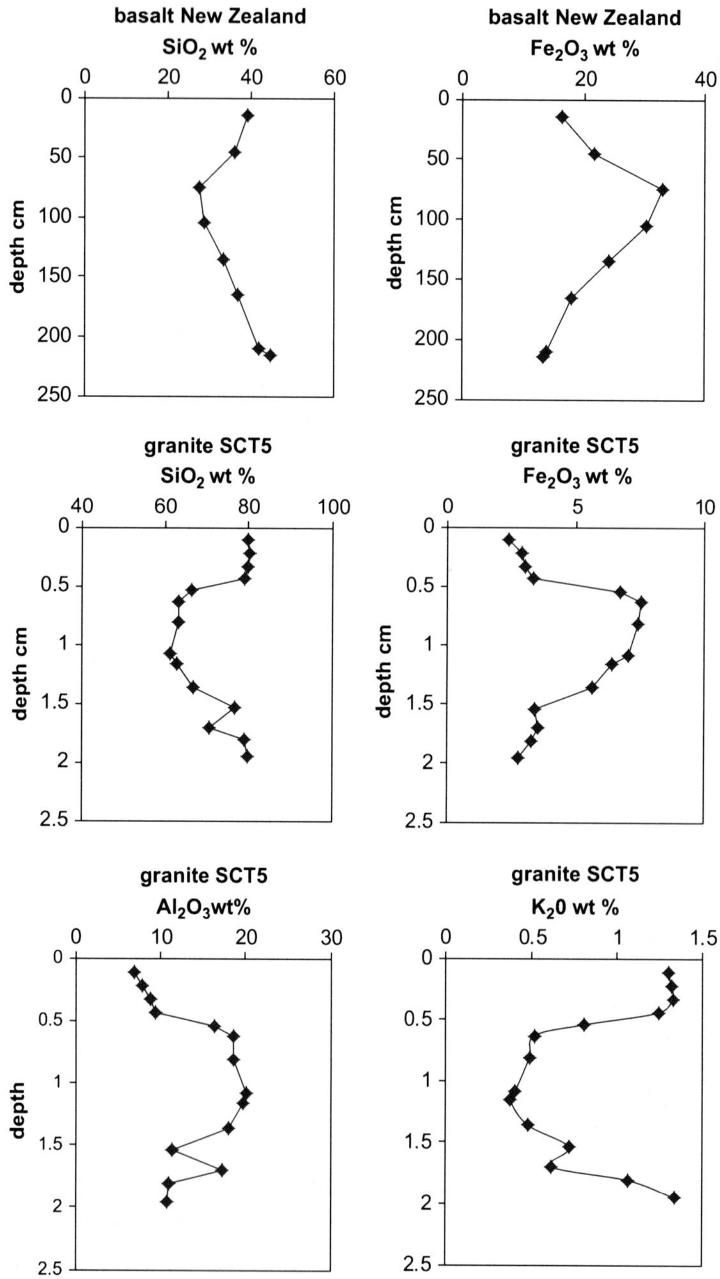

Fig. 4.9 Basalt depth trend for SiO_2 and Fe_2O_3 (Loughnan 1969) plus Al_2O_3 and K_2O in granite (data from White et al. 2008)

These soils are distributed in various zones, usually under conifer forest cover in cold climates, but also they can be found in tropical contrasted season climates near the equator [see Righi and Cauvel (1987) for various examples]. In Podzols formed under cold temperate climates there is a strong loss of iron and other elements except silica, which remains in the form of insoluble and un-reactive quartz grains. The concentration of silica (essentially quartz sand) is underlain by an accumulation of soil clays and iron accumulation forming toward the surface just below the A horizon. In some subtropical zones one can find strong laterite accumulations dominated by iron oxides, which appear to be have been disaggregated by the action of new plant regimes, which dissolve the iron and leave alumina deposits behind. This reversal of accumulation tendencies forming iron oxide deposits is striking in these regions.

In most accumulations of iron in soils, the form is dominantly trivalent oxidized iron with some divalent ions present (Schultz et al. 2010). Oxidized iron forming iron oxides and hydroxyloxides is highly insoluble in aqueous solution. However divalent forms of iron compounds are usually relatively soluble. One can postulate that the conifer forest biome on the basalt studied has produced a surface zone with a reducing environment, typical of these plants whose soil is acidic and frequently reducing. Loss of iron would be due in this case to a specific plant action on surface chemical conditions, reversing the normal oxidation process of water–rock alteration to one of reduction of iron oxide into a more soluble form of divalent iron.

In the very excellent study by Meijer and Burrman (2003) trends of potassium increase and silica increase are seen at various altitudes and hence under various biomes, fern to grass stands to tropical forest on the Turrialba volcano on Costa Rica. However iron is accumulated in the soil zone in some instances, lost in others, depending upon the type of plant regime found at different altitudes. This suggests that the presence or loss of iron in soils can be controlled by the plants and organic activity in the surface layer, which varies from environment to environment.

Fuss et al. (2011) indicate that the oxidation state of iron strongly affects its presence in soils and runoff waters in forest soils of different types of temperate climate. A clear relationship between pH, dissolved organic carbon, and iron in runoff water shows the relations of plant and microbial activity and the presence of supposed mineral elements. Iron is then an element whose presence is to a large extent controlled by biologic activity in the upper portion of alteration sequences, which determines the redox state of the materials present.

4.2.2.2 Phosphorous

In what is perhaps the most disgusting method of discovering an element, phosphorus was first isolated in 1669 by Hennig Brand, a German pharmacist and alchemist, by boiling, filtering, and otherwise processing as many as 60 buckets of urine. He isolated from urine a white, waxy material and named it phosphorus ("light bearer"), because it glowed in the dark. We have made progress in identifying and isolating phosphorus since then.

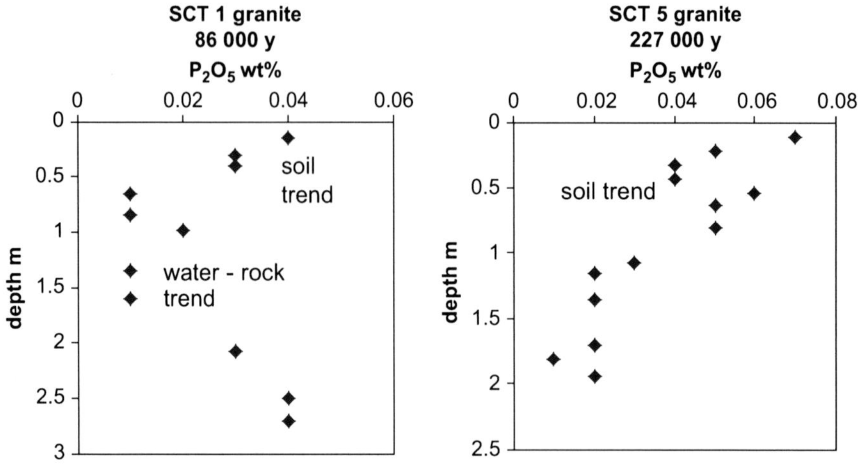

Fig. 4.10 Phosphorous content of alterites and soils in two terrace sequences as a function of depth in central California: 227,000 years (SCT5) and 86,000 years (SCT1) soil profile (White et al. 2008). Initial interactions show a decided increase in phosphorous in the soil zone, which becomes more deeply seated and increases in amount in the older soil as alteration proceeds to interact at greater depths. There is a clear tendency to lose phosphorous in water–rock interactions, below 1.5 m depth, as seen in the 86,000-year-old profile. The uplift effect of plants for phosphorous is quite evident in these soils

Phosphorous, essential to plant activity, is another element which is an obvious candidate for uplift in that it is a major nutrient for most plants. Phosphorus is essential for vegetal and animal life. As phosphate, it is a component of DNA, RNA, ATP, and also the phospholipids that form all cell membranes.

Unfortunately there is little information concerning phosphorous presence in alteration sequences. One exception is the study by White et al. (2008) on terraces on the Pacific California coast soils formed over time spans of up to 226,000 years. The younger terrace soil (65,000 years) indicates a decrease of P content in moving up the profile, due to water–rock weathering interaction, followed by a substantial increase near the surface. In the older profile there is a gradual increase toward the surface from a low level at depth. These relations can be taken to indicate an accumulation of P at the surface due to plant uplift while it is taken into solution at depth through water–rock weathering processes (Fig. 4.10). The vegetation present today in the study area is now of the Mediterranean type of grass and shrubs.

Data from Siffermann (1973, p. 76) indicate similar trends of phosphorous increase at the surface A horizons in subtropical soils in Cameroun. In general phosphorous content is less in the clay fraction than total soil and alterite materials of tropical alteration materials. The association of phosphorous and iron has been demonstrated by Peretyazhko and Sposito (2005) where the change in oxidation state and subsequent solubilization of iron oxides leads to phosphorous loss. Thus the link of P and Fe and the presence of plants seem to be obvious.

Phosphorous is a very important geochemical agent in that it can associate with transition metal and heavy elements in forming minerals and certainly will do so as an oxoanion complex in aqueous solutions.

4.2.2.3 Boron

Boron compounds such as borax (sodium tetraborate, $Na_2B_4O_7 \cdot 10H_2O$) have been known and used by ancient cultures for thousands of years.

Boron is frequently present in significant quantities in the A horizon of alteration profiles. Water-soluble boron values are variable, depending upon climate and plant types present (Aubert and Pinta 1977). It is found in various parts of plants being a fundamental minor element involved in plant structural elements metabolism. It is generally associated with organic matter in the soil where its interaction can cause isotopic fractionation (Lemarchand et al. 2005). Waterborne boron may be adsorbed by soils and sediments. Adsorption–desorption reactions are expected to be the only significant mechanism influencing the fate of boron in water.

Boron can be fixed within the silicate structure (illites) as well as fixed on clay mineral edges via chemisorption (Williams and Herwig 2002; Koren and Mezuman 1981). Thus there is a reservoir for plants on and in clay materials as it is the case for other elements in the context of plant chemical uplift. Boron can also be found as an absorbed species on oxides (Lemarchand et al. 2005). The adsorption of boron on oxides, oxyhydroxides, and organic matter in soils on variable charge sites is dependent upon the form of boron in the soil solution. It is considered that boric acid (H_3BO_3), a nonionic form, is present in aqueous solution below pH of 7–8. The oxyanion $B(OH)_4^-$ is present at pH above these values and is seen to be adsorbed on iron and aluminum oxide phases, and clays in laboratory experiments (Lemarchand et al. 2007; Aubert and Pinta 1977). Thus boron is more readily available to plants, being more loosely held on solids, at pH values below 8. There is a tendency for plants to fix boron in organic matter as constituent elements of the organic molecules, but boron is also held in common fine-grained materials in the soils resulting in a concentration of boron in the soil, A horizon from organic and mineral sources.

4.2.2.4 Transition Metals

Numerous other metal elements of minor abundance are involved in the metabolism of plants such as Zn and Cu. They occur in soils in various concentrations, some increasing near the surface and others are less present. However a major problem is the history of a soil that has been analyzed. What were the inputs such as fertilizer, and other treatments? Does the surface composition result from plant or weathering activity or human input?

For example since 1883, copper sulfate salts (Bordeaux mixture or Bouillie bordelaise) have been widely applied as fungicides against mildew on grape vines. A substantial proportion of copper sulfate sprayed annually on the vines reaches the soil, where it often remains fixed in the surface layer at much higher levels (near 1,000 ppm) than in most soils (near 40 ppm). Apparently little is moved on suspended matter in river transport according to analyses of clays found along the banks of the Gironde River nearby where values are on the order of 80 ppm Cu (Velde 2006) which is comparable to average shale values (Wedephol 1969).

For minor elements in low concentrations the analysis of data is difficult. The overall relations of minor element in soils will be treated further on.

4.2.3 Correlative Effects

In the above discussion iron and phosphorous are mentioned as being involved in the plant uplift effects. Iron it seems is often lost in the soil zone and concentrated in the lower, B horizon. Phosphorous seems to be systematically concentrated in the soil or upper horizon, being used and reused by the plant regime. Taking the data of White et al. (2008) again, one sees the increased concentration of P in the soil horizon at the expense of phosphorous in deeper zones (Fig. 4.10). It appears that as the profiles develop, toward the age of a hundred thousand years, phosphorous concentration becomes important in the A horizon, below the soil zone. The initial relations of P, found in the youngest profile, indicate a decrease in P content in the water–rock interaction zone and a subsequent increase in the soil zone. However in the oldest profile there is one trend of increase in the upper portion of the profile. Comparing these data to those for the iron crust materials in Burkino Fasso (Boeglin and Mazaltarim 1989), representing the end stages of surface alteration, the high iron concentrations are accompanied by increasing phosphorous content. One can imagine that this phenomenon is linked to a mineralogical change under tropical alteration conditions where higher iron content results in the formation of an iron mineral that is specifically capable of fixing phosphorous even under conditions of high rainfall. What is important for surface geochemistry is that the P–Fe relations of concentration will be different depending upon the iron concentration in the soil and the intensity of plant interaction.

The importance of segregation of chemical influences between plant–soil interactions and water–rock interactions can be illustrated by the results of chemical analysis of fluids reaching a stream or river. If the products of alteration in soils are moved by flowing water (either within the soil or by surface erosion, such as iron from the A to the B horizons) the materials carried in suspension in the fluids can have different concentration relations. If only B horizon material is removed, one trend will occur in analyses of the suspended material in the fluids. But if the soil zone is moved into suspension in an aqueous solution by surface runoff erosion another relation will be observed in the analysis of runoff water. Should both be moved at the same time, a strong overlap of relations will occur. This means that, depending upon the part of the alteration sequence that is removed, one will have different P–Fe relations for example which will be found in the suspended matter in rivers and in the eventual deposition of the material as sediment. Thus the elemental relations found in sediments can reflect either simple alteration causes or more complex ones. The geochemical interpretation of such material can only be conducted with caution.

4.2.4 Uplift Dynamics

The search for mineral nutrients by plants is rather efficient as can be seen from data on soil and aqueous solution content of potassium in areas denuded in a major forest biome. Figure 4.11 shows the change in soil potassium content as a function of biomass regrown on two landslide areas where the normal vegetation is deciduous forest. Both in soils and in runoff water the initial stages of alteration show significant amounts of potassium present, which decrease as plant biomass reestablishes itself. The time frame is on the order of 50 years, a rather short period for such an interaction considering that mineral transformations themselves take thousands of years under abiotic conditions. After the period of reestablishment of the biome, it appears that the extraction effect is stabilized and will reach a steady state as plant litter fall renews the potassium content of the soil in a more or less steady state of turnover.

The overall effect of chemical uplift by plants, essentially from the upper alterite zone into the A soil horizon, is that of conservation of alkalies (excluding Na) and alkaline earths through the conservation of high silica clay minerals. These clays are a major agent of exchange and redistribution of minor elements such as heavy metals at the surface of the earth. Clays are extremely important for the study of surface geochemistry. The objective of most plants is to conserve clay minerals and especially silica-rich clays that have high cation exchange capacities, by supplying silica to stabilize them chemically and by holding them at the surface by root action. Clay–organic interactions aggregate clays and bind them into coherent masses at the surface of clay materials with micro-porosity. This control of clay mineral and oxide retention at the surface has an effect of retention on the distribution of minor elements and eventually pollutants at the surface.

Prolonged alteration or intense alteration under warm wet tropical climates exhausts the silicate mineral resources that can retain minor elements through a loss of silica, despite the efforts of plants to renew silica content by the formation of amorphous silica (phytolites) at the surface. This leaves the iron–manganese oxyhydroxides and aluminous hydroxides (gibbsite) as the likely transfer and fixing agents in soils. Organic matter is important also in the soil zone through its control of oxidation state of elements such as iron through biological action and the retention of water in microporous soil aggregates.

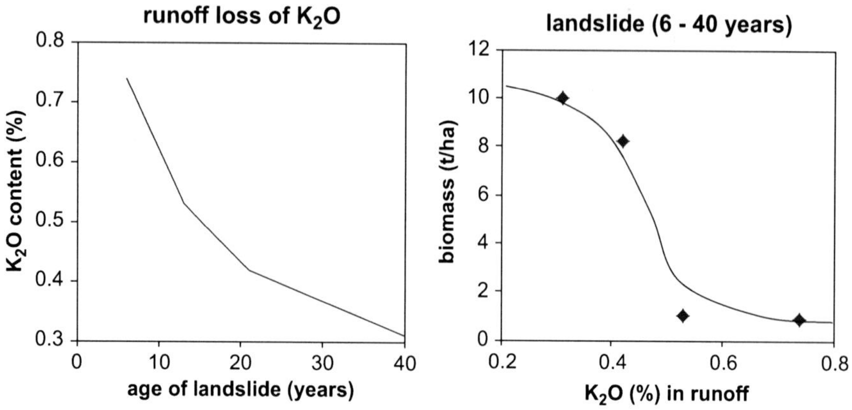

Fig. 4.11 Several studies (Zahrn and Johnson 1995; Walker and Del Moral 2003) show that a stable chemical system can be established by forest systems rather rapidly. In landslide areas runoff water potassium content has been seen to decrease as plants biomass establishes a chemical equilibrium over relatively short periods of time, several tens of years

4.3 Chemical Controls Engendered by Plants

4.3.1 *Soil pH*

The pH in the soil surface zone is an extremely important factor in the retention or release of ions in the soil zone (see Chap. 2). Hydrogen ions take the place of exchangeable cations on the surfaces of soil clay materials, silicates, oxides, and organic matter through ion-exchange mechanisms. The pH of a soil is to a large extent the result of plant and microbial action. However, one must not forget that the biome present in a given area is a function of climate and bedrock or substrate composition. Soils several kilometers apart under the same geomorphologic and climatic conditions will have different plant cover if the substrate is sufficiently different. For example plants and soils on basalts are not the same as those on granite, nor the same as those of sedimentary rocks, especially carbonate-rich types. The plant regime tends to reflect the type of pH environment of the substrate. Acid producing conifer forests are found on acidic granitic soils, and prairies of basic pH characteristics are typically found on calcareous sediments. However, there is an evolution of soil pH with time as rocks become alterites and soils develop. Overall, prairie biomes produce basic pH soils. Deciduous forests produce slightly more acidic soils, but they can give still more acidic soils depending upon the substrate. Conifer forests are decidedly acidic in their characteristics. Data from Jenny (1994) show that biomes on the same type of substrate, glacial till and loess in Illinois (USA), have decidedly different pH values at different depths depending upon the biome, prairie or deciduous forest (Fig. 4.12a).

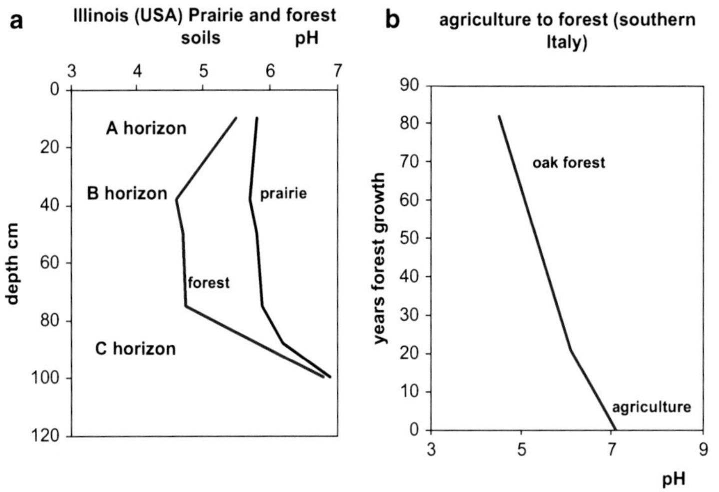

Fig. 4.12 Soil pH affected by biome type. (**a**) Illinois (USA) prairie and forest soils show different pH profiles following the data of Jenny (1994). The pH values tend to be similar at the surface A horizon but are different at different depth of alteration. (**b**) Data from Amato et al. (2004) show the effect on pH in the soil A horizon due to changing the plant regime from crop culture to oak forest in southern Italy over a period of 80 years

Deciduous forest soils take an excursion to more acidic values in the B horizon, whereas prairie soils remain largely basic in chemical character (Fig. 4.12a). Amato et al. (2004, p. 131) show that a change from agricultural land use (essentially prairie plants) to oak forest after abandoning the cultivated plot changes the pH in the soil A horizon over a span of 80 years (Fig. 4.12b) from a clearly basic soil (pH 7) to an acidic one (pH 4.5) in southern Italy. Here the plant factor is primordial, the substrate and climate remaining essentially constant. The change in pH is rather rapid, indicating that plants and ensuing microbial activity are the determining factor in determining soil pH.

Time and the evolution of the clay minerals in the soils can affect the pH also. Soils on basalt from Hainan Island (China) show a decided evolution toward acidic values as a function of time (Fig. 4.13). The change in clay mineralogy is very strong under the subtropical climate where aluminum hydroxide (gibbsite) becomes the major phase in the older samples. The low silica mineral kaolinite and especially gibbsite (aluminium hydroxide) favor acidic soil values. The shift in pH is significant in that it changes from near neutral to decidedly acidic with a decrease in pH of two units.

Climate and the resulting biome then could be expected to influence soil pH. Soils from different elevations and rainfall regimes on Hawaiian basalt soils (Teutsch et al. 1999) show a correlation of rainfall and pH (Fig. 4.14a). This of course reflects the plant regime present. At the highest rainfall zone on the island one finds a tropical rain forest and more grassland landscapes under lower rainfall regimes. Black (1957) gives data for annual rainfall and soil pH for soils across the

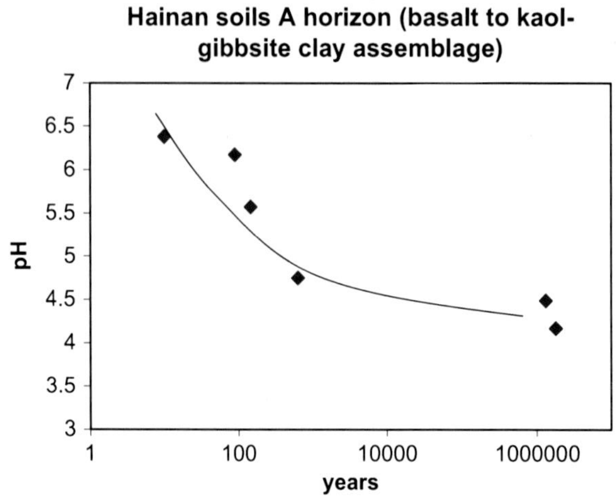

Fig. 4.13 Time–pH relations for Hainan soils based on basalt flow materials in a semitropical climate (He et al. 2008)

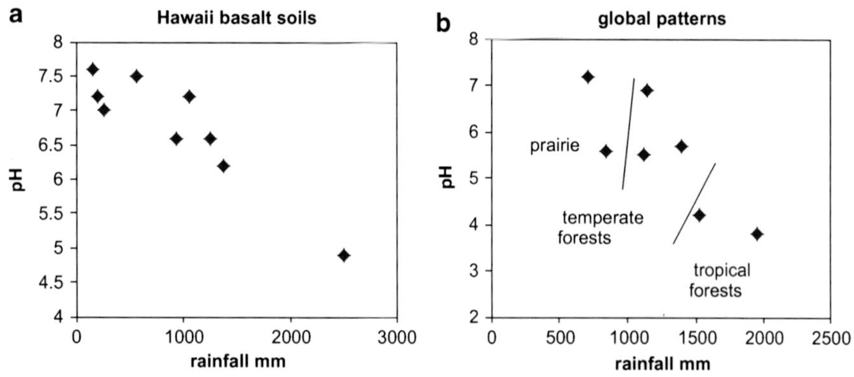

Fig. 4.14 (**a**) Correlations of rainfall and pH on Hawaiian soils (Teutsch et al. 1999). (**b**) General relations of overall rainfall and pH where the general plant regime changes as a function of rainfall from prairie to tropical forest; see Aiken (2002) for global patterns of pH and rainfall

central USA which are similar to those for basalts but slightly lower in pH for the same rainfall values. Here again rainfall does not determine climate totally nor does it describe the plant regime present. It is clear that the pH is a function of substrate and plants. In a general way one can say that soil pH is to a large part determined by plant and hence microbiological action. The plants present can be strongly influenced by the substrate materials. Climate is a factor that can determine which plants are present independently of the substrate to a certain extent. This is indicated by Aiken (2002) as shown in Fig. 4.14b. Over a range of rainfall conditions overall dominant plant types change some but pH is similar, while in the tropics where rainfall is intense, producing low silica minerals, and as a result pH is significantly lower (Fig. 4.14b).

Meijer and Burrman (2003) indicate that the soil pH in the A horizon is very similar, pH near 4, over a wide range of altitude conditions on a mountain in tropical Costa Rica where high rainfall produces low silica minerals, from forest and grasslands at 3,000–100 m which give essentially the same acidity due to the high kaolinite and gibbsite (aluminum hydroxide) content of the highly altered soils. Here the alterite mineralogy is dominant in determining soil pH and plants have a more minor effect on soil chemistry.

Plants can then influence the clay minerals present through a control of solution chemistry to a large extent under conditions of temperate climates. One major effect is the presence of amorphous silica brought to the surface as phytilites, which encourages the presence of siliceous types of clay minerals under conditions of temperate climates by increasing the activity of silica in the soil solutions over periods of much less than a million years through the uplift process. This action adds active silica to the soil chemical regime as the plants decay (phytolites, from Greek, "plant stone"). The siliceous clays (smectites and illite) conserved through high silica activity in solutions the clay minerals, which exchange basic cations in their interlayer exchange site, notably potassium, and their active surface (charge variable) exchange of cations. Soils present in areas subject to high rainfall tend to produce acidic soils due to the presence of low silica minerals (kaolinite and gibbsite) where silica activity in solution is low. The interplay of plants and alterite material is extremely important in the determination of soil pH. These variables are most important in regions of moderate climate and variable vegetal cover.

4.3.1.1 pH as a Major Factor for Minor Element Cation Retention and Movement in Soils

There are essentially three types of material present in soils which are the major cation exchangers: degraded organic matter, clay minerals (silicates), and oxyhydroxides of Fe, Mn, and to a lesser extent Al. Organic matter in a little evolved state can have a high cation exchange capacity, several times that of smectite (Sposito 1989), which decreases as the material is matured by bacterial action. The amounts of these materials in a soil will of course determine its capacity to fix or release elements in aqueous solutions. The capture or release of ionic species in solution determines the movement of elements, such as K and Ca or those elements of minor abundance such as transition metals or heavy metals.

The principles of cation exchange, structural site, and chemical attraction, which operate according to chemical parameters acting on the cations in solution, have been discussed in Chap. 2. In the following section we would like to indicate some of the effects of the principles of cation exchange as a function of the minerals and organic matter present in soils and in the end show the overall effect of cation exchange on the minor element concentration in the soils.

We would like to show the importance of pH concerning cation exchange which is often little discussed in textbooks and treatises concerning cation exchange in soils although the phenomenon is well understood and has been for some time.

It is of major importance, and frequently cited as such in current work. The pH is of greatest importance in that at the hydrogen ion activity cations are not fixed on exchange materials such as organic matter, edge sites on clay minerals, and most oxides. This is critical to the mobility of minor elements in soils systems where these elements will not find a stable mineral environment on which they can be attracted.

The three basic types of materials retaining and releasing cations and anions in soils are first of all the silicate clay minerals with essentially two types being present: smectite–illites and illite–kaolinite. Smectites have a high exchange capacity, near 120 m equivalents of charge per 100 g of dry solids (a rather complicated unit of measure but useful). These cations are little affected by exchange with hydrogen ions (pH effect). To a great extent the cations are held within the clay structure, absorbed into it. Using the same measurements non-smectite clays (kaolinite and illite) have exchange capacities of less than 10 milliequivalents/100 g. Thus the capacity to retain cations is to a large part determined by the type of clay mineral present. Further, oxides and oxyhydroxides of Fe, Mn, and Al often present in soils in the clay size fraction ($<2\ \mu$) have low cation exchange capacities, near 10 milliequivalents of charge per 100 g of material (see Chap. 2 for details of these units). However, the retention of metallic cations can be much stronger at times than that on silicate mineral clay surfaces.

Organic matter is a source of cation exchange, but since the form and chemistry of organic matter is highly variable, the capacity can be very different depending upon the type of organic matter present and its state of maturity. Organic matter tends to be a high fixation agent in its early stages of maturity in the soil but as microbial action proceeds the chemical functions that fix the cations are dispersed and oxidized leaving a much less reactive material in the soil [see Piccolo (1996) or Hayes et al. (1989) for example for a detailed description of the evolution of organic matter in soils]. In soils with immature organic matter, that which is little evolved because of climatic or local geographic conditions (bogs and swamps), the capture of cations is great but in more oxidized or evolved soils, such as those found in agricultural sites, the organic matter appears to have a smaller role in cation capture, and is frequently inferior to that of clays (silicates and oxides) as argued by Velde and Barré (2010, p. 185). Organic matter nevertheless does have an impact on cation retention and release in soils, which is important. It varies as a function of soil type, based upon the chemical–biological (microbial) parameters that promote the loss of functional groups through maturation processes. Clear relationships between organic carbon, pH, and cation exchange capacity in forest soils are demonstrated by Mareschal et al. (2010). Higher carbon content is followed by decrease in pH and a higher CEC. The relations are not linear with higher carbon content increasing rapidly at pH below 3.5. This is an indication of acidic conditions where the microbial transformation of organic matter is slowed. This results in an increase of organic carbon content in the soils and an overall increased cation exchange capacity due to less evolved organic matter.

Apparently most of the cation fixation on organic matter is similar to that which is found on clay mineral edges or oxides. Ions are present which have partially unsatisfied charges although they are covalently bonded to others in the chemical structure in which they are found. Either negative or positive charges are found in these instances. Such sites can fix anions or cations. Most studies consider the capture and exchange of cations on these sites as dependent upon the activity of hydrogen ions in aqueous solutions. The description is considered as a variable charge site.

The same situation is considered to be the case for oxides and low silica clays such as kaolinite or gibbsite. Here also the major exchange site is due to uncompleted structural bonding on the edges of crystals. The amount of exchangeable ions present on the variable charge sites is a function of pH. These general relations are well outlined in Buckman and Brady (1969) see Chap. 2.

By contrast, siliceous clays, smectites for the most part, have interlayer (internal) exchange sites, which are much less affected by hydrogen ion compensations, only below pH 2. Here the major factor of exchange and retention is the presence and the relative concentrations of cations in solution. Illite (potassium-rich clay mineral) is a special case in that the internal exchange sites are complete and non-exchangeable, where potassium is present, and neither other cations nor hydrogen has a strong tendency to displace the potassium in the mineral.

As a result, the major phases present in the soil horizon have two different behaviors, one more or less independent of pH (smectites) and the other strongly dependent on pH: oxides, organic matter, and clay minerals edges. Studies by Hooda (2010), Arnfalk et al. (1996), Holmgren et al. (1993), Ross (1994), Kabata-Pendias and Pendias (1992), Gaillardet et al. (2004), among others, demonstrate these tendencies. The concept is used to model the cation exchange behavior of different materials. It is dependent upon the chemical bonding characteristics of the cations and the chemical potential of the hydrated ions to be bound on or in exchange materials. The available exchange sites are generally called exchangeable bases when measurements are made on soils materials. The amount of bases (cations other than hydrogen) is relative to the total capacity of exchange, which is made at neutral pH values. Thus when summing up the cations present and comparing it to the inherent capacity of cation exchange for the materials in a soil the result describes the amount of hydrogen that is fixed at the expense of exchangable cations.

Figure 4.15 indicates the relations of the proportion of sites occupied by cations (exchangeable bases, or cations) and pH in soils from two biomes, prairies and forest. Here it is clear that the more acid forest soils have fewer sites of cation exchange occupied by exchange ions, while the prairies soils have significantly more sites filled by cations. There is a rapid change in site occupancy filling by hydrogen ions in the pH range of 5–6. This is the region where forest soils differ from prairie soils. It is apparent that forest soils will contain fewer exchangeable ions than prairies soils. The obvious result is that forest soils will lose much of the exchangeable cation content, alkalies and other ions, and especially important for geochemical considerations, transition metal and heavy metal cations. Prairie soils, by their inherent basic chemical character, will be richer in metallic ions.

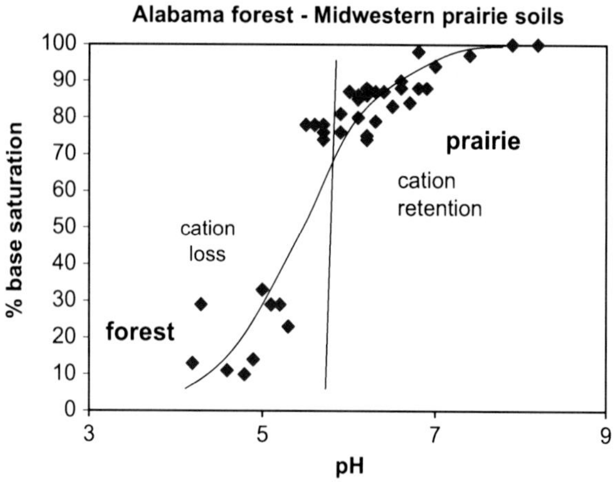

Fig. 4.15 Relationship between pH exchangeable bases for Midwestern prairie soils (Ruhe 1984) and mature forest soils from Alabama (Bryant and Dixon 1964) indicating the relatively sharp downturn in exchangeable bases in the range near pH 6. Forest soils show significantly lower exchangeable bases present due to their low pH

The retention of cations is not the same for all species, as any chemist would expect. The attraction of water molecules and the expression of residual charge is a function of ionic radius, charge, and electronic configuration. Therefore different species of cations are held more or less against the substitution of hydrogen ions on ion exchangers depending upon their electrochemical properties. Also the clays with internal exchange sites (smectites) do not behave in the same way as exchangers with sites on the surface of the crystalline or molecular matter. Different materials have very different retention capacities ranging from low values such as that of kaolinite (3–15 cmol/kg) and gibbsite to high values for smectites (60–100 cmol/kg). There is in fact a hierarchy in the retention of cations on exchange sites but very different values can be found for specific ions on oxides. The fixation capacity varies as a function of pH.

4.3.2 Modelling Cation Absorption to Describe Experimental Observation

Solid particles, especially oxides/hydroxides, develop an electrical charge on their surface when submerged in aqueous solution. Based on the molecular structure of montmorillonite cations can adsorb by either electrostatic attraction on the basal plane or in a formation of bonds with the functional groups at the surface or the broken edges (see Chap. 2). The distribution of cations in a montmorillonite system is dependent on ionic strength, pH, and the type of adsorbing ion. The ability to form chemical bonds with the functional groups is a key factor that determines if a cation will be absorbed onto the edge sites (Strawn and Sparks 1999).

Cations can interact with negatively charged mineral surfaces by purely electrostatic attraction. This interaction is called "outer-sphere complexation". Such reactions are well known for cation interactions with permanently charged clay mineral surfaces (i.e., typical ion-exchange sites). This charge is due to an isomorphic substitution within the clay structure (Al in the octahedral layers replaced by divalent cations (Mg/Fe(II)) and/or Si in tetrahedral SiO_4 layers replaced by trivalent cations (Al/Fe(III)). If the charge deficit arises in the octahedral layer, the net charge at the surface is more delocalized than if charge deficit arises in the tetrahedral layer. The isomorphic substitution is the origin of the permanent charges, which are independent of pH. The weak electrostatic nature of cation attachment to permanently charged surfaces renders this interaction readily reversible. At high ionic strength the background electrolyte ions outcompete other potentially adsorbing ions at the planar sites.

The formation of strong cation-surface oxygen atom bonds with significant ionic or covalent character results in inner-sphere complexation where parts of the hydration sphere of the metal cation are removed. Inner-sphere complexation takes place on amphoteric surface hydroxyl groups and thus varies with their pH dependent protonation/deprotonation. Other parameters affecting inner-sphere sorption include the presence of completion cations, complexing ligands, and metal ion concentration, etc.

Ionization of hydroxyl groups on the surfaces of clays, oxides, and organic matter can result in what is described as pH-dependent charges. Unlike permanent charges developed by isomorphous substitution, pH-dependent charges are variable and decrease with increasing pH. The presence of surface and broken-edge-OH groups gives the kaolinite clay particles for example their electronegativity and their capacity to adsorb cations. In most soils there is a combination of fixed and variable charge.

Anions and cations behave in an opposite manner as a function of pH with respect to their adsorption behavior (see Chap. 2). Table 4.1 indicates such relations based upon calculations using different models and parameters assumed to be determinant in ionic attraction and laboratory experiments (Hooda 2010; Arnfalk et al. 1996; Ross 1994; Kabata-Pendias and Pendias 1992; Gaillardet et al. 2004). A significant number of studies using natural materials have been done, which all corroborate these general relations presented in the table [see Arnfalk et al. (1996), for example]. Results of the modeling approach are given in the table where different models seem to converge to similar relations.

Arnfalk et al. (1996) indicate that the pH values of maximum fixation (pH_{max}) of different cations vary as a function of the soil type. Gray and black soils indicate pH_{max} values for Cd, Cr, Pb of <4 , in red soil 2–6, and in sand >5. The differences in the soil pH and the fixation of cations are due to the minerals present and their relative abundance and the activity of plant organic material. It is interesting to note that anions (oxyanions particularly) are absorbed at low pH values while they are released (desorbed) at higher pH values. For some elements their behavior as cation or oxyanion is determined by their oxidation state (Arsenic as Arsenite, As(III), (AsO_3^{3-}) and Arsenate, As(V), (AsO_4^{3-}).

The differences in pH for binding can be detected in certain sets of data for soils. For example, Pb is fixed at relatively low pH values (near 3) most likely in smectite internal exchange sites while most transition and other metal cations are fixed at

Table 4.1 Ion retention and pH (resumé of data from Evans et al. 2010)

Absorption of cations		
Element	pH for smectite	pH for oxide or kaolinite
Pb, Cr, UO_2	>2	>3–4
Cd, Zn, Ni, Co	>2–3	>5–6
Cu		>4.5
Hg	>7	>7
As(III)		>7
Desorption oxyanions		
	pH for oxide or kaolinite	
Sb(V)	>4	
AsO_4^{3-}	10	
As (V) **the same**	>8	
Cr(VI)	>7	
Mo(VI)	>6.5	

higher values (between 5 and 6) on surface and broken-edge sites ("variable charge sites"). Figure 4.16 shows this effect and the data were collected from a set of soils across the United States and from Hawaii, Egypt, and Greece. The ratio of Pb to Zn is used as a reference. The relations are similar for Cd and other metal cations. The ratio of Pb/Zn is relatively low at basic pH values, but it changes dramatically below pH 5.5. Here the retention of Pb is clearly increasing relative to transition metals. The retention of all the elements considered is strong to lower pH values for smectites but not for oxides and low silica clays such as kaolinite or for oxides.

One can expect then that prairie soils (pH $>$ 6) will retain heavy metal ions while forest soils (pH $<$ 5) will not retain them. The relative relations of pH and cation retention will determine the release of different cations in soils. Many such relations can be seen in data from different studies of soils elemental compositions.

4.3.2.1 Plant Uplift Action and Minor Elements in Soil Profiles

If pH, controlled by plant engendered chemistry, is important in releasing or retaining minor elements in soil materials, what are the consequences of chemical uplift by plants combined with the action of hydrogen ions which compete with cations on clay surfaces? Several studies indicate that there is an increase in minor elements at the surface in alteration sequences. One can always suspect recent pollution due to atmospheric input, but the presence of some chemical elements among the transition metal elements and not others would suggest that at least some of the increase in the soil zone is due to plant activity. This is particularly evident in series where some minor elements are in fact lost at the surface relative to deeper, non-soil alterite zones.

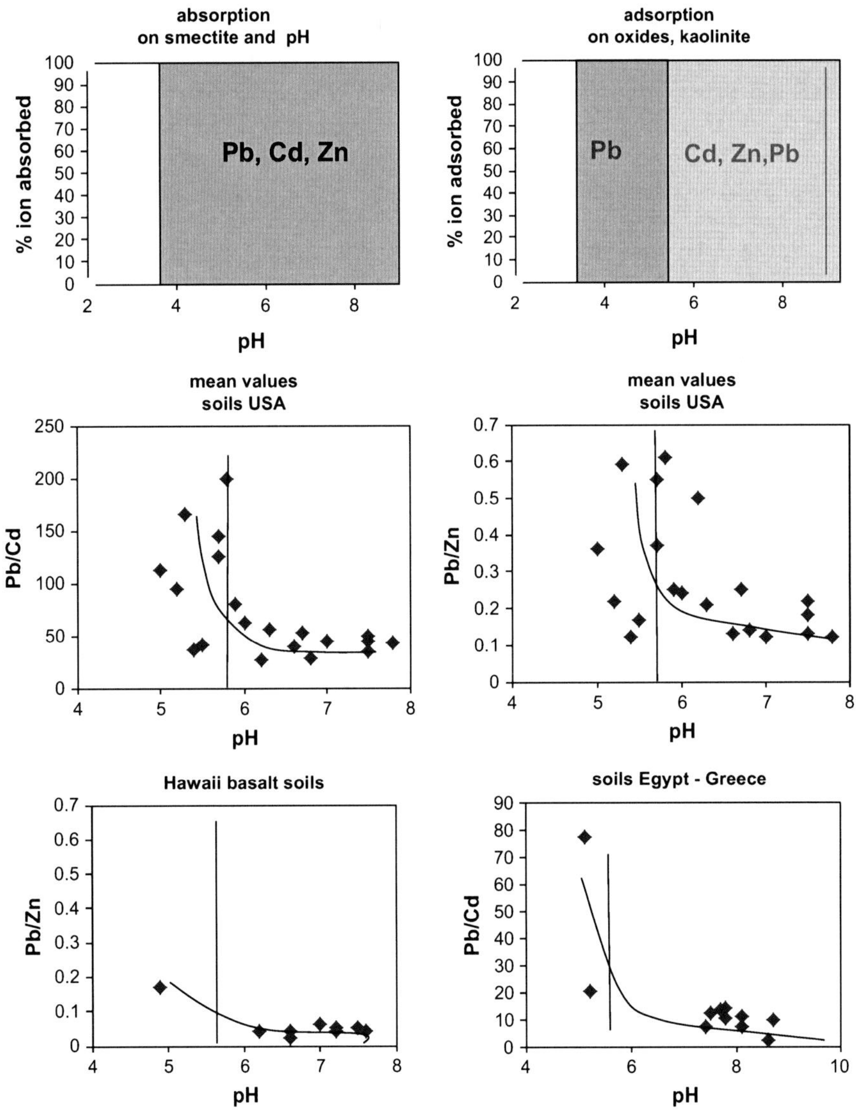

Fig. 4.16 Relations between pH and cation fixation on clays and oxides, modeled by Hooda (2010) and data of Arnfalk et al. (1996), Ross (1994), Kabata-Pendias and Pendias (1992), Gaillardet et al. (2004) and observations of element abundance in soils from Holmgren et al. (1993), Hawaiian basalts (Teutsch et al. 1999) and Egyptian and Grecian soils (Shaheen 2009). The ratios of Pb/Cd and Pb/Zn indicate relative abundance in the soils of these minor elements, which changes strongly in the pH range near 6 as predicted by modeling experiments

4.3.3 *Profiles and Uplift of Minor Elements*

- Zn, Cd increase at top but Cu decrease in Minnesota loess and tills (Pierce et al. 1982).
- Zn, Pb, Cu, Co increase at top in soil on Hainan basalt (Huang et al. 2004).
- Zn, Pb, Cu increase at top of soil on basalt Hawaii (Teutsch et al. 1999).
- Zn increase in top in most cases but Ce, Cr, and V are lost at top in a Costa Rica andesite (Meijer and Burrman 2003).
- Zn, Cu increase in top but Cd, Ni, and Cr are lost in top of German forest soil on metamorphic rocks (Huang et al. 2011).

The variability in retention or release of the same element from site to site suggests that the relations are complex, with sources certainly playing a role (availability of an element) the pH of the soil which can favor the fixation of some elements and not others, or the affinity of some plants for certain elements and not for others which results in different uplift rates.

Overall it appears that Pb is bound in soils, where it is replaced by hydrogen ions only at very low pH conditions. Also Zn appears to be retained in the surface horizons even though it should be susceptible to the pH exchange ion effect below pH 5.5 (see Fig. 4.16). The increase is seen relative to the values at depth in the alterite zone. Cu is frequently seen to increase in the upper parts of the profiles but not in all cases reported. Thus one can suspect that not all of the minor element ions are sorbed in exchangeable chemical sites due to permanent charge or on surface and broken-edge sites in the soil zones but may also be present in the structure of oxide phases.

Numerous authors have indicated the similarity of major mineral elements present in plants of different types (Broadley et al. 2004; Hodson et al. 2005; Knecht and Goransson 2004, for example). This leads one to suspect that the variations in relative content of the major mineral elements such as K, P, and Si are due to climatic or geologic (substrate) factors. However the general pattern holds in most soils under conditions of temperate climates although high weathering intensity can take away the vestiges of plant chemical uplift when plants cannot provide enough of the needed elements to compensate for high rainfall and subsequent dissolution of elements in undersaturated soil solutions.

4.3.3.1 Concentration of Mineral Elements of Minor Abundance by Plants

This subject is the central theme of a large number of published works, among them Khan et al. (2011) who give an idea of the breadth of the topic. Mineral elements of minor abundance in alterite and soil materials are found in plants, and as one would suspect in somewhat different concentrations as are major elements. Markert (1998) however proposes a reference plant composition based upon numerous analyses of plants. The similarity of concentration ratios in plants leads to the possibility of estimating an average plant composition. Mineral elements, which stand out are K, P, Si, S, Mg, and Ca present in the percent to tenths of a percent levels. Minor and trace elements are present in the ppm range such as Cu, Cr, Co, Fe, Mn, Ni, Zn, Sr, Ba, B, Al, Pb and many others are listed as being below the ppm range. Transition metals then are commonly part of plant chemistry following the more abundant elements such as potassium or silicon. Kabata-Pendias and Pendias (1992, p. 69) propose an index of bioaccumulation of mineral elements based upon average soil compositions and average plant compositions. Cd and Hg are the highest in bioaccumulation followed by B, Br, Cs, and Rb and then by transition metal elements and As and Pb. The sequence is somewhat difficult to follow, but in general one sees that transition metal elements (Zn, Cu, Cd) and some heavy metal elements (Hg, Pb, As) are found to be enriched by plants in soils. Whatever the difficulties in obtaining data for such an exercise it seems clear that there is, all the same, a tendency for biological accumulation of many elements at the surface which are now generally considered to be anthropogenic contaminants. If one finds Pb in a soil it is immediately considered to be due to recent industrial contamination. However this may not be necessarily true in all cases. The same is likely to be the case for Cd and so forth. It is very important to be aware of the complex interrelations between mineral and vegetal spheres of chemical activity. In fact the only sure way of distinguishing anthropological influences is by isotopic analysis of the elements present.

An interesting case of plant selectivity for major and minor elements can be seen in the data of Barkoudah and Henderson (2006). The subject is the change in chemistry of plant ash of species found at different distances from the influence of saltwater. The plants nearest the sea in the Middle East have the highest content of Na while those further away have a higher potassium content. K and Na form a negative correlation as could be expected in that terrestrial plants are almost exclusively potassic. Geochemically one finds that Rb is positively correlated with K. Zinc, Cu, and Ni are positively correlated with K as well as with Fe. It appears that the incorporation of transition metal ions can be associated with the potassic or terrestrial character of plants.

4.3.3.2 Elemental Concentration by Plants Exterior to the Metabolic System

The above discussion concerns the chemical concentrations of mineral elements in the shoot (aerial) parts of plants. The materials are usually recycled into the soil at the end of the life cycle of the plant. However the subaerial zone of plant activity, roots and near root areas, should be considered in that some interactions there can affect the concentration of mineral elements. In fact plants have a capacity to neutralize the presence of certain elements when they are in overabundance in the root zone and thus avoid a deleterious action of these elements if they were absorbed and introduced into the metabolism of the organism. Morel et al. (1987) indicate the role of root exudates in fixing heavy metals. Serret et al. (2007) and

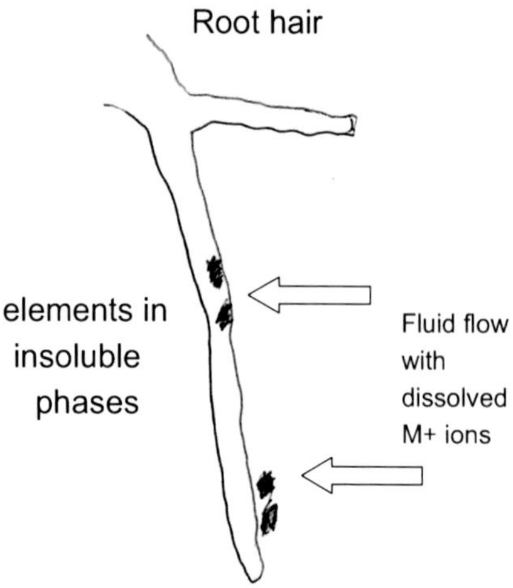

Fig. 4.17 Representation of the isolation of minor elements as insoluble phases in the root zone of plants. The insoluble material can be found on the outside or just on the inside of the root wall

Isaure et al. (2010) show that tobacco can engender the isolation and precipitation of relatively insoluble mineral grains containing Zn and Cd in calcium carbonate phases. The same type of action is found for grass root interactions (Lanson et al. 2008) where Zn is associated with manganese minerals within the roots. Copper in the metallic form (Cu^0) has been found at the root–soil interface (Manceau et al. 2008). These insoluble mineral grains are found on or in the root system. The objective of such phase isolation of metal element ions is to avoid overload in the plant system. The undesired elements are neutralized in insoluble phases in or near the roots. This capacity to fix elements in insoluble forms has led to much recent research into the possibility of bio-remediation by plants in areas of contaminated soils where the overload of contaminant elements is fixed in the soil by plants (Huang and Gobran 2005, for example).

In fact the action of plants in concentrating minor elements in specific phases is to produce a very low solubility phase, which concentrates the element outside of the aqueous system and hence keeps these elements from entering the plant metabolic system. Such an action isolates the minor element in a very concentrated form. The new, insoluble phase can be found on the outside of the root or just within the membrane of the root wall (Fig. 4.17). Such an action has been recognized recently as a possible mechanism to isolate minor elements from groundwater present due to human activity and hence neutralize dissolved pollutant elements.

The above discussion leads to the conclusion that plants can bring up elements of major and minor abundance into the shoot system or in fact concentrate them in the root zone when they are in overabundance. The two mechanisms will lead to a higher concentration of minor elements in the soils zone, either in bioavailable or cation exchangeable form (through degradation of organic material) or in insoluble form developed in or near the roots.

If plants can concentrate and use or isolate excesses of mineral elements which reside for a certain time in the soil zone, it is clear that these elements, often considered as contaminants, should be considered as a part in the balance of natural elemental transfer. The elements considered above are basically of the cation type with a more or less stable oxidations state. These can be more or less assimilated to the exchangeable bases of agricultural and soil chemists. Their behavior is one of exchangeable ions, which follow the chemical activity of different ions in the aqueous solutions of the soils, including hydrogen ions, and hence they are subject to chemical mass action principles. Other phases produced by plants are relatively insoluble: carbonates or oxides of metallic elements such as Fe and more rarely metallic forms of elements, such as copper (Manceau et al. 2008), and these elements will remain present for longer times in the soil zone and will be somewhat less indifferent to changes in chemistry determined by pH and Eh conditions.

Plant interactions for the various elements at the surface can be outlined schematically as of two types shown below:

1. Plant uplift → incorporation into the biomass → deposition as vegetal material or release through root exudates → degradation by biologic action and release in soil as exchangeable ions.
2. Root action to isolate elements → neutralization by precipitation as insoluble phases in the soil.

The presence of elements in soils due to plant action is of two sorts: one where the elements are released as soluble ions after degradation of the organic material to be reused by plants or eventually taken out of the soil by dissolution in moving rain water. Another possibility is the stabilization as insoluble minerals, which do not or only slowly react to surface chemical conditions. There is a stable and an unstable portion to many of the minor element concentrations found in soils.

Plants are part of the equation of pH balance and cation retention, but the actions of agricultural man can be of importance also. Ammonium fertilization can change the soil pH (Liu et al. 1997). Matocha et al. (2010) indicate that heavy use of ammonium fertilizer can at times change the acidity of soils to an extent which results in the polymerization of aluminum hydroxide in the interlayer of clays forming a stable, low cation exchange capacity soil chlorite mineral. Such changes in clay mineralogy can thus be important for the fixation capacity of ions in the soils, especially those that would be present in the smectite minerals at low pH values such as Pb. On the other hand, liming of soils [use of $Ca(OH)_2$ or carbonate rock] designed to increase pH values for better crop production will encourage the presence of expanding smectite minerals which absorb cations at lower pH values. Increase in pH will also increase the adsorption of metal cations on the surface sites of clays, oxides, and organic matter. Thus the various chemical manipulations of farming practice can be important for the dispersion of cations in agricultural environments.

4.3.4 Cases of Minor Elements Retention in Soils

As mentioned above the presence of plants concerns several aspects of minor element chemistry. The effect on pH and hence cation exchange is one. The effect of reducing chemical environments is another important aspect. Of course not all of the trace or minor elements are susceptible to redox effects, while others are. Several specific cases are given below to illustrate some of the major effects of chemistry on minor element abundance in soils.

4.3.4.1 Rare Earth Elements

This group of elements is very often used as a tracer of geologic history for different materials where the relative abundances of the different rare earth elements (REE) elements in the series are compared to determine different chemical actions, which are the result of different chemical processes, especially those at high temperatures. However rare earth elements can be added to the surface, soil environment through the use of phosphate fertilizers. Hence the use of rare earth element spectra to trace surface chemical interaction in modern environments can be at times complicated. In general the rare earth elements are found in important quantities in zirconium, apatite, and other heavy minerals generally found in the un-reacted portion of alterite material. If the minerals are un-reacted one would expect to find little change in elemental concentrations and hence little fractionation. However this is not the case in certain examples of weathering of forest soils (Aubert et al. 2001). Here light rare earth elements such as Nd are leached most likely from apatite, which is probably more susceptible to acid attack than zircon minerals. In rocks with less or no apatite present the change in rare earth spectrum is similar in soils compared to bedrock with a tendency to have light rare earth elements mobilized more during alteration (Daux et al. 1994). Sedimentary load of rivers in many cases reflects all the same the bedrock materials (Martin and McCulloch 1999).

By contrast Tang and Johannesson (2005) indicate that once rare earth elements are freed from high temperature minerals they will be selectively absorbed at different rates from solution onto rather inert materials such as siliceous sand, as a function of pH values in the solution. Light rare earth elements are absorbed less at lower pH values, below 5–6. In evolved soils, terra rossa on dolomite, there is a concentration of rare earth elements in ferromanganese concretions where important fractionation occurs (Feng 2010). Hence it appears that chemical weathering can influence the rare earth content of the various components of the series in the solids and by complementarity in the waters moving through the soil alteration zone.

Loss or gain of the different elements in the REE series will depend upon the types of minerals present in the bedrock, which can have differential stability to the forces of chemical weathering. The minerals in the altered material will then play a role in absorption or not of the elements as a function of pH and hence plant regime.

4.3.4.2 Zinc

This transition metal is normally found as a stable divalent ion. It is absorbed onto silicate clay minerals of different sorts: smectites, soil vermiculites, and to a lesser extent gibbsite and kaolinite, oxides (Mn or Fe), as well as being present in less abundant Zn minerals (Isaure et al. 2005; Kirpichikova et al. 2003; Manceau et al. 2005, 2004, for example). The situation in a soil of normal (i.e., ppm) abundance is that the Zn is absorbed into smectite phyllosilicates, as well as some Mn oxides and adsorbed onto Fe oxides and other phyllosilicates as well as being attached to certain organic functional groups in the organic soil materials. Relatively small amounts of Zn phases such as sulfides are present in most soils, but they can be relatively important at times. Zn mobility depends upon its type of absorption or adsorption at times, which is largely dependent on pH. Below pH of about 5, absorbed Zn ions enter into solution. Zn is more strongly held in smectites or soil vermiculites. A change of Eh can destabilize the Zn mineral, or a host mineral, such as Mn or Fe oxide. Change of oxidation state of Fe and Mn changes the mineral structure and Zn is liberated in ionic form. In certain cases sphalerite, ZnS, is destabilized by oxidation of the sulfur in the mineral and Zn is liberated.

The reverse effect can be found in soils, which are amended with sludge or sewage. Here the system is overloaded with the minor element and it tends to precipitate as a mineral (sulfide, phosphate, or carbonate) or becomes a major part of Mn minerals (Lanson et al. 2002). These minerals (birnessite for example) will hold the Zn in place until more normal soil chemical conditions pertain. The stability of Zn in soils then depends largely on the stability of host minerals and the chemical conditions that pertain to stabilize it in or on these minerals.

4.3.4.3 Copper

Cu appears to be stable as a divalent ion and is for the most part present in inner-sphere and outer-sphere water complexes which do not form multinuclear complexes on mineral surfaces at high concentrations as do Co, Zn, and Ni (Funare et al. 2005). Cu is attracted to charge variable surface sites on soil materials (oxides, silicates, and organic matter). However its presence in soil organic matter as a basic component concentrates it in the organic-rich upper soil zones where it is generally assumed to be associated with the organic residue present. In some cases Cu can be found as very small metallic particles associated with roots (Manceau et al. 2008).

4.3.4.4 Arsenic and Antimony

Arsenic and antimony are quite different cases compared to the transition metal elements in that they can change oxidation state and in doing so their ionic form and function.

Arsenic ranks 20th in abundance in the earth's crust. Arsenic is found in natural waters in both organic and inorganic forms. Inorganic arsenic occurs with two main oxidation states in natural waters, as As(V) and As(III). The toxicity of As(III) is much higher. Arsenic is mainly present in aqueous solutions as $HAsO_4^{2-}$(aq) and $H_2AsO_4^-$ (aq), and most likely partially as H_3AsO_4 (aq), or AsO_4^{3-}(aq).

Antimony has been known since ancient times. It is sometimes found free in nature, but is usually obtained from the ores stibnite (Sb_2S_3) and valentinite (Sb_2O_3). Antimony is an important metal in the world economy. Annual production is about 50,000 tonnes per year.

In the oxidized state As and Sb are for the most part oxyanions, which have a negative charge in aqueous solution due to deprotonation. This sets them apart from cations in that they will compete for very different sites on soil materials. This is evident in that they do not compete with cations for exchange sites but are found on positively charged sites instead of negatively charged ones. The adsorption of these chemical units is on positive surface and broken-edge sites. These elements can be found as cations or oxoanions. These complex ionic forms change, apparently, the absorption tendencies drastically compared to more simple ionic forms of elements in solution. Modeled absorption of As and Sb shows low coefficients of absorption of M^{3+} ions at low pH, but the absorption increases gradually at values above pH 6 (Arai 2010). The oxoanions of As and Sb have a totally different affinity compared to metallic ions in aqueous solution. These elements tend to be fixed as exchange ions on substrates at edge sites over the pH range of most soils (Evans et al. 2010) which suggests the existence of a cation form at high pH and an oxoanion form at low pH. This behavior tends to concentrate the As in the alteration zone and soil zone as noted by Melegy et al. (2011). Biological uptake and redeposition are very important in this process. The oxidation state and pH of soils then can control the type of ion present.

The strong affinity of As for iron oxide phases is well known. The various oxidation states of As and Sb do not change their absorption behavior significantly. This is especially true of the different minerals of the Mn and Fe oxides found in soils. The retention of As as a function of pH in the range 5–10 does not change significantly (Jahan et al. 2011). Hence it appears that As is a minor element which will remain in the soil zone as an absorbed species in the presence of iron oxides.

It is important to remember that different heavy metal elements can have very different chemical affinities in the surface interactions of soil environments. While most transition metal cations are not adsorbed at pH below 5–6, oxyanions are strongly absorbed at low pH values and not desorbed until pH reaches values well above 6. Hence there is a reverse behavior of the two types of ionic species in aqueous solutions. This aspect of ionic behavior is outlined in Table 4.1 above.

4.3.5 Summary

Very briefly, it seems that the heavy metal elements which can have significant range of oxidation state of high valence form oxyanion units in aqueous solutions which have significantly different absorption behaviors from transition metal oxide cations. Thus if one wishes to use a method of elimination or retention of minor element in surface materials, it is necessary to determine the chemical characteristics of the elements concerned, in that any remedial treatment which affects pH will not have the same effect on cations and oxyanions.

Concerning the overall movement and displacement of minor elements in the soil zone, one finds that cation types have various behaviors depending upon pH and the substrate, which absorbs the ions. Smectites retain elements in interlayer absorbed sites at pH around 5–6, while most ions are not held on the different substrates such as oxides, organic matter, and kaolinite at these ph values. Retention as interlayer ions can be high for certain elements such as Pb being absorbed on interlayer sites in clay minerals under acidic conditions while others are retained only under more basic pH conditions on non-smectite materials being absorbed on edge site or surface adsorption sites (see Strawn and Sparks 1999, Fig. 2). Most transition metal ions are strongly absorbed on non-smectite material at pH above 6. Given the general tendency for plant regimes to have a given range of pH values, it is important to consider the type of biome present when estimating the movement characteristics of minor elements. The elements As, Sb, Bi, and certain others will behave in a quite different manner. One would expect then that the movement of minor elements from the soil media to aqueous solutions will be a matter of the chemical affinity of each element depending upon the ambient chemical conditions of the soil engendered by the biome present.

In a general schema, the forest, especially conifers, and other acidic soils will lose most of the minor element concentrations brought up to the surface from the alterite zone. Only Pb is seen to remain in significant quantities. However there is the possibility of lead carbonate formation due to high CO_2 activity engendered by biological activity in the organic zone of soils. As and Sb will be present and concentrated in the soil zones. Since mature or highly altered soils, under tropical climate conditions, are usually of low pH, one can expect that these soils will have few minor elements at the surface. Young soils in cold climates and old soils in hot and wet climates will have similar retention properties for many minor elements. They will essentially be lost to the flux of rainwater and brought into stream flow.

However the plants present in many soils will bring the necessary minor elements to the surface to supply their own needs. This occurs through chemical uplift for minor as well as major elements. This renewal appears to compensate for the loss through loss to rainwater throughput to a large extent in that many minor elements are seen to be concentrated in the soil zone. In cases of overabundance plants are capable of fixing the element in question in an insoluble form in the soil zone outside of the active plant biosphere at the surface of roots or inside the root wall so that the end result is immobilization of these elements in the soil zone.

The effect of pH is such that conifer soils and other of low pH characteristics will contain less minor element material, especially the transition metal ions. Prairie soils will have a higher transition metal content.

4.3.5.1 Movement in Soils

The minor elements present in alterite material can have different fates concerning transport or movement from the site of the alteration process. Initially, some of the ions will be released to altering fluids in the initial alteration process and the aqueous solutions will transport them to river systems. Some will be fixed, temporarily or more permanently, to the new alterite minerals either oxides or silicates (clay minerals). Further alteration processes are intimately interrelated to plant activity. The longer alteration pertains, the further the plants are from the rock–alterite interface and the less they will influence the initial segregation of elements between aqueous solution and solids. Since plants use a significant number of minor elements in their metabolism and internal chemistry, they attempt to keep a certain quantity present within the sphere of their major chemical activity in the soil. Thus one can find changes in chemical element abundance when looking at an alteration profile as one approaches the upper soil zone. This is most notable for potassium and silica, but can be seen for boron and phosphorus also. Increase in elemental abundance at the surface has been noted for transition metals and some heavy metals, but often not attributed to plant activity in that such elements can be added to upper levels through human activity, either as airborne material or agricultural or industrial input. There is a general tendency to fix minor elements in the soil zones of temperate climate soils.

However the structure of soils under active rooted plant regimes develops and maintains porosity in the upper portions of the profile, especially the A horizon. This structure encourages the migration of clays into the lower B horizon. However, there also can be a lateral movement of clays within the A–B horizon which is gravity controlled when the soil is on a slope. This effect results in the transport of clays into the water table and into streams where they are transported out of the initial alteration area. Minor elements can be transported from the soil zones either as dissolved ions or as cations absorbed or adsorbed on solids. The material that transports the ions (solids or aqueous solution) is largely determined by the pH in the soil zone.

Figure 4.18 indicates the major zones present in such a schema and the type of material that is moved from the soil profile or within the profile.

We do not know the relative proportions of minor elements transported in soil waters or on soil particles from the different soil environments into the transportation medium of groundwater and eventually rivers. Both methods of transport are contrary to the plant soil uplift system mechanism for certain elements which attempts to conserve a useful concentration of major and minor element in the soil zone for plant sustenance. Heavy rainfall will inevitably exhaust their elemental concentrations through the effects of dilution and dissolution of the

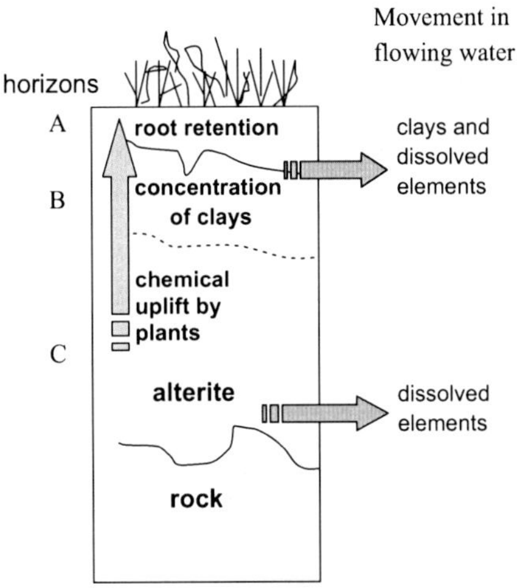

Fig. 4.18 Schematic representation of the movement of materials in an alteration profile

elements either from the surfaces of clays, oxides, and organic material or by the dissolution of these phases in the soil zone.

Soils are not only the interface between alterite and surface transport but also, of course, the interface between the biosphere and the geosphere. This intersection is important for the retention of matter at the surface and the movement of some elements in the soil and the alterite zone at depth. Certain species of plants accumulate some elements in the rhizosphere zone through precipitation of insoluble phases, as mentioned above. In a number of cases the phases formed contain, or can contain, significant amounts of minor elements such as Zn, Cr, Cu, among many others. It has been established that plants can in fact be used to fix and eventually isolate excesses of minor elements in soils introduced by human activity. In fact in many instances plants are used to develop soils on materials deposited at the surface by human activity such as garbage dumps, mine tailing deposits, and so forth. Bio-remediation for the stabilization of heavy elements is an important field of interest (see Khan et al. 2011). The interaction of the biosphere is not just a fact to be observed but also to be used to solve modern day problems of surface geochemistry.

4.4 Useful Texts

Alloway B (ed) (1995) Heavy metals in soils. Blackie Academic, London, p 368

Brody N, Weil R (2008) Nature and properties of soils. Prentice Hall, Upper Saddle River, NJ, p 965

Huang P, Gorbrfan G (eds) (2005) Biogeochemistry of trace elements in the rhizosphere. Elsevier, Amsterdam, p 465

Kabata-Pendias A, Pendias H (1992) Trace elements in soils and plants. CRC Press, Boca Raton, FL, p 364

Khan M, Zaida A, Geol R, Musarrat J (eds) (2011) Biomanagement of metal–contaminated soils (Environmental pollution), vol 20. Springer, Heidelberg, p 512

Salbu B, Steinnes E (eds) (1994) Trace elements in natural waters. CRC Press, Boca Raton, FL, p 302

Salomons W, Förstner U, Mader P eds (1995) Heavy metals: problems and solutions. Springer, Heidelberg, p 412

Sposito G (1989) The chemistry of soils. Oxford Univ Press, New York, NY, p 277

Sposito G (1994) Chemical equilibrium and kinetics in soils. Oxford Univ Press, New York, NY, p 269

Tan M (1998) Principles of soil chemistry. Dekker, New York, NY, p 513

Chapter 5
Transport: Water and Wind

The initial and fundamental chemical reaction between the elements at the surface is engendered by the contact of water (saturated with the component gases, oxygen and CO_2) and unstable silicate minerals coming from rocks. This action forms new minerals of small grain size (clays and oxides) and metastable fragments of the rocks that are not reacted. Some material is incorporated in solution by complete dissolution in the aqueous medium. Other mineral elements of lower abundance are incorporated because they are not of sufficient concentration to form a specific phase and are thus incorporated in minor amounts in minerals or on the new mineral phases produced. The material formed by alteration is then "processed" by plants in the soil zone where different chemical constraints can be imposed such as changes in Eh and pH. The fine-grained material, clays and oxides, and some more coarse parts, silt and sands, can be moved from the surface environment of alteration and displaced by either water or wind. Figure 5.1 indicates schematically the movement of materials after the action of chemical alteration at the surface. Transport by rivers can occur in various stages, interrupted at times in the movement to the sea. Some materials are moved into closed basins where the solution is concentrated to form evaporate materials, those that crystallize the dissolved load into minerals phases. Most of the dissolved material finds its way into the ocean. The suspended material can be deposited when the energy of the water movement allows it to settle out of suspension in the course of transport along stream banks or on flood plains, in lakes and so forth when the energy of transport decreases. Some material is deposited on sufficiently flat and barren areas such that wind erosion will move it from its place of fluvial deposition and disperse it as loess material, often for thousands of kilometers distance.

Interruption of movement allows interaction and reorganization of the alterite material either dissolved or particulate in nature into new mineral configurations. This is especially true when it is affected by plant-driven chemistry in soils (Fig. 5.2).

A. Bauer and B.D. Velde, *Geochemistry at the Earth's Surface*,
DOI 10.1007/978-3-642-31359-2_5, © Springer-Verlag Berlin Heidelberg 2014

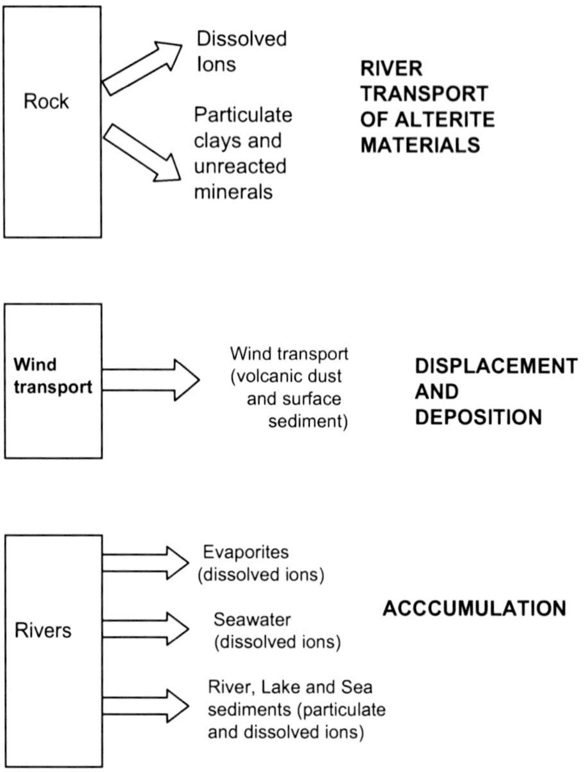

Fig. 5.1 Illustration of the transformation of rocks into sediments by different vectors. The initial stage is rock alteration and transport in rivers to sites of accumulation and a very important vector, wind transportation of sediments and volcanic dust to sites of deposition

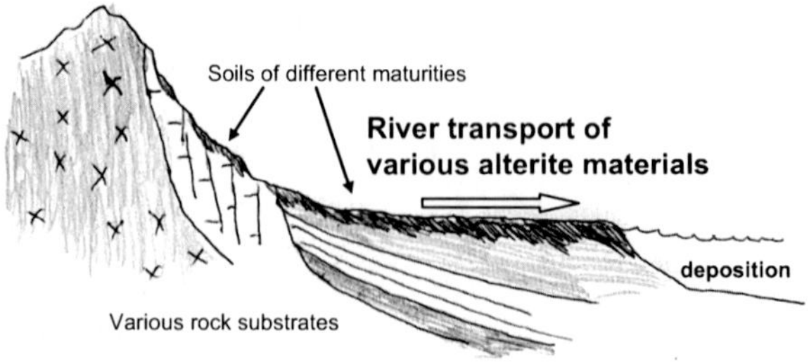

Fig. 5.2 Illustration of the various sites of alterite production which can occur in a given river basin. These include the alteration of different rock types, and the existence of different stages of alteration (young to mature) depending upon the topography of the basin

5.1 Water Transport Materials

5.1.1 Materials Present in Transport Waters

The alteration process then divides the chemical components of rocks into several groups at the atmosphere–rock interface. Several types of material are present in water transport: (1) dissolved elements which are usually in ionic form, (2) silicate clay minerals whose origin is in the transformation of rock minerals into new solid phases containing hydrogen ions, (3) oxides and oxyhydroxides, (4) organic matter which can occur in very small almost molecular units as colloidal material and (5) un-altered, high temperature minerals such as quartz. The last category is represented by a certain amount of the mineral grains of various sizes coming from the rocks brought to the surface which are not significantly chemically altered or little transformed at the surface. Minerals of very low solubility such as monazite (rare earth elements), zircon minerals, and Ti–Fe oxides can be concentrated along with quartz in the coarse fraction of transported sediments. This material is frequently the source of industrial mineral extraction for elements such as rare earths, Ti, Fe. All of the products of alteration are displaced by the action of rainwater and its accumulation in streams and rivers. Thus transport of alteration materials by moving water will consist of dissolved elements, alterite minerals of small grain size, and resistant minerals of larger grain size and also organic matter produced in the soil zone. The amount and grain size of the resistant minerals carried by moving water will depend upon the energy of movement of this liquid material. Higher energy (essentially steeper slopes over which the water moves) will displace larger grains of rock debris. As energy of transport decreases the materials in suspension will gradually be dropped out of the moving mass of water to be deposited as stream gravel and sands. Some material is moved to the edges of streams and rivers during times of high flow rate which is sedimented out when the flow rate and mass of water decrease leaving flood plain deposits. Much material is nevertheless eventually moved from the continental landmasses to the sea by further water erosion and transport. Here it is dispersed along the coastline according to the energy of transport and the importance of local coastal currents. Interaction with saline seawater creates new chemical conditions which reorganize part of the chemical forces that have stabilized the material in the solutions to form new associations. These are frequently areas of segregation of the finer particulate matter (sand) and the clay-sized fraction, which is eventually deposited after flocculation or grain aggregation in saline water.

For the most part dissolved material ends up in the saline seawater. However, at times the dissolved material can be deposited when the transport water is concentrated in evaporating closed basins in low continental areas without significant drainage to the sea. These materials upon evaporation of the aqueous solution form saline deposits which are at times combined with dissolved silicate material to form special minerals, usually magnesium-rich such as sepiolite or palygorskite. Salts of various types and compositions are usually the phases formed in these closed basins. These are concentrations of the very soluble elements Cl, Br, I, and F anions and Na, and Mg cations among others.

Another deposition mechanism of dissolved matter is that of the action of sea life, where dissolved calcium is combined with atmospheric carbon dioxide to form carbonate material, largely calcic but at times some magnesium is present and some other mineral elements of minor abundance are found such as strontium. The formation of carbonate rocks is extremely important to the balances of dissolved cations and atmospheric components such as carbon dioxide. In the nearshore environment and some freshwater lakes, abundant plant life contributes to a strong increase in organic matter in the transport medium. The major rock types developed from alteration materials are chemically inert sandstone (resistant minerals from rocks), clay sediments made of alterite silicate and oxides, and carbonates developed through the action of marine or freshwater shell bearing life. These three categories of sedimented materials will become rocks in their turn upon burial along the edges of continental landmasses and will become rocks susceptible to weathering upon being brought to the surface by tectonic movements. Very much of the surface rock material on continents which is subjected to alteration action is in fact material which has already been cycled through the alteration process to produce sedimentary rocks. Hence the alterites, which are carried by moving water today, can be largely recycled alterite material. One would expect minimal differences between shale rock and alterites of shale materials for example. However weathering intensity can change the types of alterite minerals present and change the chemistry of the resistant chemically stable alterite materials relative to the initial material. There will be different contrasts between alterite minerals and the initial rock material depending upon the origin of the rock (most likely to be changed will be volcanic or plutonic materials) and the intensity of alteration. The relative amount and type of dissolved material will depend also upon these two variables.

An important point to keep in mind is that the course of a large river covers, in most cases, areas that are not homogeneous in the substrate rocks which become alterite and in many cases a large river system drains areas of different geomorphology (mountains and plains) which furnish soil and alterite materials of different maturity (Fig. 5.1). One should not expect sampling of a river at a given point to produce a material representative of a single rock type nor alterite of a given stage of maturity. If one considers the Amazon River for example, the geomorphology of the region drained (Andean mountains to humid jungle low lands) covers a large range of alteration maturities produced under different climates. Clay minerals and alterite chemistry will be widely varied depending upon the area drained and of course the central drainage path will contain representatives of these different regions. A river such as the Mississippi in North America will represent much more homogeneous substrate materials, essentially sedimentary rocks and glacial deposits, where alterite has developed over similar periods for much of the basin and under similar climatic conditions, although small differences have been noted in the type of clays present depending upon the geographic and climatic region (see Jenny 1994). Thus river-borne material will represent in various proportions the material present in a drainage basin, which is more or less homogeneous in rock type forming the alterite and the stage of alteration of the alterite. In general one can expect a mixture of materials to be carried by major rivers to the sea.

We will look at the chemical data available for stream and river dissolved and suspended materials in order to understand the importance of the variables of alteration intensity and initial rock type material which is altered and then carried in streams and rivers. The initial distinction to be made is the physical state of the material transported by streams and rivers. Two main types are usually considered: suspended and dissolved matter. This distinction is dependent upon the size of the particles one analyzes.

Taking a sample of river water, it is easy to distinguish the silt and sand-sized material by letting the materials in the sample settle for several hours. However the finer material needs special treatment to separate it from the inherent aqueous substrate. Usually one takes the clays from the sediments by settling (either static or in centrifuges) where the <2 μm fraction is isolated and it is flocculated from suspension using an electrolyte of low concentration, such as NaCl or $SrCl_2$. The material left in suspension is considered to be dissolved or colloidal in nature. However, very fine crystallized or molecular matter can remain in the solution. Very intense centrifugation can be employed to eliminate more of the very fine particles, but there is always a line between particle, molecule, and ion which is difficult to define. In the following discussion we use the definitions of the authors who have made the laboratory determinations, realizing that some of the material analyzed might end up in another category using other methods of separation. However the definition of colloidal and truly dissolved (ionic) material is difficult to follow in analytical circumstances and the problem is a difficult one to solve [see Gaillardet et al. (2004) for a very careful discussion of the problem]. The problem of the carrying agent of minor elements lies in the definition of dissolved and colloidal in that the dissolved material can aggregate to become colloidal, forming slightly larger particles while in suspension. There are arguments for this process when freshwater from rivers encounters saline seawater. This is dealt with further in the chapter. Initially we take the definition of dissolved and suspended matter as such concerning the reported data for distinctly freshwater river transport.

The available data sets for river chemistry are not all overlapping in the type of element analyzed. Some elements are regularly analyzed, others more sporadically. This is especially true for trace elements, which are rather rarely analyzed and the elements that are analyzed are rather sporadically present over the range of potential interest.

It is possible to model the characteristics of cation attraction and fixation on various types of materials found in rivers and lakes. However the results are often disappointing when compared to field measurements (Gaillardet et al. 2004). The systems are rather complex, having several distinct types of material present whose interactions are poorly known. Thus we choose to use field observations to a maximum in our analysis.

5.1.2 Alteration Products in Rivers

We will compare major element abundance for soluble elements (Ca, K, Mg, Na) and ions forming insoluble residues (Fe, Al, Si) in rivers that drain continental shield materials in cold climates (Canada) and high intensity alterations under humid climates (Congo River Central Africa). Here we can see the impact of alteration intensity on the chemistry of alteration expressed by the material dissolved in the waters that drain the alteration zone. These relations can be compared to the observations made on the Amazon River materials, which depend upon the energy of transport (white water, lowland, and flood plain) covering a range of alterite materials. Examples of medium energy drainage basins on sedimentary rock material are given to compare with the more extreme examples.

5.1.3 Dissolved Material and Colloidal Material

5.1.3.1 Major Abundance Soluble Elements

Cations Ca, Na, K, Mg

Here we use the definitions proposed by the authors of the articles cited. Some prefer to distinguish between ionic and colloidal materials, which are of a smaller particle size than clays and associated materials. In the colloidal fraction (usually considered to be of <1 μm diameter in particle size) one can find silicates, oxides, and organic molecules [see Gaillardet et al. (2004) for a thorough discussion of the problem]. Each type of material will have different affinities for ionic elements in solution. What interests us here is the overall distribution of elements and re-distributions upon changes in the chemistry of the aqueous solutions. A common measure, which is indicative of the alteration regime, is the pH of the aqueous fluid. It can be used to compare the different compositional trends in the various rivers indicating the importance of the chemical regime of alteration.

Relations between K and Ca content and pH in river waters for the three river systems (Congo, Dupré et al. 1996; Amazon, Dosseto et al. 2006; and Canadian Shield rivers, Millot et al. 2003) indicate that in general Ca content is limited at pH below 6, with a substantial increase in Ca content above pH 7 and a similar situation can be seen for K content as a function of pH with a wider spread of change in concentration with pH. This is seen in the river waters from more or less temperate climate alteration, but in the Congo where a tropical climate dominates the relations do not hold. Nevertheless it appears that both K and Ca are affected by the pH of the alteration water that brought them into the river flow (Fig. 5.3).

The increase in K and Ca in the dissolved fraction of materials transported by rivers with increasing pH indicates a loss of the elements from the suspended materials. This displacement could be explained by the fact that minor elements

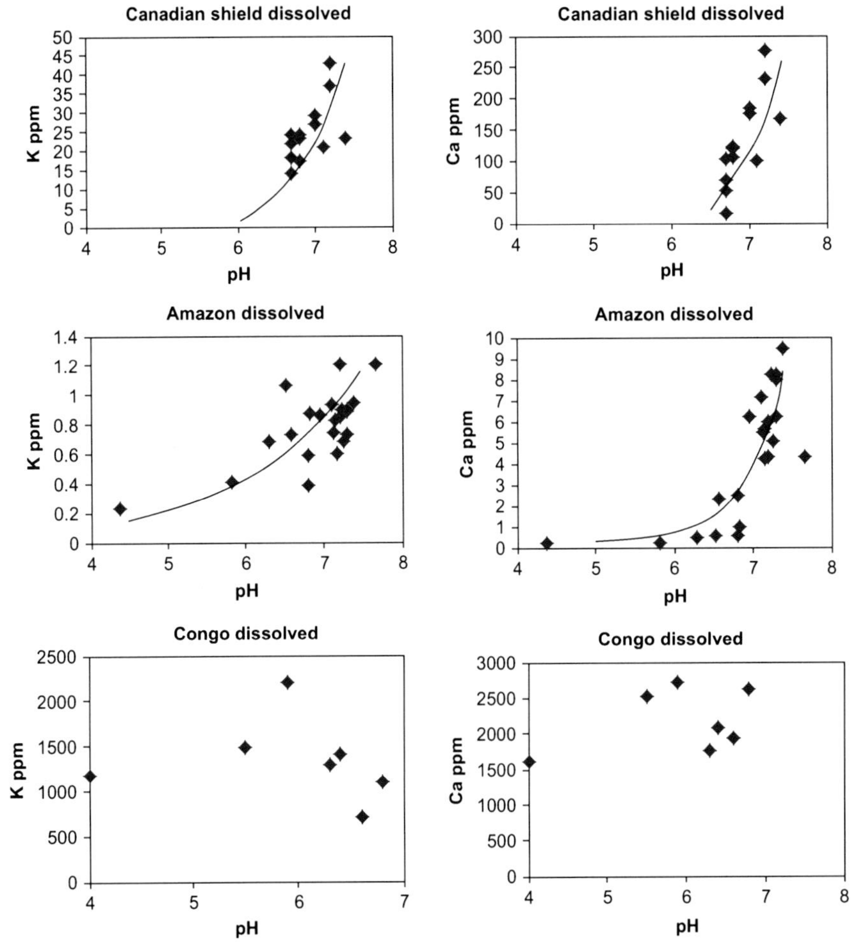

Fig. 5.3 Soluble concentrations (ppm) of Ca, K vs. the pH in the dissolved fraction of river waters for the Canadian shield (Millot et al. 2003), Amazon (Dosseto et al. 2006), and Congo rivers (Dupré et al. 1996)

are attracted to variable charge (surface) sites at pH above pH 6. Perhaps the increase in K and Ca reflects this fact or else the major elements are released from interlayer sites in clay minerals at these pH values. The data for the Congo river system show little effect with pH where the clay minerals susceptible to attract interlayer ions are almost absent. The apparent exchange of major cations from clays then would be due to the type of clay mineral present in the suspended load of the river systems.

In the river systems with less evolved clays from regions of temperate climates (lower stage of weathering giving smectite and illite minerals) it appears that there is an effect of pH on the presence of K and Ca in solutions. However in the Congo

River systems there is no clear relation. Thus it seems that in the less evolved clays, smectites, Ca, and K are held in the structures under low pH conditions but released to a certain extent as pH rises above 7. This suggests that the interlayer ion sites can be affected by pH, to the extent that other competing ions can be introduced into the interlayer sites, expulsing some of the Ca, K ions.

As one would expect, in areas where crystalline rocks are altered (i.e., where calcium carbonates are almost absent). Mg and Ca vary together in abundance, as do Na and K. The soluble elements are taken into solution in the early stages of weathering in the water–rock interface zone of reaction, and both mono- and divalent cations leave the alterite system (Chap. 3). These same relations can be seen in the data for a freshwater lake in Minnesota, USA, formed from alteration on acidic granitic rock types (Bartelson 1971). In data from rivers draining sedimentary rock basins (Moon et al. 2007) the relations of Ca and Mg are less systematic than those found for acidic eruptive and metamorphic rocks. This is obviously due to the variable availability of carbonate materials in the drainage basin, which change the Ca abundance relative to other alkali and alkaline earth elements. In fact one might use the relationship of Ca to Mg in river waters to estimate the impact of carbonate rock in the drainage basin.

5.1.3.2 Elements of Low Solubility

A different pattern of dissolved element abundance appears when looking at the relatively insoluble ions Al and trivalent Fe in the same river basins. In the rivers with acidic waters (low pH), those coming from zones of high intensity alteration, Congo and lowland Amazon basin water, the amount of alumina in solution is shown to increase in acidic waters, those where kaolinite and gibbsite are frequently the major clays in the alterite zones. In general Al and Fe content vary positively in the dissolved fraction of elements present in these rivers.

Thus, very briefly, there seems to be a segregation of dissolved major elements in river waters, Al and Fe being found in the acidic types and the soluble alkali and alkaline earth elements are found to increase greatly in abundance at pH above 7. Near neutral river waters would then appear to have a minimum dissolved major element content (Fig. 5.4). The relations of iron content in the dissolved fraction of river waters as a function of pH have been noted by Fuss et al. (2011) for soil pore water solutions in hardwood (deciduous) forests. The low pH and association with abundant organic matter where bacterial activity is important appear to be responsible for the reduction of Fe^{3+} and hence its higher solubility in aqueous solution. Again one must be careful concerning these data in that the definition of dissolved and particulate can cover overlap in particle size from one study to another. However the results seem to be internally consistent and will serve as a basis for comparison with the analyses for particulate matter.

Soluble silica is presented in river waters as a function of pH between pH 4 and 8. This is quite in contrast to the presence of dissolved Al and Fe where high solubility is found at low pH values. The relationship to pH is striking. The strong

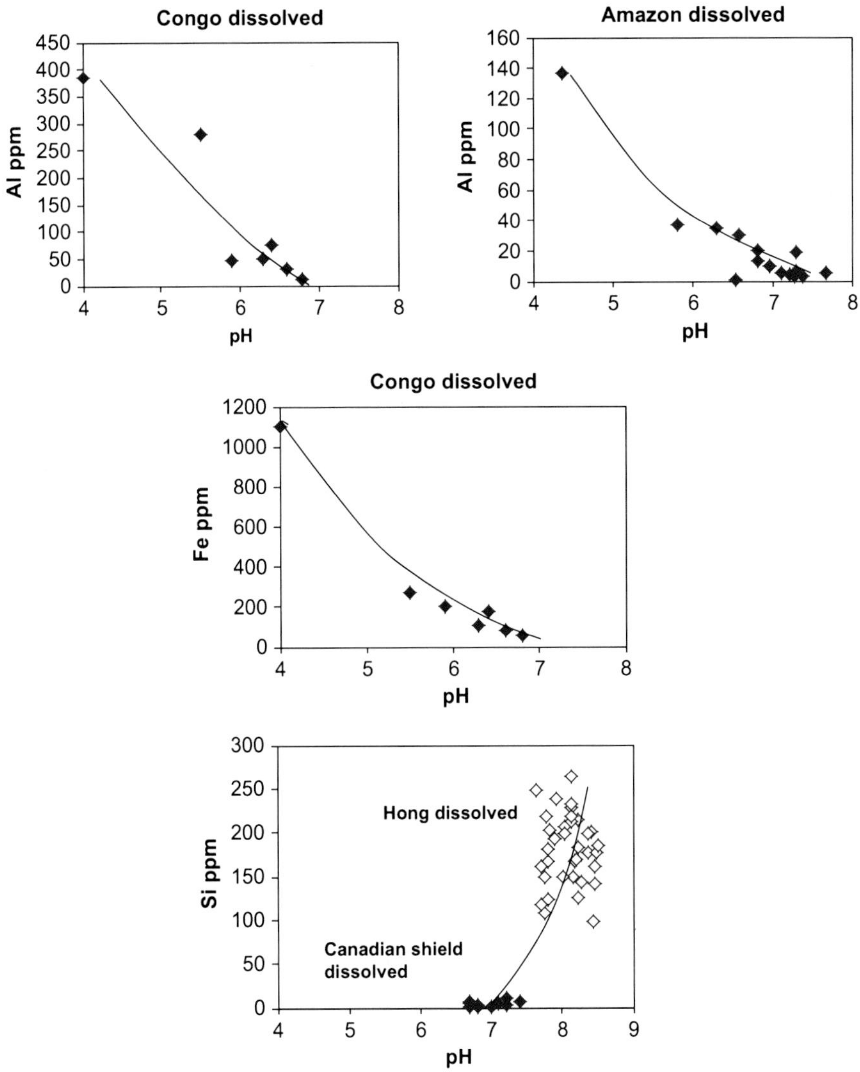

Fig. 5.4 Major element concentrations in dissolved fraction of river waters for less soluble elements (Al, Fe, and Si) as a function of pH. Data from Millot et al. (2003) (Canadian Shield), Dosseto et al. (2006) (Amazon), Dupré et al. (1996) (Congo), and Moon et al. (2007) [Hong (Red River)]

presence of silica at pH near 8 is slightly lower than the laboratory solubility data summarized by Degens (1965) where significant solubility of silica as quartz does not occur until values of near pH 10 are reached. Quartz is of course present under most conditions as a metastable mineral, but it does not control silica activity in solutions. The dissolved silica present due to mineral dissolution and recrystallization is characteristic of early stages of rock alteration, which would be expected to be dominant in the data for Canadian Shield rivers. Here the values are relatively low at pH 7. However the data set includes measurements for continental farmed and prairie areas of China. The dissolved silica content of river water is significantly higher than for the lower pH, immature forested Canadian Shield rivers. In the Chinese area the plant regime is dominated by grasses and grass crops. Here the deposition of phytollites (amorphous silica) from plant activity at the surface appears to contribute to the significantly higher silica content of river waters. Since phytollites are essentially amorphous silica, highly soluble, the resulting high silica content of rivers draining these regions is to be expected.

The concentrations of dissolved Fe and Al in soil waters reflect the solubility of the oxyhydroxides of the elements (Dosseto et al. 2006). Certainly the prevalent oxyhydroxides of Fe and Al commonly found in tropical soils would not give low pH values to drainage waters since the range of pH for their precipitation (Lide 2000) is above 7 for Al and Fe hydroxides. The insoluble elements Al and Fe are of significantly higher concentration in acidic waters due to the instability of these phases at low pH. In the river waters discussed here a reciprocity in dissolved element abundance for Fe and Al exists compared to alkali, alkaline earths, and silica as a function of pH. Where Fe and Al are stable in solids other elements tend to enter into solution to a greater extent.

5.1.3.3 Soluble Elements, Major and Trace Abundances

Horowitz (1985, pp. 5–11) discusses the distribution of minor elements between dissolved material and that associated with suspended particles in rivers. Certain elements are systematically present in large part and are associated with the dissolved fraction, such as Cl, Br, and S. Other elements are associated with both dissolved and suspended materials such as Na, Ca, Li, and Sr. Transition metals and heavy elements such as rare earths are dominantly associated with suspended matter. This reflects the types of alteration products present in the river transport material, where relatively insoluble high temperature oxide minerals carry transition metals and rare earths, while clay materials fix different types of material on their surface as do oxides produced during weathering processes.

Soluble minor elements are commonly associated with soluble major element cations; the pairs Ca–Sr and K–Rb show associated presence in the dissolved ion content of the Congo and Canadian river waters (Fig. 5.5). This is to be expected in that these elements are frequently co-related in abundance in solid phases because of strong similarity in chemical characteristics and when released through water–rock interaction they will be present together in solution.

Minor element data for dissolved species in river water, especially for Sr and Ca, indicates that these elements follow one another in abundance as would be expected by the similarity of chemical characteristics of the two ions. Overall it appears that the highly soluble alkali and alkaline earth elements are more present in river

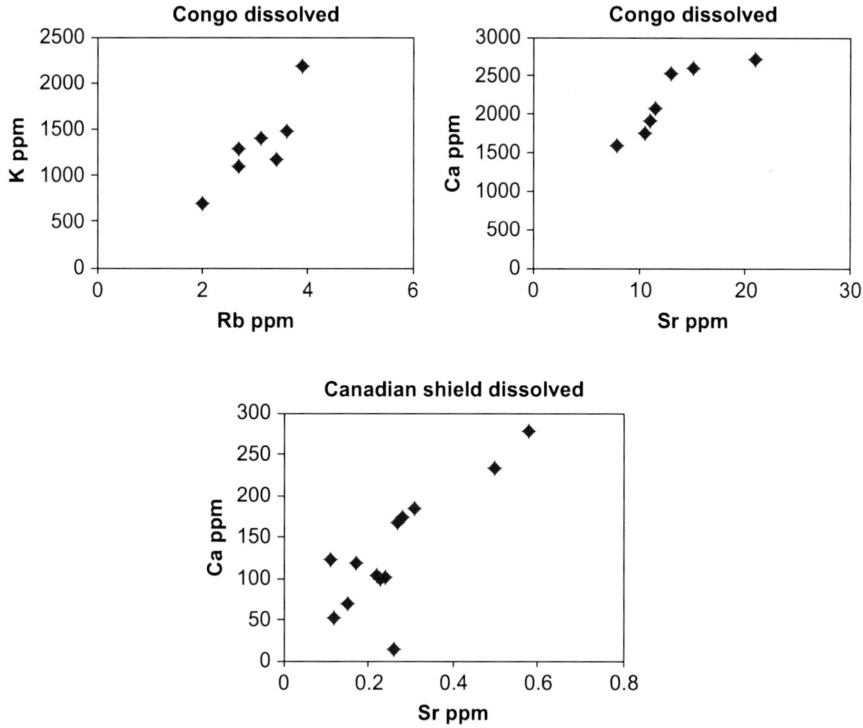

Fig. 5.5 Dissolved element concentrations for the major–minor element cation pairs K–Rb and Ca–Sr [data from Dupré et al. (1996) for the Congo and Millot et al. (2003) for the Canadian Shield]

solutions above pH 7, and hence the terminology of alkaline waters is respected. The importance of pH cannot be too strongly emphasized.

An illustration of the problems in comparing data sets can be seen in the study of Kerr et al. (2008) where material in streams coming from inland wet lands show relations between ions in the solutions and the abundance of dissolved organic matter measured as carbon content (DOC). Metals (Pb, Zn) and rare earth elements (Ce and La) are associated positively with carbon content, while Ca, Mg, Na, Sr, and Ba are negatively associated. This suggests that in fact alkali and alkaline earths are not attracted to the charge variable sites on organic matter while the metals are. The organic matter is present in molecular chains in the water, but in the very fine size fraction. It is not truly dissolved and the elements associated with it are not in a true dissolved or ionic state. However this material will remain in suspension for great distances and will eventually be attracted to larger sized materials in zones of deposition. One finds the same results in comparisons of three different river types (Dupré et al. 1996) where rivers with high organic content show higher "dissolved" ion content of rare earth elements and heavy metals.

5.1.3.4 Insoluble Trace Elements

Rare earth elements (REE) are often used in geochemical studies as tracers of provenance and chemical change because they are contained in phases of low solubility and hence reflect their geologic origin of the initial materials. However, some of the rare earth elements enter into solution and can be used as tracers of source materials and chemical change of the solutions that have affected their chemical history (Johannesson et al. 1997).

Ma et al. (2011) deduce that the weathering of different high temperature minerals in altering rocks that contain REE can determine the content in stream waters and at times effect a fractionation of middle weight REE for example. However, Hagedorn et al. (2011) come to the conclusion that the REE in streams and rivers in the Australian Alps are due to several factors, such as water–rock contact time (weathering intensity in a profile) and less dependent upon rock lithology or river discharge. Due to the low content of these elements in stream and river waters, it is probable that the effects of weathering at a specific site (rock type and mineralogy, stage of weathering, etc.) which can result in changes of REE relative concentrations producing anomalies in their relative distribution can be effaced through the mixing of waters as streams and rivers incorporate more and more material of different origins into their dissolved load. Nevertheless it is apparent that the chemistry of alteration produces differences in the overall composition of REE in solution. This is most likely due to the selective dissolution of different minerals which carry REE in the rocks. For example phosphate and carbonate minerals carrying REE will become unstable under different pH conditions than do oxides. Since each phase selectively incorporates minor elements according to the chemical affinities due to its structure the overall distribution of REE will be changed when one phase becomes unstable and others remain stable.

5.1.4 Suspended Matter

5.1.4.1 Major Elements

The term suspended matter can cover several types of material in river transport systems. The most evident is the silicate material of fine or larger grains, clays, silts, and sands. Basically the mineralogy and composition of the suspended silicate material changes with grain size where the clays in the fine fraction are of low alkali and alkaline earth content and of a relatively low silica content. A second type of material is more difficult to define in that it does not respond readily to X-ray diffraction identification. This is the oxyhydroxide material dominated by Fe and Mn but accompanied by Al forms. This material is highly reactive chemically but very difficult to determine in a quantitative way. Complicated chemical extraction methods are necessary in order to determine the state of crystallization for the different materials present. A third substance, common in materials derived from soils, is organic matter which can be present in small molecules (of low density) which also do not respond to X-ray diffraction identification. Some studies take into account the different types of suspended material and others do not. We will attempt to define the limits of our knowledge in the following discussion using available data.

The relative amount of suspended matter in rivers and streams is dependent upon the energy of water movement: the higher energy the water movement will carry more coarse material and carry it greater distances than rivers with low gradient and lower energy. This variable then is dependent upon the geomorphology of the drainage basin of the river. The higher the slopes, the more un-altered and coarse-grained material will be found in the streams and rivers. Hence the composition of the suspended material depends not only on the type of source rocks, or the intensity of alteration that has formed the alterite material but also on the topographic structure of the drainage basin. As is evident, mountain streams will move a relatively high amount of suspended material, in early stages of alteration. It will be sedimented when the stream slope changes at the base of the slopes and much of the suspended load is deposited. The larger rivers tend to have more slowly moving water masses and less coarse material suspended in the water. In a general way, the slower the water moves the more mature alterite material it will carry. Chemical analyses of suspended and bed load materials in the Congo river (Dupré et al. 1996) indicate that quartz dominates the suspended material in that the chemical analyses show above 92 % SiO_2 for the samples investigated. This is to be expected for materials coming from very mature landscapes with low relief and thick alteration zones. Few non-altered primary minerals will be deposited by this river.

Bouchez et al. (2011) present some very interesting data and give conclusions concerning the size of the grains transported and the composition of the material for the Amazon River. The Amazon River drains parts of a very high mountain chain and lowlands, which have been subjected to intense tropical alteration. There is a strong contrast between the alteration on the slopes of the Andes and the forest of the Amazon basin. There is a strong correlation of grain size and Al/Si ratio indicating that most of the coarse material is quartz-rich which suggests an overall strong alteration leaving only the metastable quartz fraction. Hence the coarse particulate material will not show strong variations in chemical characteristics. Such material will not reflect the rock initial compositions of the alteration areas. In the Amazon River suspended materials reaching the mouth of the river the separation has been made between alterite and minor bedrock relicts and quartz, which is the most robust and chemically resistant of the rock minerals.

In most cases of suspended materials in large rivers there is a good positive correlation between Fe and Al content, which indicates that oxides and oxyhydroxides are associated with the alterite silicate minerals as seen in alteration profiles. This is true for sequences in specific rivers of high alteration intensity and those of more temperate climate conditions. Amazon river data (Dosseto et al. 2006), data from Gironde river in France (Jouanneau and Latouche 1981), East coast of the USA (Horowitz 1985), Congo river basin analyses (Dupré

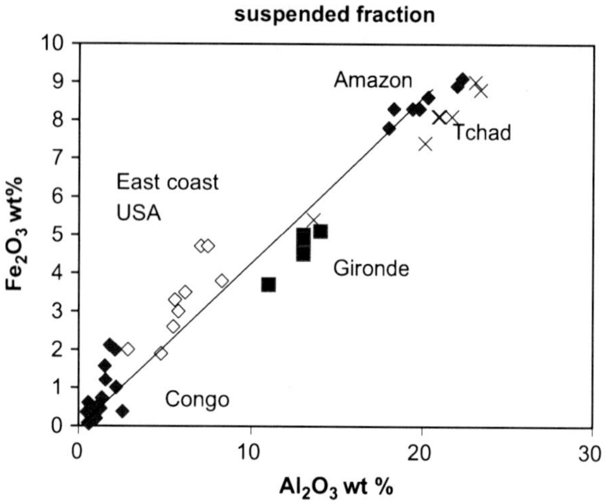

Fig. 5.6 Relations of relative concentrations of Al and Fe in the suspended material in several rivers from various climate and topographically configured river basins. Amazon river data (Dosseto et al. 2006), from Gironde river in France (Jouanneau and Latouche 1981), East coast of the USA (Horowitz 1985), and the Congo river (Dupré et al. 1996), rivers feeding Lake Tchad (Gac 1980)

et al. 1996), and rivers feeding Lake Tchad (Gac 1980) indicate a rather constant trend in the relationship between Fe and Al, which are the elements that remain present longest in surface alteration processes (Fig. 5.6).

Using the sets of data for rivers from different areas of the globe draining soils developed under different climates, it appears that there is a general relation of Fe and Al, which suggests a relation of near 3.5 Al atoms per Fe atom in the suspended matter. The Al/Fe atomic relation is much lower in average shales (near 2.3) but almost identical to the average sediment value given by Mason (1966). These significant differences between present day sediment composition and geological examples could suggest a difference in weathering regime or a gradual change in source material, sedimentary rocks instead of volcanic material as continents become more mature and form sedimentary rocks which enter into the tectonic cycles.

The parallel association of Fe and Al makes it difficult to distinguish with which alterite phases, silicate or oxide, the minor elements will be associated with in the suspended material of rivers. This being the case, we will treat the data by associations of minor elements, with no attempt made to distinguish the chemical associations of the elements to a substrate.

5.1.4.2 Minor Elements

The question asked concerning minor elements in rivers and river sediments is: how much contamination is there these days? This is difficult to establish unless a base line of normal conditions can be established. One can try to indicate that a given river is pristine or subject to only very low anthropogenic input, by industrial, agricultural, or urban input, but there are in fact few areas where these influences will pertain in a whole river basin. Extreme cases of climate make the polar regions ideal study sites, but the normal alteration processes are very weakly expressed and they can hardly be taken as examples for tropical weathering cycles for example.

Some data are available for single samples from different rivers and some samples from different sections of the same river. What is interesting is to see if the minor element concentrations diminish or increase along the course of the river and which elements are correlated. It is unlikely that pollution in different areas will give the same elemental distributions. We will look at several cases, where the elements are of similar chemical types.

The element pair Pb–Zn is one of typical anthropic pollution in industrial areas or those of old mines. Data for 12 world river systems (Gaillardet et al. 2004) indicate no specific correlation for rivers from different basins and stages of alteration. However in looking at several specific river systems (Gironde in France, East Coast rivers in the USA, Lake Tchad in north central Africa) one sees that the concentrations of Pb and Zn are related in much the same way from one region to the other, in industrial USA, in less industrial southern France, and in non-industrial Tchad. This suggests strongly that lead contamination does not enter significantly into the suspended loads of these rivers. The relation of about 1/3 for Zn/Pb indicates a general relationship due to rock sources whereas local pollution would change these relations significantly. Gaillardet et al. (2004) indicate that the amount of Zn transported in or on the suspended phases in rivers is roughly ten times that of the amount in the dissolved phase. They show that the amount of Zn in the dissolved phase decreases strongly from pH 6 to 7 in several rivers (Fig. 5.7).

An interesting study by Bourg (1983) shows the differences in minor element attraction by suspended matter in two different river systems in France, the Rhone and Gironde rivers. Adsorption of Cd, Cu, and Zn as a function of pH are compared for the sediments of the rivers compared to silica and alumina. The strong increase in adsorption characteristics in the Rhone river sediments is seen at a differential of three pH units less than that in the Gironde sediments. The silica and alumina relations are similar in the sediments of two rivers. This suggests that although of similar bulk chemistry the Gironde sediments are composed dominantly of variable charge type attractors, losing adsorbed ions more easily while the Rhone river sediments are more dominated by interlayer clay mineral type attractions where the interlayer ion populations are less affected by pH (see Chap. 2). In the Meuse River Gaillardet et al. (2004) find a strong Cu adsorption at pH 3 while that of Zn and Cd occurs at pH 6. Although Cu, Zn, Cd can be in the interlayer absorption sites of clays they are not preferentially fixed compared to Ca, Mg, or Na hydrated ions. Hence they are found mostly on the charge variable sites of absorbers.

These studies seem to indicate that it is not possible to predict the behavior of suspended materials in rivers without determining the type of absorbers present in the sediment material. Each river system will have its specific suspended matter, which depends upon the sources of the materials and the extent of alteration creating clays and oxide materials along with the organic matter charge of the rivers.

If the absorbers are of different types and abundances in river-borne suspended sediments, can one make generalizations concerning the presence of certain minor

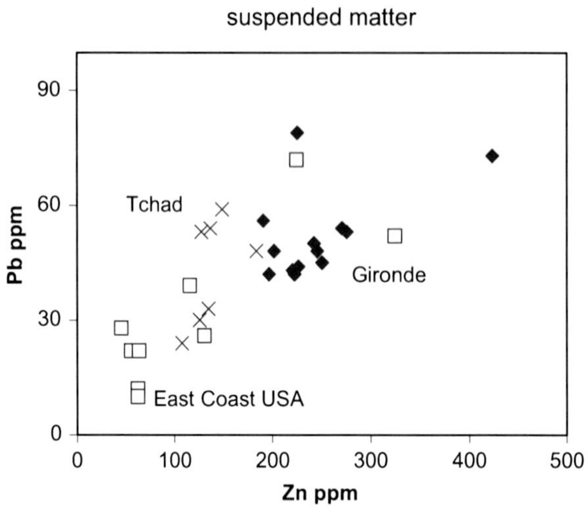

Fig. 5.7 Relation of Pb vs. Zn concentration in the suspended material for the rivers around Lake Tchad, East Coast of the USA and the Gironde river in France (data from Gac 1980; Horowitz 1985; Jouanneau and Latouche 1981)

elements? Transition metal elements such as Ni, Co, and Cu, among others, seem to be associated in relative abundance in suspended matter in rivers in a systematic way. This would suggest that each river basin although it has a different chemical structure and distribution of these minor elements gives similar interrelations of transition elements. One must remember that the sites of attraction and retention on suspended matter in rivers are variable; organic matter, clay minerals, oxides, and the different proportions of these phases will respond to the changes in pH in the river waters. Selectivity on surface sites, silicate, and organic and oxide materials, is especially dependent on pH and for each element the attraction is found to be varied in a different manner at different pH values. In general when the pH is above 8 most minor element cations are fixed to a maximum on suspended matter, but the oxyanions such as As are fixed at low pH and released at high pH. Förstner (1986) shows the dramatic effect of pH on lead and cadmium dissolved ion content in Swedish lakes where a change in pH from 4.5 to 5.5 decreases dissolved ion content by a factor of six such that these elements are probably fixed on the solids of the sediment fractions.

Some studies do however show specific relations between host minerals and minor elements such as transition metals. Suspended sediments in a fresh water lake in Michigan, USA (Lienemann et al. 1997), indicate that the reduction of manganese in the water column fixes cobalt in the manganese oxide structure. Cu and Fe are also associated with the manganese oxyhydroxide phases. Lead can form a very fine particulate material in these freshwaters. Gueguenat (1985) indicates that the capacity of fixing transition metal cations is highly variable according to the mineral species of the same element (Fe) suspended in a freshwater river near an industrial influx. The values of elemental fixation for Zn can vary by factors of 10^3 for different forms of iron oxide and hydroxide, as do those for Co and Mn.

The assembled data in Gaillardet et al. (2004) for rivers in various parts of the world indicate that there is a somewhat systematic relationship between Co, Cu, and

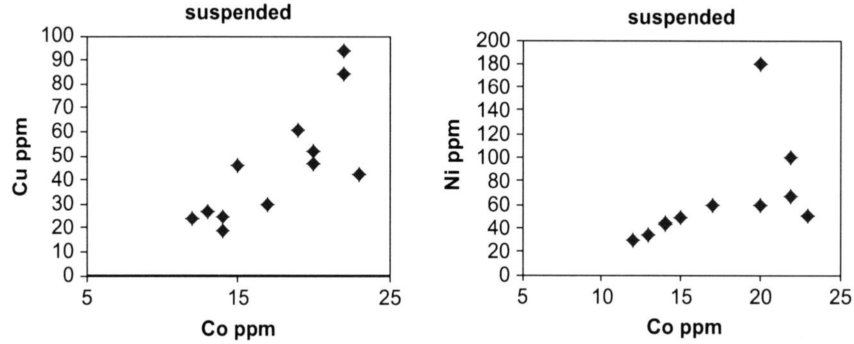

Fig. 5.8 Relation of Cu vs. Co and Ni vs. Co concentration in suspended matter in various river systems around the world (data from Gaillardet et al. 2004)

Ni abundances in suspended matter (Fig. 5.8). Such relations are to be expected in that the chemistry of these elements in the normal reduced state is quite similar. In general these elements are related to the abundance of Fe in the suspended matter.

Given these complex relations it is not surprising that correlations between major elements and minor elements present in suspended matter in rivers can be but are not always quite variable. The different carriers and their mineral structure can determine the relative amounts of minor element fixed on their surfaces. Since the different phases are a part of the alteration history of the rocks and the alterites in the basins drained by the rivers, one can expect to find a variety of relations of minor element abundance in the suspended load of rivers.

5.1.5 Comparison of Dissolved and Particulate Matter in a River

5.1.5.1 Soluble Elements

The soluble elements, such as Na and K, can be compared in suspended load and dissolved element concentrations in the cases of two rivers the Congo (Dupré et al. 1996) and the white water or areas of high slope and immature soils in the Amazon (Dosseto et al. 2006) and various rivers over the globe (Probst 1990). In the two examples with multiple samples for a given river system, K/Na relations show that there is more K in the suspended load material and higher Na content in the dissolved phase. This is to be expected in that Na is essentially not fixed on or in any of the phases in soils and alterite minerals whereas K is found in the clay fractions for the most part fixed in the form of illite minerals and in illite layers in various structures. The potassium is largely in a non-exchangeable state. There should be limited chemical equilibration between these clays and solutions concerning these two elements. However there is limited exchange of K from the

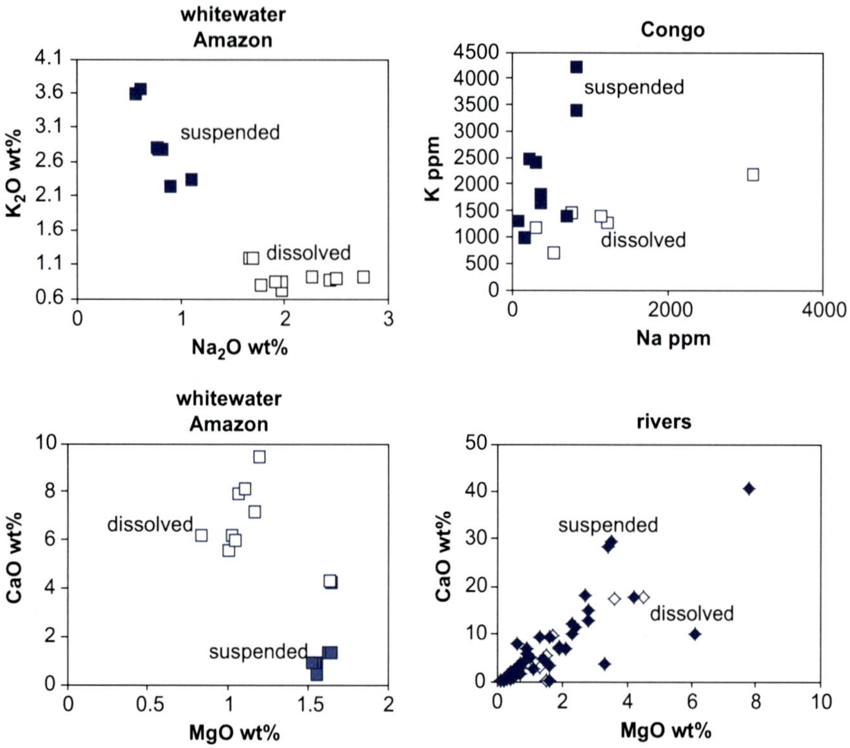

Fig. 5.9 Comparison of some soluble elements in dissolved and suspended fractions in the Congo (Dupré et al. 1996), white water (strong relief) Amazon (Dosseto et al. 2006), and a variety of world rivers (Probst 1990) rivers. Trends for soluble and suspended material for Na and K are discernable for the Congo River system and Amazon rivers

clay fractions. This is seen in the Congo River data where the solids show much more K than Na, but in the dissolved fraction another trend is discernable where Na is of higher concentration in the solutions, being less attracted to the clay surfaces. Hence the manner in which the element is absorbed (incorporated in the structure) into the different mineral phases can be determinate for the relations between elements in dissolved and suspended matter states in river transport (Fig. 5.9).

5.1.5.2 Elements of Low Solubility

Al and Fe

We will consider the major elements of low solubility, i.e., those which remain to a large extent in alterite residues. Taking the major residual low or insoluble elements Fe and Al, in the different river samples, it appears that Al is more abundant in the bed load and suspended materials (see Fig. 5.6). There is a reasonable continuous

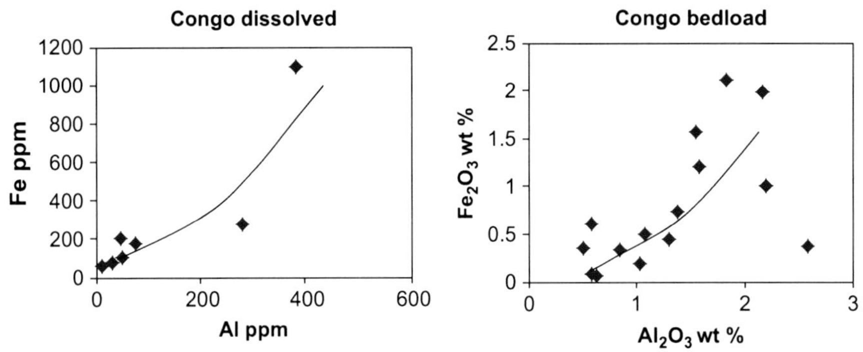

Fig. 5.10 Fe–Al dissolved and suspended fractions in the Congo River (Dupré et al. 1996)

trend following the different river system data indicating a roughly fourfold difference in atom percent. For different rivers the Al_2O_3/Fe_2O_3 ratio is near 3 and in the Congo River it is near 1.

In the Congo River samples dissolved Al is less abundant than Fe, indicating that the Al oxide solubility is significantly lower than that of Fe phases in the solid soil materials even though the solubility product constant (Ksp) for Al hydroxide minerals is 10^{-13}, while that for divalent iron hydroxide compounds is 10^{-14} and the oxidized form, Fe^{3+}, is only 10^{-36} (Hodgeman et al. 1959). The same relations are seen in soil interstitial waters in China compiled by Hong et al. (1995). One would expect then that the suspended and bed load materials contain significant amounts of alumino-silicate minerals, such as kaolinite, which would have a lower solubility than the iron hydroxides. There appears to be a major non-equilibrium between dissolved and suspended load values for Al and Fe (Fig. 5.10).

5.1.5.3 Minor Elements and Insoluble Elements

Co–Fe

Iron is relatively insoluble during the alteration process, ending up as one of the concentrated elements in highly weathered materials. Co is often associated with iron in high temperature minerals in minor quantities as well as in low temperature alteration products. It is present in minor quantities in the major phases present. Co is assumed to be an element, which is fixed on surface sites on oxides in alteration processes. If one considers the abundance of Co and Fe in Congo River waters (Dupré et al. 1996) Fe can be present at near 1,000 ppm, while Co occurs at near 0.5 ppm levels. This reflects the low solubility of iron oxides and hydroxides as noted above. In the data for numerous rivers suspended loads reported by Gaillardet et al. (2004) and the East Coast USA rivers reported by Horowitz (1985) the relative amounts of Co and Fe are similar. If one compares the relative abundances of the two elements as being near 4 ppm Co to 70,000 ppm Fe oxide (see Fig. 5.11) the

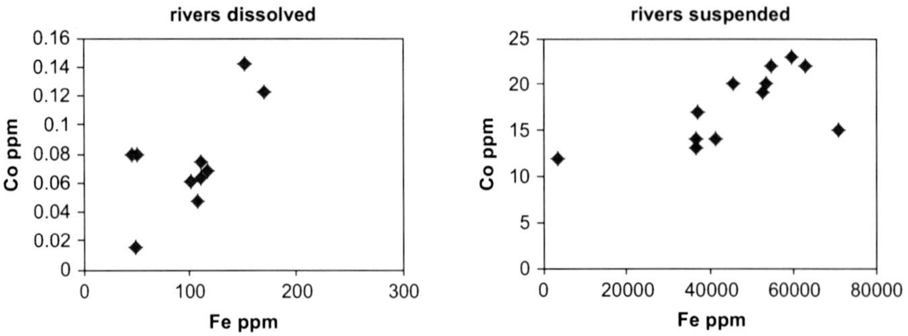

Fig. 5.11 Relation of dissolved and suspended Co and Fe element abundance for various world rivers (Gaillardet et al. 2004)

dissolved content of Congo River waters is quite similar to that of suspended material in other rivers. The ratios of Co to Fe oxide in altered basalt (Huang et al. 2004) and studies cited in Chap. 3 for highly altered soils show similar values of Co to Fe concentration (oxides) in the altered materials of various stages of alteration. One might assume then that the Co/Fe ratio in particulate matter in rivers is inherited from alterites and that it is in chemical equilibrium between the suspended and dissolved fractions of the river materials. Here of course the Co is fixed on surface sites on the Fe minerals and is thus subject to the chemical forces of concentration in solution. Thus in this case the elements adsorbed on the surface of oxide minerals and the oxides seem to be in equilibrium with the river waters that carry the materials from sites of alteration to an eventual deposition as solids. The higher the proportion of Fe present in the solids, the more Fe in solution and the more Fe present the more Co is present.

It appears then that the material transported by rivers is of several sorts. Essentially dissolved major element species reflect the relations of alterite produced solids where the more mobile elements are not retained in the new minerals. The more insoluble materials form products of alteration producing new phases, which interact with minor elements dissolved in solution. The minor elements become attracted to the insoluble matter, which can be transported from the sites of its formation. Chemical equilibrium is attained in solutions between the dissolved species and the insoluble materials during the alteration process, but often these insoluble materials do not come to equilibrium with the river transport solutions. Thus minor element equilibrium on surface exchange sites of insoluble phases can occur, but the substrate materials do not react readily to come into equilibrium with the transport solutions. The suspended matter in rivers will then reflect to a certain extent the source areas but to a lesser extent an equilibrium between dissolved and particulate matter of materials moved by a river.

5.1.6 Rivers and Seawater: The Deltas

The inevitable end of the transport cycle is the encounter of river flow with the sea. Several things happen in this environment due to the slowing of water flow and diminishing energy of transport and the change of overall chemistry as river, freshwater, material encounters the saline chemical environment. Chemically the change is one of relatively low concentrations of soluble elements encountering a system of high dissolved ion content, principally that of sodium cations and chlorine anions. However this is not the only part of the change in environment. The slowing of water flow, and the dispersion of the river water out over a larger surface area, subjects it to the interaction with plant life in the form of algae and other smaller units of vegetal life. Here there is a strong influence of chemical activity directed to the sustenance of microbial life. This can affect the chemical properties of the solutions such as changing the Eh and to a certain extent the pH of the solutions.

The clays and other suspended matter exchange cations in and on the particles for Na, at least in part, as salinity increases. Also there can be an exchange of hydrogen cations due to changes in pH from river water to seawater conditions. This is outlined in Fig. 5.12.

A certain portion of the suspended material is coagulated into floccular structures, which gain overall mass allowing them to sediment out of the water medium. This material is often fixed to the active organic matter in solution. When it sediments it forms a strongly bound assemblage of clay, oxide, and organic materials which, when buried and conserved, become a source for petroleum reserves.

One of the major vehicles for the transport of minor elements in river-borne material is generally considered to be iron oxide (Allard 1995). In these waters, of increasing salinity, the loss of iron from the dissolved or colloidal fraction has been clearly noted (see Gaillardet et al. (2004) for a review). However the interaction of new organic matter forming in the estuarine environment and iron hydroxyl-oxides is such that it is difficult to establish which of the chemical agents is responsible for the accumulation of trace elements (see Gaiero et al. 2003, for example). The state of iron, mineralogical or structural, is very important concerning its ability to attract and fix cations of minor elements. Gueguenìat (1986) indicates that there is 100-fold difference in the ability of iron hydroxides (di- or trivalent) to fix Zn, Co, or Mn compared to the more structured oxyhydroxides and yet another hundred fold difference with the pure oxide phases. Amorphous hydroxides apparently have a significantly higher number of active sites on their surfaces. Hence the hydrous phases of iron attract cations to a much greater extent than the oxides themselves under equivalent conditions of pH. Loss of iron from the suspended matter in river-borne material through flocculation can be of great importance concerning the content of minor elements in this material. Similar relations have been noted for Mn hydroxyoxides also in laboratory aqueous systems (Lienemann et al. 1997).

Organic matter in colloidal or particulate form also can be a vehicle for major and trace elements. Kerr et al. (2008) indicate fixation of major elements (Al) and minor elements such as transition metals, Pb, and rare earth elements. Hg can be

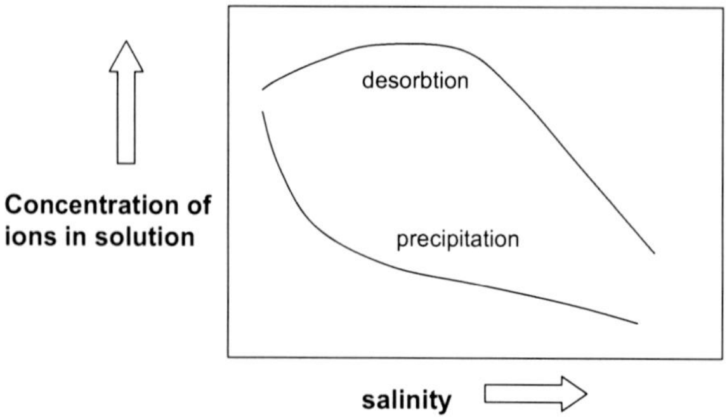

Fig. 5.12 Illustration of the relative concentrations of minor element and other dissolved ions in solution as a function of solution salinity when river water comes into contact and mixes with seawater. Three major effects are seen, (1) desorption of ions by mass action of Na ions in solution, (2) and precipitation of certain elements such as iron which fixes other elements onto its surface which decreases their presence in the aqueous solution (adapted from Millward and Turner 1995)

found to be associated with colloidal organic matter as well as oxyhydroxides (Merritt and Amirbahman 2007).

A summary of trace metal loss or increase from the dissolved fraction in estuarine waters as they become more saline (Millward and Turner 1995) indicates that there are various patterns of element loss or conservation depending upon the river concerned. In summary it appears that only Zn is systematically lost while some elements such as As, Sn, Cd, Cu, Ni can be found to increase in the dissolved element fraction. Conflicting patterns of elemental abundance in the dissolved and/or colloidal fraction in rivers can be observed as a function of salinity. For example silica is conservative, i.e., follows a linear relation with salinity (salt content essentially) while in certain cases iron content decreases rapidly in river water from the Merimac river estuary for example (Troup and Bricker 1975) and less so, but in the same manner one finds a drop in Fe in the dissolved fraction of the Gironde and Loire river systems of France (Kraepiel et al. 1997; Frenet 1981). Iron is likely to form oxyhydroxide particles and then it enters into the particulate fraction. Other minor elements can be incorporated into the iron hydroxides and follow iron and thus they show a decrease in the dissolved fraction, such as Mn and Pb which are probably associated with the formation of iron oxyhydroxide particles. However in the same system, increase in salinity initially increases the dissolved content of Cd, Ni, Zn, Hg, Pb, and Cu. Some indication of the provenance is seen in a slight decrease in the content of these elements in the particulate fraction of river-borne materials. This could well indicate the replacement of these elements on cation exchange sites, which are replaced by the Na or Mg ions in the sea water solution (Kraepiel et al. 1997).

Particulate matter minor element content remains rather constant until a high degree of salinity is reached where one can assume that flocculation of the particles occurs and some of the material is sedimented out of the estuarine water system. Similar observations have been made for some minor elements such as Cd in

Mississippi river delta sediments (Shiller and Boyle 1991) but in their study iron appears to be conservative, i.e., no loss above mixing dilution values, as is Ni but one finds a continued increase in Mo, Cr, and V in the dissolved fractions of the estuary waters. The desorption of ions due to exchange for Na and Mg from solids is clearly expressed in these samples.

Rare earth elements are of low abundance in the fine-grained colloidal or dissolved fractions of aqueous transport systems. One interesting chemical feature of rare earths is the tendency for the elements to differentiate in relative abundance according to atomic weight, light to heavy rare earths. Laboratory experiments by Coppin et al. (2002) give the parameters, which can produce such changes in relative abundance. The tendency for rare earth elements to become enriched in light elements on clays is a function solution concentration of sodium where saline solutions will show enrichment. Low solution concentrations do not favor a relative change in heavy or light REE elements. Data summarized by Gaillardet et al. (2004) for REE concentrations in the ultrafine fraction of materials in different rivers over the globe shows various types of relative ionic concentrations, enrichment or not. Such behavior seems to be adequately explained by contact between fresh river water and somewhat saline waters in estuaries. Similar observations have been made by Nozaki et al. (2000) for rare earth elements in the Chao Phraya river delta, Thailand, and by Sholkowitz (1993) for Amazon River estuary waters.

One can propose several scenarios for minor element abundance in the dissolved fraction of river water in estuaries as salinity increases (Fig. 5.12). Ions that are replaced on their exchange sites in the different types of matter present (colloidal organic, suspended oxides, and silicate clay minerals) will show an initial increase in relative abundance as salinity increases in the initial stages of mixing. High sodium and magnesium concentration exchanges for absorbed and adsorbed ions from suspended solids. So-called conservative behavior is that where the elements are not displaced from their substrate by the increasing abundance of exchange ions in the form of Na and Mg ions in seawater. Loss of ions in solution can be explained by precipitation from a dissolved form into a solid form, being for the most part incorporated into the suspended matter present in the estuarine water. This is likely to be the case for iron in many cases for example. As dissolved iron precipitates it brings minor elements with it from solution.

Suspended matter in river waters is in part kept in suspension due to the low ionic strength of the in solution. High ionic concentrations (seawater) tend to flocculate small particles, which settle only very slowly. They are kept in suspension often by the energy of moving water, which continually puts them into suspension by turbulent movement of the water mass. Cations tend to join particles together where strong local charges are present, especially on the edges and the surfaces of clays or on oxide surfaces. Here a cation can satisfy partially negative charges on two particles joining them and forming a larger particle. The juncture of several particles increases the relative mass of the particle, which allows it to descend in the water column. Figure 5.13 illustrates in a schematic way cation exchange and flocculation phenomena. The sodium ions enter the interfoliar spaces

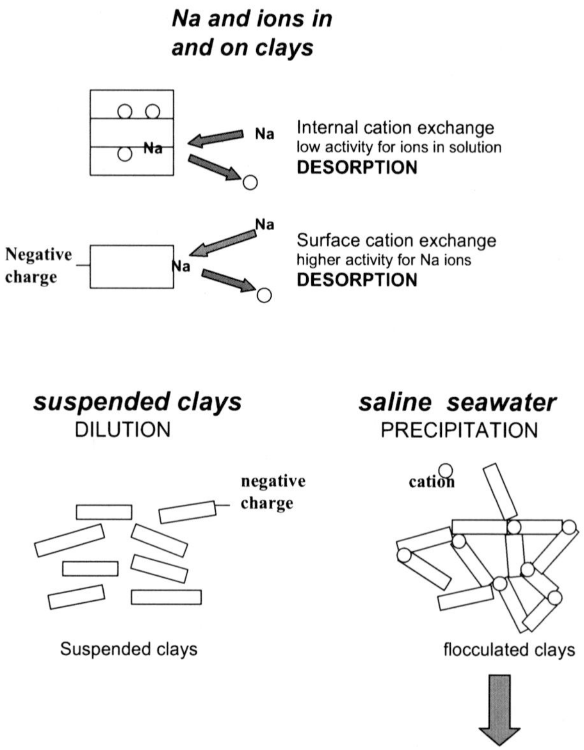

Fig. 5.13 Illustration of the effects due to the introduction of suspended material into seawater. A reaction of desorption of elements on and in the clays occurs; suspended materials are assembled into flocculated particles which sediment out of the aqueous suspension

in clay minerals and exchange the hydrated cations fixed there due to a high chemical activity of Na in the saline solution. However, the cation selectivity affinities of most ions are higher than that of sodium. Hence the internal cation exchange will be low and most of the ions carried by the clay particles in the riverine water will remain in the clays. However, the sodium ions in seawater (more concentrated solutions) will eventually exchange off much of the adsorbed ions on the clays where they are fixed by local negative charges. Here less selectivity occurs, apparently from observations of estuarine water chemistry, and the normal laws of mass action apply more directly.

When river and seawater is mixed, the ionic strength of the solution with suspended material increases. In general, each charged surface is charge-compensated by a cloud of (oppositely charged) counterions to satisfy electro-neutrality. Due to the permanent negative charge of clay minerals cations are attracted to the clay surface, and their concentration is diminished with distance from the surface until it reaches that of the bulk solution. The distance over which this occurs depends on the electrolyte concentration, e.g., ~1 nm at concentrations of ~1 M and ~100 s of nm at concentrations of ~10^{-5} M. This charge arrangement around the particle is called the diffuse electrical double layer (electrostatic repulsion) and opposes aggregation by the universal attractive van der Waals force, which acts to bind particles. Adding salt (seawater) to a colloid suspension in river water causes the double layer to shrink around the particles; this is known as double-layer compression and forces the agglomeration/flocculation of clay particles.

The flocculated particle assemblages are ultimately of greater mass than the individual clay particles and will tend to sediment out of the solution. These clay particles will be associated with iron and manganese oxides, all of which can fix some minor elements at the surface of or within the particles.

The larger particles formed of many smaller ones then settle toward the bottom of the water column and form sediments which are generally moved parallel to the sea coast by currents and tides. This material contains aggregates of fine particles of silicate and oxide material with highly active surfaces, which attract and fix elements from solution. Such matter becomes land material, when enough sediment has accumulated or when human activity fixes seaward boundaries as polders, which in effect increases the land surface relative to that of the sea. Poldering and use of estuarine deposits has been a standard method of increasing land surface for agriculture.

5.1.7 Summary: River Transport

Rivers move by far the greatest amount of surface materials which have been affected by weathering. Their activity brings together the dissolved and suspended materials of water–rock interaction and the new minerals formed by this alteration. Also the rivers can move associated organic matter. The materials moved in rivers in general will give an indication of the overall chemical activity and stage of alteration within the drainage basin. This incorporates the various types of material from different source rocks often in different stages of alteration. Alterite zones of slopes will be less mature than those on basin floors where alteration has progressed more due to the low amount of surface erosion. However, all this material from different types of rock alteration stages ends up in the same zone of deposition when it forms sediments at the interface of river and seawater along the coasts of continents. Some elements will stay in suspension or in a dissolved state, such as sulfur and phosphorous or sodium, calcium, bromine, and chlorine while others are fixed to iron hydroxyoxides, manganese oxides, or clays. However significant adsorbed ion loss occurs as the salinity of the waters increases. Overall the seawater concentrates the soluble elements, those of highest ionic character. Some cations of these types can be systematically fixed in clays, such as potassium remaining in the sediment materials.

The process of displacement in water is one of solution chemistry, surface chemical activity of particles, and their concentration in solution. In most rivers the materials transported are of different origins, stages of soil maturity, plant regimes, and bedrock sources, which will lead to a mixture of materials. It appears that chemical equilibrium is not reached between the solid phases and solution as the concentration of these elements in solution indicates. The concentrations of Fe and Al are good examples. However, the concentrations of adsorbed ions (transition metal and other trace elements) suggest that the surface site chemistry responds rapidly to the chemistry of the aqueous solutions.

The role of very small molecule organic compounds as carriers of adsorbed ions such as heavy metals and rare earth elements is very important to consider. The high oxidation state ions (3+ and higher) are strongly attracted to surface variable charge sites at pH above 5, and can be significant carriers of these minor elements into the zones of eventual deposition and sedimentation. The pH and the type of carrier materials determine which minor elements are dissolved or absorbed on solids.

Overall the transport of major and minor elements occurs via two types of materials, those suspended in the solutions and those in the dissolved state. The sites at which the different elements from the dissolved state can be fixed on suspended matter are a function of the pH, the composition of the transporting solutions, and the types of sites to which the elements are attracted. Chemical equilibration between elements in solution and those adsorbed on or absorbed in solids will depend upon the intensity of chemical bonding of ions onto the materials and the types of sites, relatively low energy absorption or high energy adsorption. Even on the same types of attraction sites, charge variable adsorption, the intensity of chemical bonding can be different depending upon the cation in the structure of the solid. Kulik et al. (2000) indicate different bonding intensity for Si or Al sites on clays which is seen in the retention of ions on the surfaces of clays as a function of pH. This bonding intensity will probably be different for different types of clay minerals with different structures. Hence heterogeneity can be expected in river-borne materials where different clays from different sites of variable weathering intensity are brought together as the river moves materials to sites of sedimentation.

5.2 Wind-Borne Materials

Particulate material when it settles from wind-borne air masses in significant amounts is called loess by geologists. The term "Löß" was first described in Central Europe by Karl Cäsar von Leonhard (1823–1824) who reported yellowish brown, silty deposits along the Rhine valley near Heidelberg. According to Pye (1995) four fundamental requirements are necessary for its formation: a dust source, adequate wind energy to transport the dust, a suitable accumulation area, and a sufficient amount of time.

This is material for the most part of fine grain size, several tens of microns being the largest material. Of course the size of particles is dependent on the energy of the air mass moving them and hence that material transported for great distances will be of smaller particle size than that due to local movement. Wind transport is a function of energy, wind velocity, and the altitude at which the air masses move. Low-level movement, hundreds of meters, is of higher energy over short periods whereas strong vertical movement moving air masses into the stratosphere allows material to move over significantly greater distances but grain size is smaller and this material is commonly called dust. Low-level transport usually deposits the suspended grains through slowing of the air mass and loss of energy giving what is largely "dry" deposition. High-level movement, in the stratosphere, necessitates rainfall to bring down the material in suspension, which is usually called dust after it is deposited. Usually the term loess in present day vocabulary terms indicates movement to low atmospheric levels and transportation of hundreds of kilometers. Very high-level transport gives materials that are to a large extent not considered to contribute to loess but to dust fallout. Wind-transported materials under glacial retreat conditions were often moved shorter distances and in greater abundance than the material on the thousand kilometer scale. The large loess deposits of China today are moved to a large extent at less than stratospheric levels. The differences in transportation distance and amount transported are due to the geological situation in which the fine-grained materials are exposed to wind movement (Keilhack 1975).

Initially one must consider that the prerequisite for dust or fine particulate transportation is that there is little or no vegetation at the surface. American farmers found this out in the period of the 1930s when intensive farming on land in rather arid regions was taken into air suspension more easily than under the initial prairie conditions. This created the great dust bowl famous for the period of exceptionally dry weather at that time. Normal geological conditions during the past several million years create such situations under two circumstances which give rise what are referred to as "hot" or desert derived and "cold" or glacial derived loess deposits. A third and more variable source of air-borne material is from volcanic eruptions. We will use the term loess to distinguish fine-grained material from eroded rocks, largely of sedimentary types, that from volcanic origin, which is designated as volcanic ash.

5.2.1 *Types of Loess*

Loess refers here, for the most part, to materials of fine particle size dominated by clays. The clays are generally chlorite–illite types. They are minerals based upon abrasion of rock material that is little altered by surface water–rock interaction processes but still largely phyllosilicate and fine grained in nature coming from sedimentary rocks or low-grade metamorphic materials that have been physically broken into fine grains. Low chemical alteration is due to little contact with the plants in the soil zones where they are deposited. The reasons for this low alteration state can be seen in the description below.

5.2.1.1 Cold Loess

Cold loess conditions occur when glacial retreat (usually on a continental scale) occurs and a large amount of fine-grained material is deposited by the retreating glacier, usually called till. The ensuing rainfall on such unconsolidated material

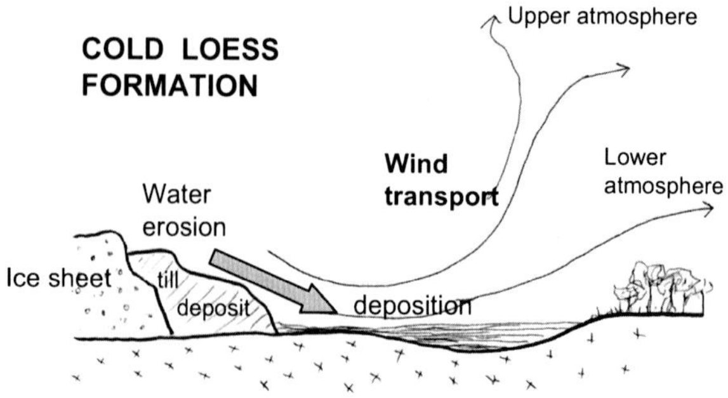

Fig. 5.14 Illustration of the origins of cold loess, i.e., that which has its origin in glacial deposits, usually on a continental scale. Alterite and rock material is deposited by a retreating ice sheet, which is then washed onto lower lying areas. It is deposited on outwash plains. Subsequent drying of the fine-grained materials allows them to be uplifted by wind currents as dust. Most of the material is deposited by low altitude wind currents at distances of hundreds of kilometers from the ice front. However other materials can be transported by more energetic wind currents to high altitudes and deposited at much greater distances

transports it locally by surface runoff erosion to be deposited nearby in local flat areas. Here the important and continuing influx of material inhibits the growth of plants, or at least enough so that the recent deposits when dried are subject to wind transport (Fig. 5.14 cold loess).

The dimension of transport distance has been demonstrated by Fehrenbacher et al. (1965) who show the relations of thickness of loess deposits as a function of distance from the edge of glacial deposition. The range of transport distance measured was several kilometers to several hundreds. The loess thickness is not a linear function of distance dropping off significantly at distances of several kilometers. Local accumulations can be of tens of meters or more in certain cases. However thinner layers of loess, tens of centimeters, can be found several hundreds of kilometers from the last glacial depositional fronts on the several continents. They have been deposited in southern South America, the central part of North America, the center of Europe, and an increasingly wider zone on the Eurasian continent. A typical characteristic of periglacial loess deposits is that the clay minerals present are of soil types, where smectite/illite minerals were quite abundant but have been produced by alteration in soil zones by biological action after deposition. Loess deposited directly is of an illite–chlorite mineralogy. The smectite/illite material is most likely the result of weathering of the wind deposits by plants under normally moist climates. Thus the hallmark of glacial loess deposits where plants could cover the material eventually is that of a typical soil clay mineralogy derived from the low temperature rock minerals illite and chlorite.

Cold loess deposits, formed in periglacial situations, are frequently found to cover glacial till deposits as might be expected in very short-distance transport. Thus the retreating glacier leaves unconsolidated material behind which furnishes loess while further retreat leaves till material still further from the initial front which is reworked to form loess again. Loess is essentially very mobile, once deposited and can be displaced easily again this time by water movement. Accumulation of loess along riverbanks is typical where the thickness is multiplied greatly.

Cold loess deposition has essentially stopped since about 9,000–10,000 years (Catt 1988; Ruhe 1984) depending upon the continent, North American or European.

5.2.1.2 Hot Loess

Hot loess is of a different origin where fine-grained material is deposited on low, enclosed continental desert basins from incoming river sources which drain local surfaces that have been affected by surface alteration or glacial phenomena. Due to the climate plant life is rare. Most often the fine-grained material reaching the internal sedimentary zones is little altered from its original state of rock mineralogy. The material transported is fine grained in size and of a phyllosilicate mineralogy, that of illite–chlorite for the most part in these materials. They are the most frequent phyllosilicate minerals in sedimentary or metamorphic rocks, which make up the largest part of continental material, which has been altered over the last several millions of years, as is still the case today.

However some loess materials, originating from surface deposits of dry saline lakes, can contain secondary sedimentary minerals such as sepiolite or palygorskite which have precipitated from concentrated solutions in the closed basins and thus have an entirely different origin from the fine-grained detritus of sedimentary or metamorphic rocks. Thus hot loess phyllosilicate minerals can be of two distinctly different origins: products of alteration and erosion or products of precipitation from saline solutions in the desert basin itself (Fig. 5.15).

Loess materials transported from desert environments are of two broad categories, low altitude transportation which can displace material for hundreds of kilometers where the grain size can be above 20 μm and another type of movement which occurs when the fine-grained material is taken up to high altitudes, kilometers where it is displaced by high-level atmospheric air mass movements for large distances, often from continent to continent. This high altitude material is largely deposited by raindrops and is very fine grained, in the clay size fraction of near 2 μm. At this level materials of different origins can be mixed in the uplift from several areas of significantly differing geographic origins, which can furnish fine-grained material that can move together over long distances. Also material can be deposited in the same area when it comes from several different high altitude wind directions of transport bringing material from different geographic sources, at times different continents, to be deposited in the same place at different times. High altitude transported material of present day origin often contains components of industrial origin or those coming from other anthropogenic origins. Loess today is largely the result of wind-borne movement of desert sediments. The major source basins of silicate clay materials are in desertic central China, the Sahara area of

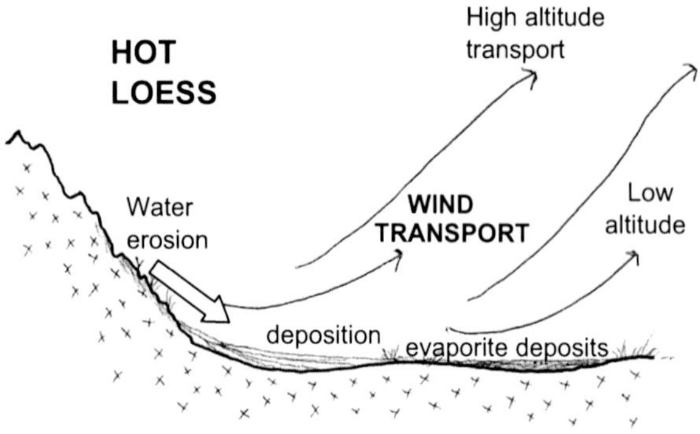

Fig. 5.15 Illustration of the origins of hot loess, i.e., that formed from desert or arid zone deposits. Alteration of the rock masses is only rudimentary but fine-grained material can be moved to outwash plains by periodic violent rainstorms typical of desert climates. Subsequent action by strong winds moves the material over different distances depending upon the energy of the winds and the altitude to which it is brought. This material can frequently be followed on trans-oceanic and continental scales using satellite photographs (see http://earthobservatory.nasa.gov/ for example)

Africa, central Australia, and some more minor areas along the Rocky Mountains in North America and the Andes in South America (see Fig. 1.8, Chap. 1).

Prospero et al. (2002) give a very thorough account of available information on loess (atmospheric dust) sources and movements over the globe, which is very useful to understand the dispersion of these present day events. The sourcing of loess dust deposits can be made using clay mineralogy for example, where different local sources in a general area of dust resources can be seen to contribute to long-range transport. One example is the Sahara region where a very large area is involved which can contribute to movement of materials in just about all directions from the sources. Clay mineralogy shows that a specific part of the northwestern source zone in North Africa is responsible for movements toward and above Europe (Molinaroli 1996). The key mineral is palygorskite, a magnesium silicate formed in evaporate basins, which is found today in Spanish soils and as far north as Scotland (Bain and Tait 1977).

An example of dust mineral composition which can be used to determine the different sources of material transported at altitude and deposited locally can be given for materials found in low-level atmospheric dust sampling and soils found on Bermuda, a rather isolated island in the Atlantic Ocean. A description has been given by Church et al. (2010) and Fontaine et al. (in press). It is generally accepted that the major source of Aeolian dust in the Caribbean area is from the Sahara with key minerals being illite–kaolinite–chlorite (Glaccum and Prospero 1980; Caquineau et al. 1998). Also it is known that palygorskite (clay mineral) is found to be transported from the areas of the northwestern African area (Molinaroli 1996; Goudie and Middleton 2001). Another rather special mineral, alkali zeolite, has been identified in the atmospheric dust input on Bermuda also. This mineral is not typical of the Saharan mineral cortege but can be traced to a desert of North

American origin. The origin of palygorskite formation is a precipitate from basic solutions in shallow water evaporate systems while zeolite forms from solutions reacting with acidic composition volcanic ash material in arid climates. Thus the mixture of two very different materials of different chemical origins is evident in this Bermudan dust material. Further the presence of boussingaultite, a hydrated sulfate of ammonium and magnesium, indicates input from an industrial area such as the eastern coast of the United States. Here the mixture of minerals from different source areas which are deposited together is certainly a highly chemically metastable state. Herwitz et al. (1996) indicate that the materials in paleosols can be traced by REE analysis to origins in the North American loess sources as well as Saharan origins. Their conclusions are that the present day variable sources of dust input have persisted well into the past. Muhs et al. (2007) come to the conclusion of multisource and multicontinent provenance for soils on Barbados, the Bahamas islands, and in Florida. The result is that the soils of these essentially calcareous (coral reef) islands are of an origin quite far from the island itself. Thus dust transport of hot loess, derived from desert areas, over long periods has affected the surface geochemistry of Bermuda for tens of thousands of years and longer periods.

Although high altitude dust transport is less abundant than lower level transport material, nevertheless this material can in certain circumstances have a strong geochemical impact on surface chemistry when it is deposited on the surface. Use of detailed geochemical observations can identify the origins of the different components of the dust material that is deposited, using such determinations as rare earth elements (Muhs et al. 2007) or Pb and Sr isotopic estimations (Simonetti et al. 2000) and hence geochemical determinations can be of great use in the study of dust and loess origins and movements in the present and the past.

5.2.2 Volcanic Ash

Volcanic ash consists of fragments of pulverized rock, minerals, and volcanic glass blown into the air by an erupting volcano. The volcanic ash is largely glass, and as such extremely chemically reactive in the surface alteration environment.

Ash deposits of volcanic origin are of course well known, and studied by atmospheric scientists as well as geologists. At times ash deposits have been so thick as to form local sedimentary deposits of tens of meters thickness. These materials, once buried and subjected to diagenetic conditions of temperature and pressure, produce a material called bentonite. This diagenetically transformed material is essentially a smectite mineral with very special physical and chemical properties. Its use in industry is for drilling mud in producing deep holes to extract petroleum, as more or less impermeable barriers against chemical migration in waste deposits: industrial, chemical, and nuclear, and as a thickening agent for paints, and for cosmetics and as a filler in hot dogs. However bentonites are not physically stable at the surface as they are easily eroded and carried into streams and rivers. The volcanic ash deposits can be found as deposits on soils, and they are rapidly altered to clay minerals and fuse into the local mineralogy, most often overlooked by soil chemists and agricultural specialists. Although unnoticed when in thin layers, some regions, for example, near active volcanoes such Mount Etna or Vesuvius in southern Italy, form a very prominent part of the surface chemistry and agricultural substrate.

The range of transport distances is great; some volcanic ash clouds can be traced for transcontinental distances or others found at the foot of the volcanic mountain source. Overall the addition of volcanic material that forms smectite, a chemically reactive clay mineral, is an enriching factor for the soils under agricultural use. If the dust is in the air and modern jet aircraft fly thorough it, the motors tend to wear out rapidly and this forms a danger for long-distance travel stopping flights for days or longer. In general volcanic loess is of a bulk composition similar to igneous rocks, usually silica rich and hence containing a relatively high alkali content, sodium and potassium. Also the crystallization of smectite from the ash releases silica, which can stabilize other silica-rich clays or perhaps form them from low silica types such as kaolinite or gibbsite already present in soil substrates. This action is most likely important for soils in tropical regions such as those of southeast Asia where volcanic eruptions are rather frequent. Volcanic ash can very well be an important factor in maintaining fertility in such soils. However inhabitants nearby usually fear the abrupt and massive production of loess by volcanic eruption preferring to see it from a distance (Fig. 5.16a, b).

5.2.2.1 Volcanic Input

Volcanic ash can be disseminated over great distances. Unfortunately there is little direct data concerning this material at distance from the sources. It is known that the ash materials can travel great distances but since such events are sporadic, it is difficult to sample the material in the air, and more difficult to trace its origin in soils present thousands of kilometers from the source. Nevertheless over long periods of time volcanic dust can make a significant contribution to soil materials. In going closer to the sources, near volcanoes, more information is available concerning the geochemical interactions of such material with the soils under biotic transformations and water–rock interaction. One can expect that the volcanic material will react rather rapidly in that the major material present is a silicic glass which is highly unstable chemically and will transform into clay minerals and oxides rapidly. One characteristic of the mineral reactions in the transformation of ash to minerals stable at the surface is that there is a strong tendency to produce metastable, cryptocrystalline silicate materials (imogolite and allophane). The relations of mineral transformation and chemical stability in soils are discussed in greater detail by Velde and Barré (2010, pp. 150–154). One important aspect of the metastable phases formed from the alteration of volcanic ash such as imogolite and allophone is their chemical reactivity. First they attract and stabilize organic matter produced in soils. Second they retain a significant amount of water in the

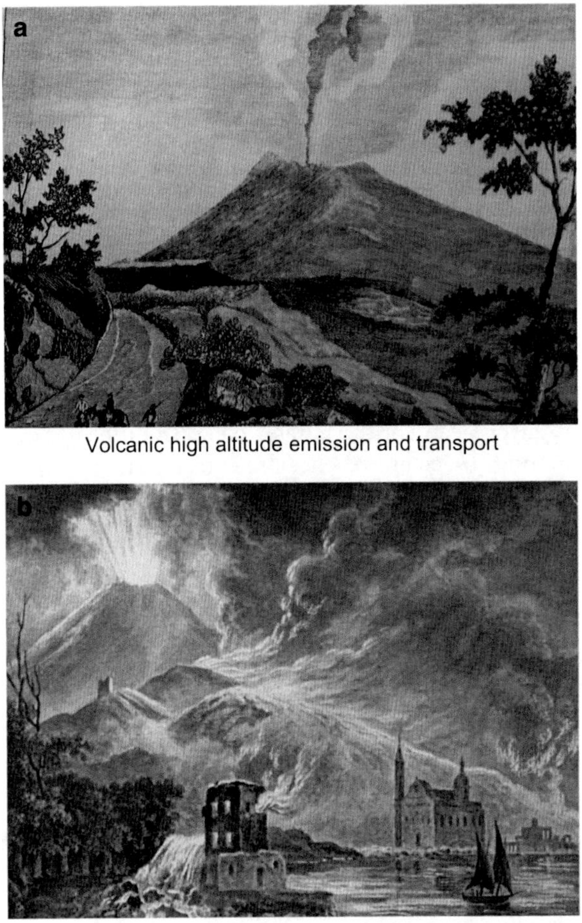

Volcanic high altitude emission and transport

Volcanic low level emission and transport

Fig. 5.16 (**a**) Figure illustrating volcanic high-level eruption which will disperse ash material into the higher atmosphere (photograph, collection of BV). (**b**) Figure illustrating a violent volcanic eruption which produces large amounts of ash that is deposited around the volcano (photograph, collection of BV)

clay–organic complex. This high chemical activity and the presence of organic matter in abundance give these soils a special quality in agriculture, as the Romans and earlier civilizations found out to their benefit. Volcanic soils, especially those based upon volcanic ash, are especially fertile. Thus volcanic wind-borne material has a significant agricultural importance.

The geochemical impact of volcanic ash deposition is treated further in this chapter.

5.2.2.2 Wind Transport of Special Elements

Nitrogen

Nitrogen is an essential, fundamental building block for life. Without an adequate supply of nitrogen, crops do not thrive and fail to reach their maximum production potential. In many ecosystems, nitrogen is the limiting element for growth.

Natural accumulations of nitrogen are known, especially in the Chilean Andes on arid plateaus. The transport vector is assumed to be mist from seawater, coming up the mountains from several thousands of meters below. The nitrates would come from oceanic organic matter dissolved in surface waters and transported in water droplets which are deposited on the dry high plateaus and to become nitrate minerals (Garrett 1998). The precise mechanisms are only conjectural but as far as anyone can tell the nitrates come from the sea. Thus relatively short-distance airborne trajectories concentrate material of a specific geochemical character, which forms commercial deposits of a very valuable element.

Mercury

Mercury is familiar to most people as the silver-colored liquid, which expands and contracts in a thermometer to show the temperature. But perhaps you have read Alice's Adventures in Wonderland and you remember the Mad Hatter from the Mad Tea Party. In the eighteenth and nineteenth century mercury was used in the production of felt for hats. People who worked in these hat factories develop dementia caused by mercury poisoning. Thus the phrase "Mad as a Hatter" became popular as a way to refer to someone who was perceived as insane.

Mercury is a highly toxic element that is found both naturally and as an introduced contaminant in the environment. Major sources of mercury pollution include coal-fired power plants, boilers, steel production, incinerators, cement plants, and gold mining contribute greatly to mercury concentrations. Mercury is released into the air through the smokestacks, so atmospheric deposition is the dominant source of mercury over most of the landscape. Once in the atmosphere, mercury is widely disseminated and can circulate for years, accounting for its wide spread distribution. The toxic effects of mercury depend on its chemical form and the route of exposure. Mercury is strongly affected by microbial actions (Ehrlich and Newman 2009). Bacteria that process sulfate in the environment take up mercury in its inorganic form and convert it to methylmercury [CH_3Hg] through metabolic processes. Methylmercury is organic and soluble in water.

This interaction is frequent in coastal and salt marsh environments, where it enters the atmosphere and is absorbed by fog (Weiss-Penzias et al. 2012). When the fog moves onto the land, it collects on the vegetation and drips onto the ground, depositing significant amounts of mercury onto the land. This material is easily moved in wind currents, which can disperse and redistribute the Hg. This is especially well known in the near marine environments (Fitzgerald et al. 1981). Mercury is thus not only deposited but redeposited by wind transport of the volatile methylated materials.

In these two instances, special cases to be sure, specific elements are moved by air transport in marine and near marine environments. These transportation modes are to a large extent independent of carrier materials, particulate materials or the necessity of being dissolved in aqueous solutions.

5.2.3 *"Human Loess"*

Airborne materials due to human activity come of course from various areas, large cities or areas of high-density inhabitation and areas of industrial and mining-smelting activity. These materials are of lower overall mass than materials coming from deserts, but they can contain relatively high amounts of minor elements and can contribute to the acidity of rainfall deposits. Elements such as As, Pb, Hg, Cd have been seen as being transported and deposited by atmospheric transport (Perkins et al. 2000; Holmes and Miller 2004; Shotyk et al. 2005; Simonetti et al. 2000 among many others). Chemical behavior of minor and major elements can be modified by the interaction of solar radiation between the elements in the particles within the raindrops in the higher atmosphere. The material of anthropogenic origin appears to behave differently from that of surface rock outcrop materials (soil and sediment materials). Chester et al. (1996) indicate that Pb, Zn, and Cu for instance have very different apparent solubilities in rainwater depending upon its origin. There is a high "solubility" in the sense of concentration in the aqueous solution, in materials of anthropogenic origin. This rainwater is acidic (pH 4–5) where hydrogen ions will take the place of exchangeable ions on mineral surfaces. By contrast the loess of rock origin in rainwater droplets tends to be more basic in character (pH 6–7) and hence in cases of more basic solutions more of the minor element cations will be attracted to particulate material. This effect has been noted by Guieu and Thomas (1996). Thus the availability of these elements to surface waters is to a large extent determined by pH values which will be modified when the rainwater comes in contact with the soils of the area of deposition. If the solar radiation interacts with solids in the rain drops, the pH could change through redox reactions and the balance of elements associated with the suspended matter could well change. Mercury and methylmercury exposure to sunlight (specifically ultraviolet light) has an overall detoxifying effect. Sunlight can break down methylmercury to Hg(II) or Hg(0), which allows mercury to leave the aquatic environment and reenter the atmosphere as a gas.

5.2.3.1 Industrial Input

Materials other than those of alterite origin can be found in high-level airborne matter, such as industrial ash material and other material borne by burning different materials.

Sulfur

Sulfur, an element usually found in the dissolved ion part of surface aqueous solutions, can be found in the air also. Sulfur dioxide in the air is essentially of industrial or heating origin coming from fossil hydrocarbon combustion (coal and petroleum). Some SO_2 comes from mineral smelting practices in that the easiest way to separate many ore metals (Pb, Zn for example) is to heat them so that the sulfur in the sulfide ore is oxidized to leave a rather clean metal deposit in the crucible, but SO_2 is released directly into the air.

In the atmosphere sulfur is usually oxidized in the air from organic sulfur or elemental sulfur to sulfur oxides like SO_2 and SO_3 ending up as sulfate in sulfate salts $M(II)SO_4$, $M(I)_2SO_4$ or sulfuric acid H_2SO_4. This reaction occurs when seawater is dispersed into the atmosphere carrying cations that react with the sulfate ions (see Usher et al. 2003). The sulfate compounds dissolve very well in water and come down again with the rain, either as salts or as acid rain.

Through various reactions the sulfur compounds are formed and form coatings on dust particles in the upper atmosphere. Movement of dust particles upward or downward brings the particles into contact with gases where redox reactions can occur, forming coatings on the particles which are carried to different levels of the atmosphere and transported by wind currents over oceanic masses to the continents. The deposition of these particles leads to acidic water solutions that affect the materials on which they are redeposited, plants or soils.

Organic Matter

Organic compounds in the atmosphere can be of biogenic or anthropogenic origins. Ocean sprays can move organic molecules from bioactivity into the atmosphere and on continents reasonably complex organic compounds can be released from plant activity and find their way into the atmosphere. Of course anthropogenic actions of burning release organic compounds in the smoke, which can be moved into the atmosphere to different levels. Depending upon the geographic regions and the origins the organic compounds in the atmosphere will vary greatly. It appears that hydrogen bonding is the main chemical fixation process of attachment of organic compounds to silicate and oxide solids that can be found in atmospheric dust (Usher et al. 2003). The release of such compounds from silicate-oxide and organic bonding upon return to the earth has apparently not been given much attention up until now.

Some minor elements can be associated with anthropogenic carbon input into atmospheric dust. Such input can be traced on a seasonal basis in the Florida Everglades where As is correlated with C in winter dust input (Holmes and Miller 2004). Many other cases have been found for specific elements transported over shorter distances. But for example Hg and Pb identified on the Faroe Islands indicate movement of anthropogenic materials over significant distances (Shotyk et al. 2005).

5.2.3.2 Gases and Dust Particles: Reactions During Transport

Iron is a very important constituent of desert dusts. It is usually in the oxidized form and most often anhydrous (hematite) as indicated by Mackie et al. (2008) for Australian desert derived dust. The action of solar radiation in raindrops changes the mineral state from hematite to ferrihydrite, a ferri-hydroxide mineral which has a relatively high solubility compared to the other forms of iron oxide present in soils. Takahashi et al. (2001) indicate that the oxidizing effect can be extended to the chlorite minerals present, where iron is dominantly divalent at their formation but which are oxidized during their trajectory at high levels in the atmosphere. The dissolution of ferrihydrite indicates that a major substrate for minor element adsorption is lost and the elements fixed on it will be readily released in aqueous suspension as dissolved ions upon their arrival on the soils of deposition. This effect renders the iron in the particulate matter bioavailable which has a large impact on ocean waters where algae, plankton, and other plant life depends upon this source of Fe for its metabolism in making chlorophyll. The impact of this biogeochemical effect on iron release is sufficiently important to show up on satellite maps of the ocean surfaces where green patches occur after significant dust storms have deposited material in a given area.

Nitrogen is another element, which is important to airborne dust chemistry. For the most part one can identify nitrogen as nitrate associated with solid materials in water droplets (Usher et al. 2003). Seasonal variations of nitrate presence indicate the influence of plant activity on the land masses (Keene et al. 2002). Nitrogen sources are vegetal, through mineralization of plant matter by bacteria, and industrial where coal is used as a fuel and some gas production of nitrogen compounds can occur in different industrial chemical processes. Also burning petroleum can give significant amounts of nitrogen compounds and gases. Extensive work has been done on the absorption of nitrogen gases on different mineral substrates. It was determined that little gaseous HNO_3 was introduced into exchange sites in clays and that most of the material was adsorbed on various substrates in roughly the same proportions. However important amounts of ammonium nitrate, formed by reaction of NH_3 and HNO_3, are found associated with aerosol particles, especially of 100–300 nm size. In fact the nitrogen-bearing compounds produced form crystals, which coat the dust (silicate and oxide) particles in the droplets of rain at high altitude. Gaseous NO_2 can be attracted to mineral surfaces and eventually bonded as NO^- complexes with covalently bonded cations such as Al. This is another reaction, which is important to troposphere chemistry. Other reactions of nitrogen compounds in aqueous media fixed to mineral substrates are possible [see Usher et al. (2003) for a discussion].

5.2.3.3 Radioactive Fallout

An example of the importance of man-made distributions of materials in the atmosphere can be given by Cs, which has been identified as a serious contaminant in surface environments.

Through atomic bomb testing since 1945 and the reactor incident in Chernobyl, radioactive fission products have artificially entered and spread worldwide throughout the atmosphere. Of the remaining fission products in the long term only the isotopes Strontium-90 and Cesium-137 have a significance: Both isotopes behave physiologically like the important elements for organisms: calcium and potassium. The interactions of radioactive isotopes of Cs with soils are of concern in environmental studies, due to the high transferability, wide distribution, high solubility, long half-life (half-lifes of Sr-90 = 28 years and Cs-137 = 30 years), and the easy assimilation by living organisms. The spatial distribution of radiocesium in the soil depends in particular on its adsorption on soil minerals. The vertical rate of spreading of radiocesium in undisturbed soils is relatively low (with the exception of sandy soils and tropical laterites). The cause of this low depth penetration is the selective sorption of cesium by the crystal lattices of clay minerals (especially smectite and illite). Cations with low hydration energy, such as K^+, $NH_4{}^+$, Rb^+, and Cs^+, produce interlayer dehydration and layer collapse and are therefore fixed in interlayer positions. According to concurring results from several authors, at least 80 % of Cs activity remains in the upper 15 cm of soil (ANPA 2000; Kühn 1982; Ritchie and Rudolph 1970; Squire and Middleton 1966; Zibold et al. 1997).

5.2.4 *Summary*

This brief review indicates the complex systems that occur when earth material in a fine-grained state is lifted by wind into the atmosphere. In this state geologic (silicate and oxide) material is combined with atmospheric materials, water of course but atmospheric gases of nitrogen and sulfur and with organic materials that are taken into this complex chemical system. Rocks contain the volatile surface elements carbon, sulfur, and nitrogen (to a limited extent) which are released either into the alteration waters of surface water–rock interaction or transformed by biochemical action into gases (N_2, CO_2). The cycle of sulfur is less direct coming apparently from gases released from anthropogenic activity into the atmosphere. But sulfur dioxide and carbon dioxide are the most common gases released in volcanic eruptions (following water). One can however consider that the dust cycle reunites some of the dispersed elements that were separated from their initial state in rocks by aqueous chemical reactions into an assemblage of elements similar to the initial state though not in the initial proportions.

"Cold loess," i.e., that material transported over shorter distances, does not show the remixing of initial components to such an extent as high altitude dust materials do. Cold loess tends to retain the chemistry of the mineralogical content of its original source materials.

5.3 Geochemical Alteration of Loess and Volcanic Materials at the Surface and the Effect of Plants

The situation of wind-borne material in the alteration cycle is rather unique in that the little altered material (formed by the rapid erosion of rocks (illite–chlorite and volcanic ash) in an essentially a-biotic environment and deposition) is successively brought onto the plant–soil interface renewing the stock of new material that is unstable in the surface environment. Rainwater, unsaturated in mineral components, is brought into contact with the wind the high temperature deposited materials, and this material is the substrate for plant interaction where alteration occurs and certain elements are taken into the plant soil system by rainwater action.

Alteration of loess is a renewable process in that as long as the loess accumulates the surface A horizon is renewed in fresh, un-altered material. The plants installed in the A horizon use the fresh material and begin to change its chemistry for their use. At the same time water–rock interaction occurs. Figure 5.17 illustrates this process. The A horizon plant zone is superficial and renewed. The B horizon is where the altered material accumulates, as in the normal water–rock alteration sequence where the bedrock alters to C horizon material. However in loess deposition, the surface A horizon is the source of "bedrock" or unstable silicate material. Bedrock alteration occurs over periods of hundreds of thousands of years whereas loess alteration occurs over periods of thousands of years or less (see Velde and Barré 2010, p. 145).

One can pose the question: can plants significantly influence the initial chemical actions of rock–water interaction which produce the alterite material stable at the rock–surface interface? Are these interactions significantly different from water–rock interactions? In loess deposition the chemistry of the plant–soil systems at the surface acts on the unstable minerals in the transported materials as well as does the unsaturated rain water which comes into contact with the minerals in the materials deposited at the surface. If one considers the possible origins of wind-transported materials, the typical loess materials are formed from low-grade metamorphic or sedimentary rocks which generally have a largely illite–chlorite mineralogy. This material is largely phyllosilicate in nature and should be closer to thermodynamic equilibrium with surface chemical conditions than magmatic rock minerals.

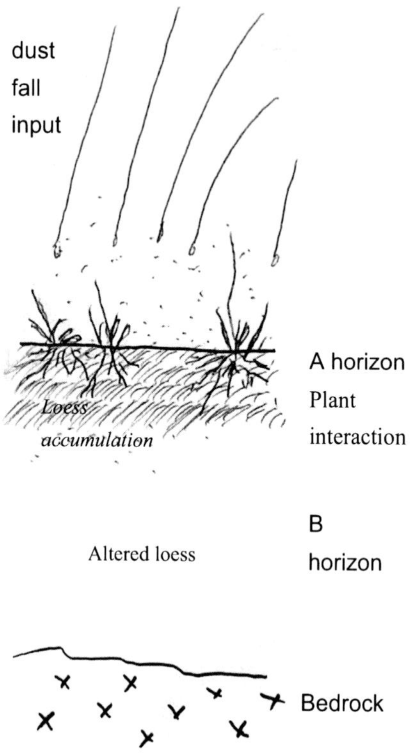

Fig. 5.17 Schematic representation of the development of soils on wind-borne materials (loess or volcanic dust)

The alteration of loess should be less intense as far as dissolution and recrystallization of the initial products compared to that of volcanic ash from magmatic materials. Volcanic dust, the second major category of wind-borne material, is highly unstable under surface conditions being formed of glassy material and high temperature minerals. One would expect high alteration rates of such material as it is affected by surface chemical conditions. Thus one can consider two major types of alteration interaction operating at the same time: (1) water–rock where magmatic materials are affected by rain water, as described in Chap. 3, and (2) plant interaction with un-altered materials where the surface chemistry is controlled by plant interaction which would be the case for loessic materials, either illite–chlorite or volcanic airborne deposits.

Data provided by Fabio Terribile (University Ferderico II, Naples) and described in Mileti et al. (2013) give us an insight into the problem. Soils sets from the Apennine mountains north of Napoli were sampled in the A (soil and plant interaction horizon) and the B horizons, usually considered to be that of the accumulated surface alterite material. Two types of material of different geological origin are present in different sites: volcanic ash deposits and loess (illite–chlorite cold loess) deposits. The materials are clearly differentiated by the clay minerals present in the A horizon, where the volcanic material changes to smectite and illite/smectite mixed layer mineral–kaolinite clay assemblages in the B horizon and the loessic illite–chlorite material which is less altered except for the formation of some hydroxyl interlayered (Al–OH) clays. The initial mineralogy in the two sets of materials is quite different where the volcanic material changes greatly in its mineralogy responding to surface chemical potentials. However the illite–chlorite loess mineral assemblage, common to much of cold less materials, even though of low temperature origin can be altered over relatively short periods of time as well, such as in temperate climate salt marshes where the chlorite component is lost being replaced by mixed layered minerals in a matter of tens of years (Velde and Church 1999). This reaction is significantly slower when biologic activity is decreased as a function of surface temperature, being accomplished on the order of 4,000 years under northern climate conditions of Finland (Righi et al. 1997). In the Apennine samples the chlorite is partially altered at the surface and more altered in the B horizon, as would be expected given the younger average age of the A horizon material which is periodically renewed by dust transport. The two different types of loess or airborne deposits are of different initial mineralogy, but both are significantly reacted in their new environment.

The investigation of Apennine wind-borne sediments forming soils presents a very interesting case to study the chemical effects of this alteration processes. In the instance of volcanic materials one has highly unstable high temperature material, often with a significant proportion of glass present, which will react to the water infiltration as its chemistry is controlled by plant activity. The tendency to recrystallize and form new, low temperature minerals is very great losing some elements to dissolution processes. A major reorganization of major and minor elemental components has been observed. In contrast, the illite–chlorite loessic material represents minerals that are in fact very similar to those found in alterites and soils, where the minerals are very closely related to soil clay minerals. Here one can expect minimal change in chemical interaction.

5.3.1 Major Elements

Our first step is to compare the chemistry of the plant-affected A horizons of the wind deposited material to the B horizon of altered material. Here one expects to see the impact of alteration seen as a comparison of the more altered B horizon to that of the less altered A horizon where new, unaltered materials are periodically deposited. The B horizon mineralogy and composition will essentially represent the transformed material that has been initially deposited at the surface. The A horizon, reacted by plants, becomes further buried by successive deposits of transported material. The A horizon then indicates various stages of the initial transformation and the B horizon the transformed material. Information has been presented in Chap. 3 indicating the chemical changes that happen when a rock is in contact with surface waters that produce the alterite, which is the water–rock interaction. It is useful to observe changes in major element abundance in order to compare the relative impact of water–rock alteration compared to plant alteration of materials.

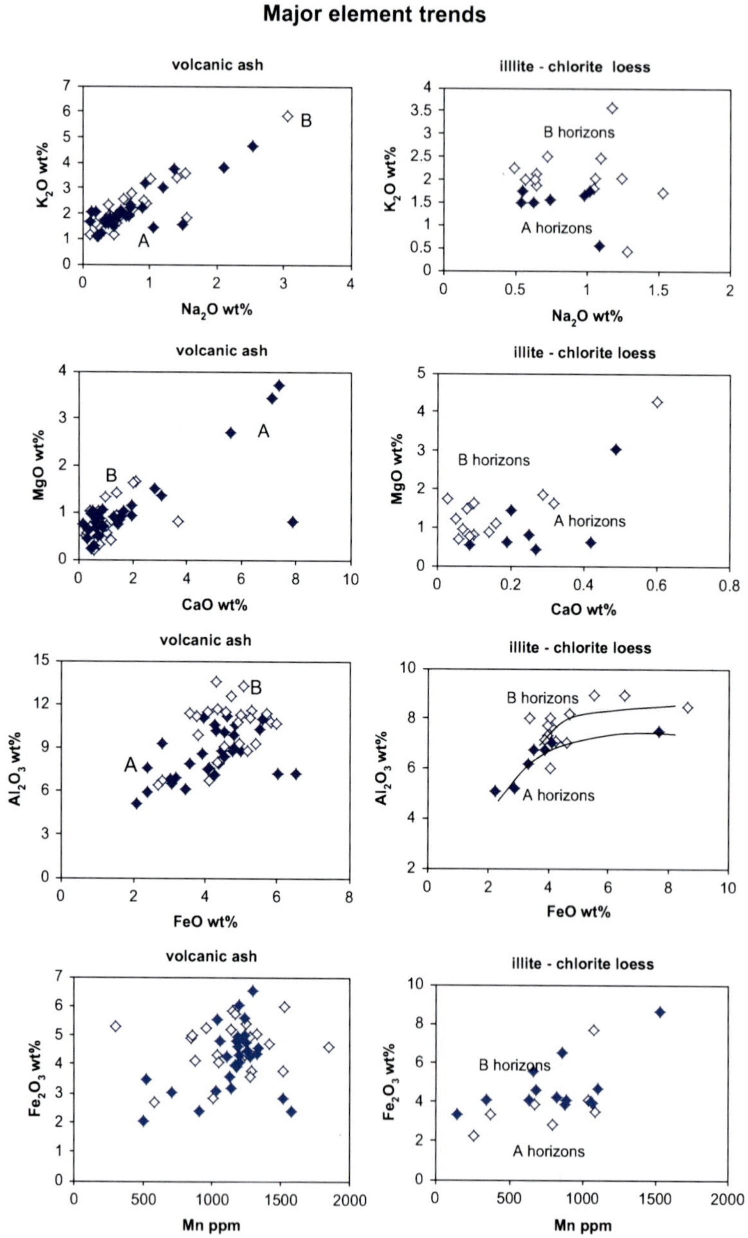

Fig. 5.18 Relations of major element abundance in volcanic ash and loess deposits in soils on the Apennine Mountains, Italy (data by Mileti et al. 2013). A horizon (*closed symbols*) and B horizon (*open symbols*) are compared. The A horizon shows relations for the initial deposited materials while the B horizon shows relations for elements after alteration by water–rock ("dust") interaction and above by chemical changes due to plant alteration. *Solid lines* indicate possible element trends for each horizon. *Solid lines* indicate trends for both A and B horizon elemental abundance

Table 5.1 Summary of elemental distribution under different types of alteration situations: water rock and plant alteration of new materials (illite–chlorite loess and volcanic ash)

Alteration	Water–rock	Plant	Plant
Soil zone	Rock to alterite horizon: igneous rocks	Volcanic ash A horizon to B: "rock" to alterite	Loess soils A horizon to B: "sedimentary rock" to alterite
Element			
Na	Loss	No change	No change
Mg	Variable	No change	Increase
Al	**Increase**	**Increase**	**Increase**
K	Variable	No change	Increase
Ca	Variable	No change	Loss
Mn	Increase	No change	Increase
Fe	**Increase**	**Increase**	**Increase**

Data from Mileti et al. (2013) and Table 3.1, Chap. 3. Similar changes in all horizons for Al and Fe are shown in bold

The relative loss and gain for major elements is given in Fig. 3.1 for major elements. Trends for major elements in the illite–chlorite loess and volcanic materials from the Apennines are shown in Fig. 5.18 for major elements. We can use this as a basis for comparison to indicate the influence of plant alteration as shown in Table 5.1 below. Here the gain or loss from rock to alterite (water–rock "dust" alteration) is indicated for crystalline rock data previously shown in Fig. 3.1.

In the soils developed and developing on wind-borne materials, one can pose the question: which trend will the chemistry follow, water–rock or plant–soil interaction? The initial deposited material can be considered as being a type of bedrock. In going from wind deposited A horizon to B horizon one has the same effects of initial alteration as going from bedrock to C (alterite) horizon in a classical alteration sequence (see Chap. 3). Looking at the volcanic ash materials as well as the illite–chlorite loess deposits, it appears that for certain elements their relative concentration pattern is different from that of the igneous rock–alterite relationship. It appears that the chemical pattern of plant–soil interactions is stronger than that of water–rock interaction, which should occur at the surface along with the plant interactions.

These data sets indicate that the major element trends of loss for some elements and subsequent gain for others in the water–rock action process in the alteration of crystalline rocks are not maintained in the alteration of wind deposited materials by plants in the upper soil zone except for the elements Al and Fe, those elements that are relatively less soluble under alteration conditions. In the graphical presentation of the data for alteration of wind deposited material shown in Fig. 5.18 which compares A horizon (deposition zone) to B horizon (post-interaction of plants and deposited material) the immobile elements Al and Fe follow roughly the same trends but with a tendency toward Al enrichment compared to Fe. There seems to be no specific change in Mn content related to Fe in the materials altered by plants unlike that found for water–rock alteration.

The mobile elements K and Na follow different trends depending upon the materials present as influenced by plants as do the elements Mg–Ca. Ca appears to be lost in loess materials compared to Mg in the plant zone as in water–rock interaction, but no significant change is seen in the plant-altered volcanic materials. In volcanic ash deposits Na–K relations seem unaffected by plant action in contrast to water–rock interaction but for loessic materials where there appears to be an enrichment in K in the B horizons due to plant cycling of potassium to the surface. There is a distinct increase in K after plant interaction for loessic illite–chlorite materials as is seen in many soils developed on different materials where the plant zone shows an increase in potassium at the surface. Differences between A and B horizon compositions seem to be greater for illite–chlorite loessic materials than for volcanic ones.

Overall, from these observations reported in Table 5.1 which compares water–rock interactions on various rock types to the effects of plant alteration on volcanic and illite–chlorite materials, it appears that the effect of plants is to conserve elements more than in the case of water–rock interaction. The volcanic ash material appears to be less affected by the surface interactions than the loess of low temperature mineral origin (illite–chlorite).

5.3.2 Minor Elements

Relations of minor element abundance in the Apennine profiles have been determined for a large range of elements. Unfortunately this set of data cannot be compared with rock–alterite data in that such detailed information is not available. Here we will look at the differences in minor element abundance in series of soils developed upon volcanic and loessic illite–chlorite mineralogy. Looking at these data it is evident that certain elements are closely associated. In many cases they form linear trends of relative abundance in alterites of a type of source material while they are related in only a general way in others.

The linear correlation indicates that the two elements are associated in the same mineral phases, most likely by the attraction to the same substrate, clay or oxide or their presence in unaltered detrital phases. Changes in general abundance, seen in the displacement of the average and extreme compositions, often in a nonlinear relationship, indicate a trend of change not related to a single substrate material which attracts the different elements but a multiphase system. Here overall multiple mineral changes are effected with a gain or loss of certain elements.

Minor element pairs such as Ni–Co, Sb–As, Ca–Sr, Li–Al, Ce–Zr are quite evident in the volcanic dust derived soils (Fig. 5.19). Such inter-element relations are less evident in illite–chlorite loess soils. This suggests that the presence of major phases in the alteration of the volcanic materials remains effective throughout the initial stages of alteration into the more mature stage, below the soil zone of plant–soil (rock) interaction shown by materials found in the B horizon of the alteration of wind-deposited materials. The volcanic source material in the example given is

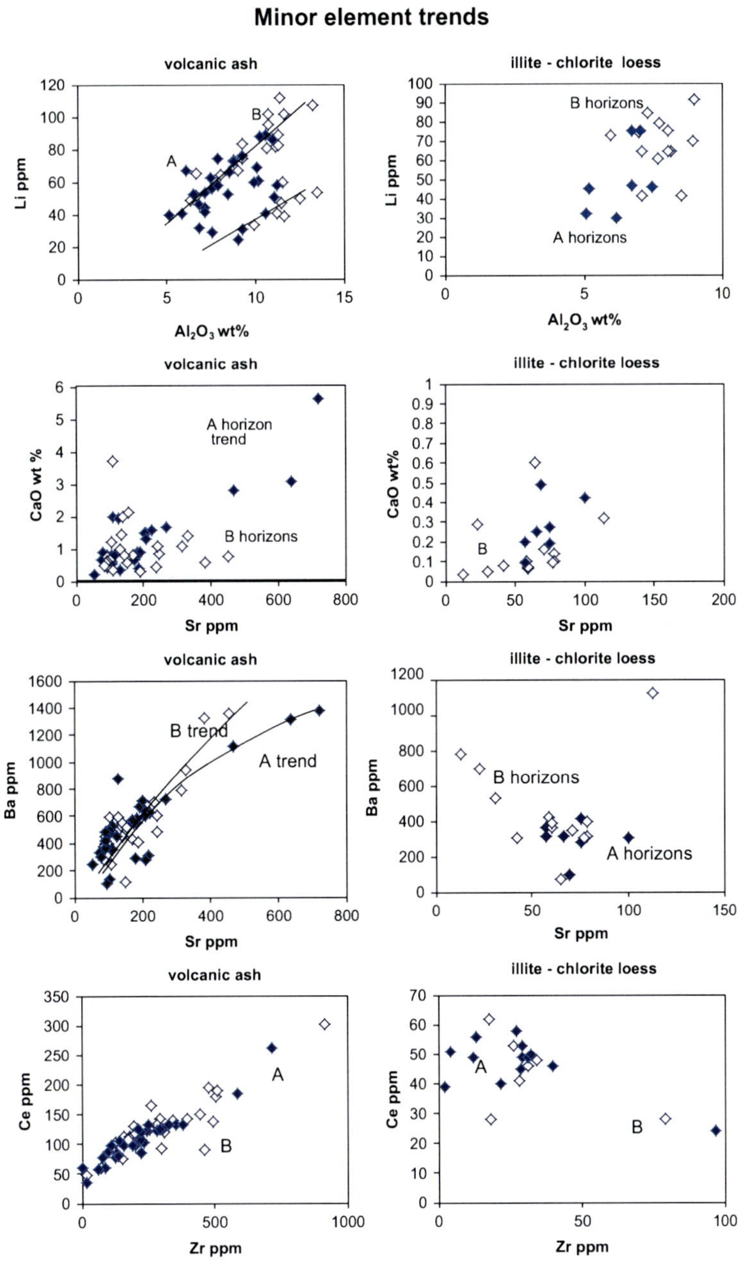

Fig. 5.19 (continued)

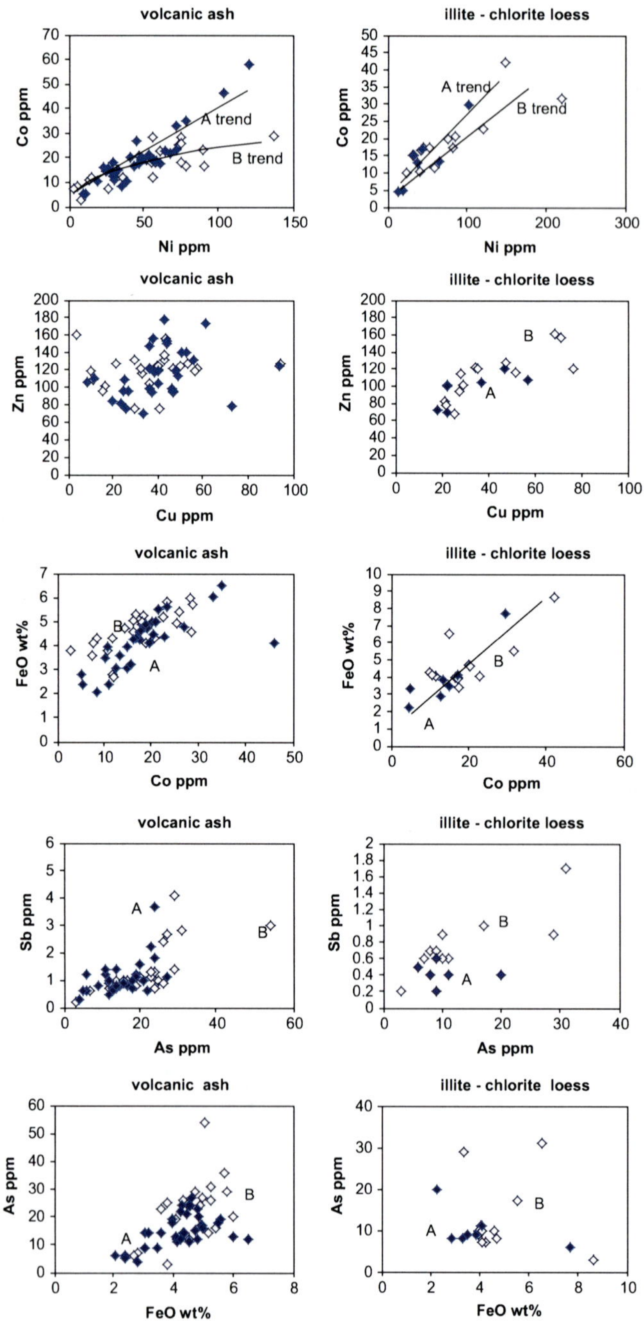

Fig. 5.19 Minor element abundances in A horizon materials for wind-borne deposits of volcanic materials and loess sedimentary rock in the Apennine mountains (Mileti et al. 2013). *Solid lines* indicate possible element trends from the different horizons. A horizon material is indicated by *closed symbols* and B horizon material is indicated by *open symbols*. The two trend series found in some elemental relations of the volcanic series suggest different source materials where the relative elemental abundances are not affected by the alteration processes under plants

likely to be coming from the same southern volcanic region of Italy, and hence one could expect systematic inter-relations to occur. The initial formation of minerals and the associated minor elements remains effective into the lower portions of the profiles. However it is clear for certain elements (As, Li, Co, Ni) there are two trends in the volcanic materials suggesting two major sources of volcanic ash. Parallel behavior of the elements, especially evident in the Li–Al, As–Fe, and Co–Ni elemental pairs, suggests that the same chemical processes and attractions are operative for both source materials and their alteration products.

In the samples studied the illite–chlorite loess materials are multi-source, largely from local materials of low-grade metamorphic origin or perhaps materials coming from similar rocks either from glacial period loess deposits or from materials in North African deserts. There are fewer linear relations between the element pairs in these materials, which confirm this multi-sourcing. The effects of alteration should be seen as being of the same sort in the plant-starting material relationships of materials in the A horizon (starting materials) and B horizon materials (result of plant alteration).

5.3.3 Soluble Elements

Lithium can be associated with alumina content in that it is often found in clay mineral structures or fixed strongly to surface sites. The relations of Li to Al seen in Fig. 5.19 indicate that there is some change in Al–Li abundances with a strong covariance. It appears that there is a tendency to have more Li present in the B horizon materials in both volcanic ash and illite–chlorite loessic materials as Al content is greater. In both cases there is more Li in the B horizon material and hence a concentration during the alteration processes by plants.

Strontium is often associated with Ca in surface minerals, on or within the structures. In Fig. 5.19 one sees that the relations are only vague in altered volcanic ash and in the chlorite–illite loessic materials as they are altered by plants. Further no selection of one element with respect to the other is apparent in the B horizon material after alteration.

An interesting case of element covariance in the two different types of wind-deposited materials is that of Ba and Sr (Fig. 5.19). In the alteration of volcanic loess, there is a positive relationship, which is maintained from A to B horizon with a possible slight increase of Sr in the B horizon. However in the illite–chlorite loessic materials there is an negative correlation, which is more strongly developed in the B horizon materials which appear to have a higher Ba content than the A horizon materials. Thus in both types of materials it appears that the Ba content is increased by plant alteration, but there is a relative loss of Sr in the illite–chlorite loess whereas there is an increase in the volcanic ash materials.

5.3.4 *Transition Metals and Heavy Elements*

One can take the example of the element pair Zr–Ce (Fig. 5.19) to assess the impact of plant alteration on insoluble elements such as rare earths and zirconium. Ce can be taken to represent the REE elemental series and Zr the host mineral (in many cases) which has a very low solubility. The volcanic deposits show a tight linear relationship where A and B horizon elemental trends overlap and the illite–chlorite loess a much more dispersed relationship which does not show significant differences between A and B horizon materials. This suggests that for both types of materials the plant regime does not affect the stability (insolubility) of the zirconium-bearing minerals.

The minor elements related to iron (transition metal elements) indicate similar behavior in volcanic and illite–chlorite loessic alterations. Co–Ni ion pairs show trends in both volcanic and illite–chlorite series where there is a slight loss of Co in the B horizon materials for both series.

As a point of comparison, one can consider Fe and Co relations, where the minor element Co should be associated with the major element Fe. In Fig. 5.19 it is clear that the two are associated in both A and B horizons of the altered materials. There is perhaps a slight relative increase in Fe over Co in the volcanic materials with weathering.

One can consider the heavy multi-oxidation state elements As and Sb in the alteration series also. As and Fe do not appear to be affected specifically by alteration, neither in loessic materials nor in volcanic ash alterites. In fact the different series of volcanic ash compositions are seen in the As contents, in both the A and B horizon materials, which indicates little redistribution by plant alteration. No significant change is observed in the Sb–As relations in either series. Thus the heavy elements appear to be little affected by the alteration processes in the surface horizons of soils.

Copper and zinc are not linearly related in the volcanic materials but are related in the illite–chlorite loess soils in both A and B horizons where they essentially overlap in abundance (Fig. 5.19). The linear relationship of Cu–Zn can be found in illite–chlorite loess in the Apennine materials as well as in materials of very different origins, such as North Africa, USA, and China. Given that the Apennine soils have not been cultivated, and that there is a general trend in various parts of the globe, one can assume that the Cu–Zn relationship is not one due to surface contamination of human activity. The concentration ratios of 1:3 for the Cu–Zn element pair is similar to that found in laterite materials in Cameroon (Fig. 5.20) for example as well as in the illite–chlorite loessic Apennine alterations. This suggests a fundamental abundance relationship that is not changed during alteration processes, either water–rock interactions or those induced by plant activity.

In this comparison of the two types of wind-borne loess deposits, volcanic and illite–chlorite type, affected by plant alteration at the surface one sees that the plants change the general chemistry of the normal water–rock alteration process significantly for some major elements and, depending upon the type of secondary

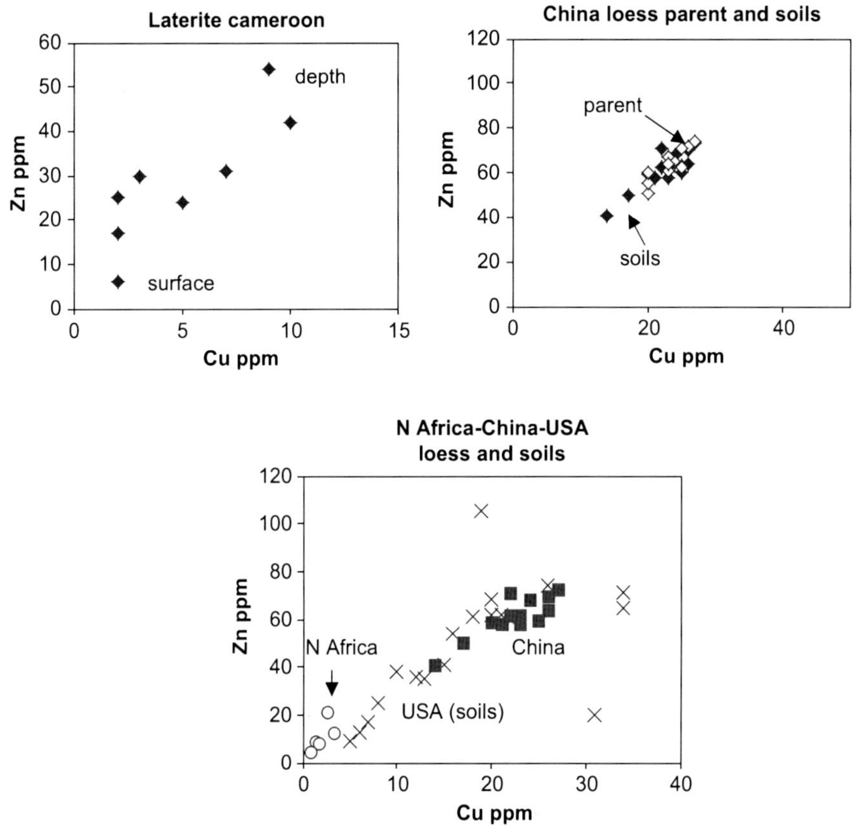

Fig. 5.20 Cu–Zn relations in laterite deposits in Cameroon compared to those for various loessic soils (Temgoua 2002; Tungsheng 1985; Pye 1987; Holmgren et al. 1993)

minerals formed, for some minor elements also. In this example the initial material is deposited, un-altered, into the or on the soil zone. Here the first alteration process is that of some water–rock interaction but above all the interaction of plants which control the chemistry of the soil solutions that are the agents of mineral alteration of the new un-altered wind-deposited material. Thus it is clear that the influence of plants changes the chemistry of the alteration process for certain elements. The transformation of rock minerals to low temperature phases stable at the surface when comparing the chemistry to water–rock action on initial substrates of higher temperature origin.

In Table 5.2 we see that the transition metals show little change in abundance during the plant alteration of illite–chlorite loess or volcanic ash. Low solubility elements Zr and Ce (REE element) show no change (Fig. 5.19 for example). However the elements Li and Sr are affected in the illite–chlorite loess alteration. Heavy metal oxyonions such as Sb, As show increases during the process in most cases. Lead is interesting in that one would expect a concentration in the A horizon

Table 5.2 Summary of minor element abundance due to plant alteration of ash and chlorite–illite loessic materials comparing the A horizon (plant interaction with incoming materials) to the B horizon where altered material is accumulated

	Volcanic ash A to B	Illite–chlorite loess A to B
Element		
Zn	**No change**	**No change**
Cu	**No change**	**No change**
Co	Loss	No change
Ni	**No change**	**No change**
Ce	**No change**	**No change**
Zr	**No change**	**No change**
Li	No change	Increase
Sr	No change	Loss
Sb	Loss	Increase
As	**Increase**	**Increase**
Pb	No change	Decrease
Ba	**Increase**	**Increase**

Data from Mileti et al. (2013). Bold emphasis shows that the same effect is seen in both contexts

due to human input which seems to be apparent for the illite–chlorite loess but not the volcanic alterite. Since both soil types were deposited by similar processes and experienced similar histories, one must assume that the differences were not due to human input.

The summary of relative abundances of minor elements in wind-deposited and altered material indicates that in many cases there is little effect of plant alteration on elemental abundance in illite–chlorite or volcanic materials. It is interesting to note that in some cases for volcanic ash, Li abundance versus Al and Sb versus As for example, there are two compositional trends in the initial deposited material which are still found in the B horizon alterite. This occurs even though there is a slight tendency to gain arsenic in the B horizon.

Wind-deposited materials are very important to human activity in many areas in that they form the basis for agricultural soils. Wind and rain deposits now contain materials other than the traditional loessic or volcanic input. Human components must be considered and the reaction of added minor elements to the material is of great importance. In the zones of deposition of wind-transported material most often the incoming material is affected by the chemistry of soil solutions that are determined by plant and biological action.

5.3.5 *Summary*

Transport of surface alteration products can be classed into essentially two means of displacement: that by water (essentially streams and rivers) and that by wind at low or high altitudes. In the instances of water transport there is little change in the physical state or chemical structures of the solids. The suspended matter seems to change little and in fact to little affect the chemistry of the transporting aqueous medium.

Wind transportation, which deposits loess and volcanic ash materials, is quite a different matter. In low altitude and short-distance transport, the solids are little affected by the movement cycle. However in very high altitude transport one can find some important changes in oxidation state of the transported materials and a change in their chemical affinities. Oxidation of solid materials can change the balance of surface absorption for the solids, creating new adsorption sites, and release some material into solution as pH changes when the materials reach the surface of the earth. Complex chemical rearrangements between organic and other materials can create new phases and new chemical affinities during the transportation phase. What is of greatest importance of course is the reaction of these materials as they become deposited at the surface, in soils for instance. Much of the wind-transported material is affected or integrated into the biological cycle of plant–soil interaction where the more loosely bound ions can interact readily with the biological materials that they come into contact with.

The interaction of plants with the relatively un-altered wind-borne materials indicates that their major element concentrations are more or less maintained compared to the initial phases of water–rock interaction where the major change is the loss of soluble elements such as the alkalies and alkaline earth elements and the retention of the less soluble elements Al and Fe. Most minor elements, transition metals especially, are not affected by plant alteration of the initial deposited materials in the soil zones. As and Sb are however for the most part concentrated in the alterite materials compared to the input deposits. It appears that the stability and change in phase of the initial material is very important in determining the concentrations of the elements in the deposited materials of wind-borne transport.

In general one can say that the altered materials formed in the initial cycle of rock–atmosphere chemical change are little modified during the transportation cycle. However the elements of minor abundance, which do not form a stable mineral phase but are adsorbed onto those stable phases present, are relatively easily affected by changes in general chemistry of the transporting solutions or media. These relations are very important for considerations of environmental chemistry in that the transport cycle is very important to the biocycles at the surface of the earth.

5.4 Summary: Transport by Wind and Water

The two major vectors of transport for altered material at the surface of the earth are water and wind. For the most part, the physical and chemical state of the solids transported remains much the same during transport except for some of the iron in materials that reach very high altitudes in wind-borne materials, which tend to be oxidized. Essentially wind-borne material reaches the point of deposition, contact with the earth, in much the same state that it was in before transport. Given that most of the area where it is deposited on continental massifs is land, and hence covered with plant active soils, this deposit will be transformed according to the chemistry that is imposed upon it's by the vegetal zone. Mineral transformations will depend upon the relative instability of the phases present. For example volcanic ash will alter rapidly to form a clay mineral assemblage whereas the materials from sedimentary rocks and sediments derived from there will be closer to chemical equilibrium at the surface and will change much less in their mineral and chemical characteristics. The wind-transported material will find its way into the water transport cycle as rainwater moves it into the streams and rivers of the continents.

Water transport by streams and rivers combines the two types of alterite material, dissolved and suspended solids, into a single aqueous system mixing the components that had been previously separated by alteration chemical forces. The major effect of transport is in fact to mix alterite materials, dissolved and solids, from different chemical alteration sites and to create a non-homogeneous system where the different components will be affected by the different concentration of each component. Clay minerals from low alteration regimes, such as illite and hydroxyl-interlayered minerals, can come into contact with dissolved ionic species formerly in equilibrium with highly altered materials such as iron and aluminum hydroxyl-oxides. A wide range of clay mineral types and bulk compositions can occur depending upon topography and local climate effects (Liu et al. 2012b). These new phases can be much more reactive and capture transition metal element ions to a greater extent than silicate clay minerals for example. The heterogeneity of dissolved mineral–solid material relations in rivers indicates the variety of materials present in transported materials that have not come into chemical equilibrium in the aqueous environment of transport. Basically the transport of various materials, solids and dissolved ions, to areas of sedimentation creates a chemically unstable assemblage of geologic materials, which must readjust upon sedimentation and burial.

These principles are very important and are used to interpret observations on chemical analyses of river-borne materials for instance, and they are used as a basis for calculations and models of transport, which is extremely important in problems of pollution (see Westrich and Förstner 2007 for example).

5.5 Useful References

Drever J (1982) The geochemistry of natural waters. Prentice Hall, Upper Saddle River, NJ, p 436

Keilback K (1975) The riddle of loess formation, pp. 47–50. In: Smalley I (ed) Loess: Lithology and genesis, Benchmark papers in geology, vol 26. Dowden Hutchinson and Ross, Stroudsburg, PA, p 429

Pye K (1987) Aeolian dust deposits. Academic, London, p 334

Salomons W, Förstner U (1984) Metals in the hydrocycle. Springer, Heidelberg, p 349

Sly P (ed) (1986) Sediments and water interactions. In: Proceedings of the 3rd international symposium on interaction between sediments and water. Springer, Heidelberg, p 517

Westrich B, Förstner I (eds) (2007) Sediment dynamics and pollutant mobility in rivers. Springer, Heidelberg, p 430

Chapter 6
Sediments

6.1 Introduction

Sedimentation is the end stage of surface geochemistry. At this point the geological materials that have been affected by surface chemical influences are deposited and become essentially closed systems isolated from the surface chemical effects of air and rainwater. The sediments will be subjected to changes in pressure and temperature due to burial on the edges of continents or more rarely on continents. The basic context of surface geochemistry is that of element migration and change under different chemical conditions whereas the regime of sediment burial is one of more or less constant chemical conditions where new minerals can be formed due to changes in the physical conditions pressure and above all temperature. The early stages of burial (diagenesis) allow some migration of geo-materials, water, but usually on a relatively limited scale. The loss of material comes from within the sediments, which change phase and release some elements into the superfluous aqueous fluids that constitute a part of sedimentary material. In diagenetically affected sediments, producing diagenetic rocks, the fluids expelled from the sediment are the sedimentary fluids that circulate from rock layer to rock layer at times forming new minerals, or transferring substances such as hydrocarbon materials from source rocks to reservoir rocks. Overall the sediment environment, from the time of initial deposition, is one of a more closed chemical system than that of alteration and transport. The fluids in each layer are close to chemical equilibrium with the minerals present. The importance of sediments concerning the transfer and sequestration of certain elements can be illustrated by the fact that carbon is incorporated into various sediments in the sediment forming stage. Ehrlich and Newman (2009) indicate that carbonate rocks, formed in sedimentary environments, contain 1.8×10^{22} g of C and other sediments (shales) a similar amount while soils contain 3×10^{18} g and living matter 8×10^{17} g whereas the atmosphere contains only 6×10^{17} g. Therefore in the last stages of the surface geochemical cycle, large amounts of material can be transferred from the different

A. Bauer and B.D. Velde, *Geochemistry at the Earth's Surface*,
DOI 10.1007/978-3-642-31359-2_6, © Springer-Verlag Berlin Heidelberg 2014

realms of the surface, atmosphere, living matter, and soils to sediments which become rocks.

The initial stages of sedimentation can be affected by biological activity, which influences pH and Eh. Major actors in the control of elemental oxidation state and the form (dissolved ionic, particulate) that elements have in the surface transportation and especially sedimentation process are biological and especially microbial agents. The oxidation state of Fe, Mn, and S are an especially important source of energy for many forms of bacteria and vice versa. The chemical state of these elements depends upon the activity of these elements in the sedimentation environment (Ehrlich and Newman 2009).

There are three general environments where material is deposited from transport fluids: (1) ocean or peri-ocean sedimentation (the most common and most abundant sedimentary formation process) where the sediment is dominantly particulate, (2) freshwater lake sedimentation where again the material is dominantly particulate, and (3) arid closed basin evaporitic concentrations which provoke the formation of solid material from that dissolved in the transporting fluids. Each sedimentary environment has its specific chemical constraints and the resulting sedimented, solid material reflects these constraints. In each context the distribution, capture, or release of a large portion of minor elements is determined by the chemical state of dissolved elements in the depositing fluids. They can form either oxoanion complexes that can change oxidation state due to bacterial action to produce new minerals such as sulfur or phosphorous, or elements that are affected by bacterial action which changes their oxidation states such as Fe and Mn. In the initial stages of sedimentation, concentration and release or migration of elements in aqueous solution in sedimentary processes is often largely determined by the chemical potentials imposed by bacterial action as bacteria extract energy from changes in oxidation state of different elements in the geological materials.

A special chemical environment is encountered in evaporite deposits, where the aqueous phase is reduced and essentially lost, and where the highly soluble elements (anions such as Cl, Br, and the oxoanions such as CO_3^{2-}, NO_3^-, and SO_4^{2-}) are successively brought out of suspension by loss of aqueous matter concentrating the ions which form solids by combining various anions with the concentrated cations. The evaporite minerals are composed of salts, sulfates and carbonates. However at times silicate minerals are produced, usually strongly magnesian, which completes the sequence of minerals that are the result of final concentration of all dissolved material in transport waters.

In the initial stages of sedimentation there is a significant chemical flux due to the specific consolidation and biological activity of the environment. These effects diminish as burial proceeds reducing the chemical variability of the system.

6.2 Freshwater Sedimentation: Lakes and Streams

In freshwater lake sedimentation, the alteration products of water–rock and plant–soil interactions are brought together in essentially their initial state of alteration. Deposition is more or less local, i.e., close to the source materials and hence the sediments represent a small variation in source, with minimal mixing of alteration types due to topography or climate. Of course in eventual freshwater–seawater mixing zones at the end of surface transport, there is a maximum of different origins, concerning bedrock type, climate and biome influences, and soil age. In lake sediments there is often a strong influence of biologic activity due to the configurations of a semi-closed system with low input of water and sediments, which controls the pH and Eh of the sediment and its pore water. Here the major influences are chemical constraints imposed upon the pore water of the sediment by biological processes. Two elements dominate in the retention and sequestration of minor elements in such situations: Fe and S. Depending upon their chemical state, dissolved or solid phase, they fix elements either within the solid phases or on the surface of the solids. Since lake water is somewhat mobile, often participating in the flow of ground water to river outlets, the state of retention of minor elements on different phases is very important to the movement of elements in these instances.

Bourg (1995) outlines the solubility trends of heavy metals in solutions as a function of the Eh and pH of solutions. Essentially heavy metals are strongly affected by the solid phases present in the sediment load. Initial increase in Eh destabilizes sulfides through oxidation, releasing heavy metals. Continued increase in Eh oxidizes metals forming oxyhydroxides that capture the heavy metals in solution. Increasing pH at constant Eh is responsible for inner-sphere complexation at amphoteric surface hydroxyl groups of the minerals present (see Chap. 2). This effect also decreases the presence of heavy metals in solutions. Thus at high Eh and pH heavy metals will be fixed on the solids. The key phases are oxyhydroxides and sulfides.

6.2.1 Fe Effect

Two of the elements that are of low solubility under conditions of surface oxidation in soils or rivers are iron and manganese. Iron is much more abundant than Mn and frequently is associated with the less abundant Mn. As such it is difficult to distinguish the associations of minor elements with either Fe or Mn. In the discussion below we will consider only the iron oxides present and dissolved iron in sediment and pore waters.

The elements Fe and Mn usually form hydroxy material that is amorphous or poorly crystalline, but which is at times crystallized in regular crystalline structures. In any case very much of the surface of the solids is highly reactive and captures by cation and anion adsorption many of the elements in solution. Ferric (Fe(III)) hydroxy-oxides are frequent in particulate material in rivers and hence in sediments in freshwater deposits in lakes. This particulate material strongly attracts minor elements such as As, Zn, Pb, Zn, V, Mo, Ni, and Cu, for example (Belzile and Tessier 1990; Johnson 1986; Balistrieri et al. 1994). Transport of dissolved iron in sedimentary fluids of freshwater sediments has been observed and modeled to establish the dynamics of iron movement and precipitation in deposited material. The results indicate a stronger absorption of elements by freshly precipitated iron material than that already in suspension.

Let us look at the iron effect in the deposition environments by steps: (1) settling in the lake water and (2) interaction of elements under changing chemical conditions during burial below the sediment–water interface. In the first case the redox zone is within the water column and in the second the redox zone is within the sediment.

6.2.1.1 Water Column in Freshwater Sedimentation

Depending upon the example investigated, one can see changes in Eh and pH in each of the environments. Within the water column, one can identify a redox interface where iron plays a very important role in fixing or releasing minor elements by surface interaction. At the interface Taillefert et al. (2000) identify the formation of hydrous iron oxides, which scavenge minor elements on their surface, notably lead. Balistrieri et al. (1994) show that not only lead is concerned but also the elements Co, Cr, Mn, Ni, Zn, As, Mo, Ni, Zn, Cu, and V among others can be concerned by release from iron hydroxides as they are dissolved at lower oxidation states or by fixation of the elements in new sulfides formed after the reduction of the oxidation state of iron. Usually the presence of sulfur is of minor importance in freshwater lakes but under conditions of intense human activity such as mining or smelting or the introduction of urban waste into the water system, sulfur can become an important actor in the chemistry of minor elements. The critical zone is that of change in oxidation state where ferric iron hydroxides are dissolved to a certain extent and where minor elements fixed on the hydroxide surface are released to the freshwater solutions. The redox boundary is the result of organic activity, bacteria which usually drive reactions to more reducing environments. Below this boundary one can find sulfides forming. These minerals are largely Fe_2S phases.

Iron then occurs in the suspended matter in the lake waters and it is present as well in dissolved form as water enters lakes and these materials move out of suspension and solution to be deposited in the freshwater sediments. The amount of minor elements present associated with iron in a solid oxyhydroxide form depends upon the pH of the solutions, being higher above values of 6–7 than in more acid solutions (Johnson 1986). Iron is found to precipitate from sediment pore water during the early stages of sedimentation where it interacts with dissolved ionic trace elements. The Fe dissolved in transport fluids is largely in the divalent state compared to that in the particulate phases, which is essentially trivalent oxidized iron. Oxidation under surface conditions tends to form iron hydroxide in the trivalent state, which precipitates on the particulate matter in suspension. For the most part iron in freshwater sedimentary material is in the ferric state initially.

6.2.2 *The Ferrous Wheel*

The effects of changing redox conditions in freshwater sedimentation are indicative of the movement of iron ions in solution into and out of the redox zone. As an illustration we show a general movement of ions by migration due to changing chemical potential within the water column. The redox zone is shown within the water column where sedimentary material is introduced into a freshwater lake in which bacterial action creates a lower zone of reducing conditions (Fig. 6.1a). The redox boundary separates the oxidative upper water zone from the more reducing environment below the redox zone. Input from river transport is essentially oxidized iron in the form of particulate matter, either as individual particles or as material adhering to clay particles. As it traverses the redox zone, a portion is reduced and dissolved in the lake water, ferrous iron being more soluble than ferric. This is the bottom of the "ferrous wheel" circle, where ferrous iron can combine with reduced sulfur to form pyrite, which descends to the sediment surface. The ferrous iron oxide material in the sediment is attracted to the reducing solution through the lower iron content present due to loss through the formation of sulfides.

Above the redox boundary the oxidation of aqueous iron occurs in solution impoverishing the solutions in the reduction zone of iron. This attracts iron from the surrounding sediment to the solutions below the redox zone. Iron oxidation above the redox zone eventually forms oxide phases which descend into the redox zone to be reduced or descend into the sediment at the bottom of the water mass. Overall an accumulation of iron sulfide in the sediment occurs due to the reduction of iron and mobilization of ferrous iron to form in soluble iron sulfide.

6.2.2.1 Freshwater Sediments

The ferrous wheel mechanism of iron migration can operate then in the water column of sedimentation or in the sediment column itself. In the first several tens of centimeters of fresh water sediments one finds oxidized materials, which can be slightly reduced by organic action. In freshwater lake sediments one finds a strong effect of organic matter and bacterial action which, upon sedimentation, tends to reduce the oxidation state of the iron oxyhydroxides. Since Fe^{2+} is more soluble than Fe^{3+}, the iron present is dissolved in part during its reduction and significant amounts of adsorbed ions of minor elements are released to the interstitial fluids. Belzile and Tessier (1990) show that one often finds a strong increase in dissolved iron and associated minor elements in the first tens of centimeters of sediment

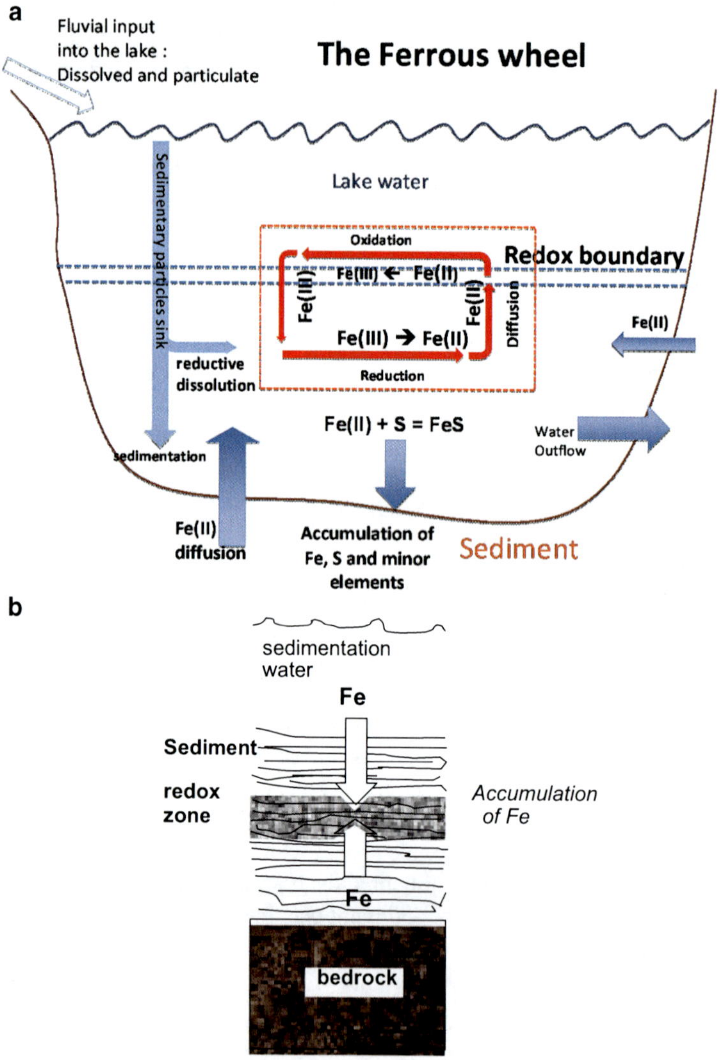

Fig. 6.1 (**a**) Schematic representation of the oxy-reduction layer and element migration in a fresh water sedimentary situation where a redox zone affects iron in dissolved and particulate states in the water column. Here the major chemical effect is reduction where Fe and S can be combined to form a solid phase. (**b**) Schematic representation of redox zone within the sediment column where oxidation is a major chemical effect on iron compounds and dissolved species. The forces of chemical oxidation and ionic migration occur within the sediments where the redox zone occurs in the sediments themselves below the surface

where reduction occurs. The iron present in the solid state in fresh water sediments can be closely associated with organic matter (Taillefert et al. 2000).

It has been observed that there is a zone of iron concentration in freshwater sediments several tens of centimetres below the sediment–water interface. This zone contains ferric iron hydroxides. In fact there is a diffusion of dissolved ferric material downward and ferrous material upward to the level of redox action in the sediments. An important effect concerning iron oxyhydroxides in freshwater sediments is the dissolution and precipitation of matter in the near surface zone. Iron can be reduced at depth as sedimentation proceeds and the change from trivalent to divalent ion species brings about a destabilization of the iron lower in the sediment column. This iron can migrate upward to the oxic zone near the surface where it is precipitated in a trivalent state (Belzile and Tessier 1990). Thus there is a tendency for iron and minor elements to migrate towards the oxidation zone of sediments in freshwater lake deposits giving a zone of increased concentration near the surface of the sediment column.

A close association of arsenic with these oxyhydroxides indicates the formation of Fe–As (AsO_3^{3-}, AsO_4^{3-}) oxoanion associations (Belzile and Tessier 1990). Such material is very stable under different conditions and the Fe and As associations are stable together unlike other anions fixed in variable charge sites on oxides, which can be replaced by hydrogen at low pH values. Here we have an example of a stable phase of iron and a minor element (As) forming in the sedimentary environment. Johnson (1986) reports the case of adsorption of Cu and Zn on amorphous iron hydroxides, which are affected by changes in pH where higher values produce more adsorption of Cu and Zn, which confirms the more frequently encountered case.

Thus one finds two types of associations of minor elements with Fe in freshwater sediments. One as a distinct chemical compound of iron acting as a cation and As forming an oxoanion which combine to form a stable phase. Here there is absorption of As. The other is the more classic adsorption of transition metals on iron oxyhydroxide matter, which are more labile and are affected by exchange potentials with hydrogen cations. There is a redox zone in the lake waters as well as one just below the surface of the sediment–water boundary. This indicates that much of the particulate matter does not react to the reducing conditions and reaches the bottom of the water column in an oxidized state.

6.2.2.2 Sulfur Effect

Sulfur is basically solubilized during the water–rock interactions of the initial stages of alteration. For the most part sulfur is present in sulfides, cation–anion minerals, which form at high temperature and pressure in sedimentary, metamorphic and igneous rocks. This sulfur is readily oxidized in rainwater and forms sulfate oxyoanions, which are present in usually sufficiently small quantities to allow the element to be moved out of the system as a dissolved species. Data from sources cited in Chap. 5 indicate that there is usually more sulfur in river transport than chlorine.

Sulfur is an element that can have several oxidation states: −1, 0, +2, +4, +6, but in surface environments the −2, 0 and +6 oxidation states are the most common (Ehrlich and Newman 2009, p. 439). The versatility of sulfur in an electronic sense allows it to play many roles in natural processes. One is that it is a source of energy for living organisms such as bacteria, which derive energy from the change in oxidation state. Also sulfur is essential to several metabolism processes in plants and is therefore almost omnipresent in active biological zones.

Bacteria frequently use the sulfate–sulfide redox reaction as a source of energy, when enough sulfate is present in the ambient solutions where bacteria reside. Some material is transformed in the process of plant decomposition. It can be important in producing the organic material that is active in fixing and converting sulfur through bacterial action (Brüchert 1998). Most sulfur is moved in ground water solutions. In the case of lake deposits where water is not moved rapidly, one finds that sulfate ions diffuse to sites where the redox process occurs. The redox process favorites the formation of iron sulfides, and in doing so depletes the concentration of sulfate in solution, hence the diffusion of sulfur ions towards the sediment surface. In this process of sulfide formation, a certain number of trace elements are fixed in the crystal structure. Huerta-Diaz et al. (1998) indicate the preferences for As, Ni, and Cd in the iron sulfides or at times forming independent phases with these cations whereas Co, Cu, Mn, Pb, and Zn are less strongly attracted to a sulfide phase. These reactions are reflected in the pore water compositions of the sediments. Here we see that there is a selection of elements by sulfides for internal incorporation that is not the same as that found for cation adsorption on surface sites where all of the elements mentioned would be fixed on an appropriate surface, iron hydroxide, for example.

Reasonably often, in looking at compositional profiles of sediments, one sees a peak of Fe and S in the near surface, about 10 cm depth (Carignan and Tessier 1988 for example). The correspondence of the two element concentration maxima suggests that this is not a pollution effect but a surface geochemical phenomenon. Sulfate reduction creates the FeS concentrations found to accumulate at the surface of marsh sediments (Lord and Church 1983). The zone of Fe–S concentration is one where redox phenomena engendered by bacterial action occur, and Fe^{3+} is changed to Fe^{2+} the latter of which combines with the reduced form of sulfur oxyanion or H_2S to form a sulfide. Carignan and Nriagu (1985) propose that trace elements such as Cu and Ni can diffuse from the overlying water above the sediment to form a part of the element concentrations associated with the Fe and S accumulations just below the sediment surface.

Gröger et al. (2011) report similar phenomena in acid sulfate soils developed in recent river sediments in Vietnam. Here the concentrations of Fe and S are deeper, near one meter, but they are associated with minor elements such as Pb and As. However the minor elements appear to form their own sulfide phases instead of being incorporated into the FeS phases. In these sediments the pH and Eh curves cross in the zone of oxido-reduction where Fe and S are concentrated.

The presence of sulfides in the deposits of early sedimentation in freshwater–saltwater contact is not uniform over the climatic variability of the earth. Aller and Michalopoulos (1999) indicate that the amount of sulfide of the ferrous iron component (FeII) of such sediments can be a function of the climatic conditions generating the sediments. In tropical areas (Amazon and Gulf of Papua) the proportion of divalent iron bound in sulfides is on the order of 10–20 % whereas in similar materials from Long Island Sound and Eel River delta (USA) the proportion is between 60 and 80 %. This suggests that the overall oxidation potential (Eh) is significantly different depending upon the climate of the weathering and transport mechanisms and this affects sulfur accordingly.

These observations suggest that the affinities of S and Fe are determined by several factors in the zones of mixing of freshwater and saltwater sediments and the geochemsitry of minor elements will be affected by the interplay of overall oxidation potential and action of the activity of organic materials which are determined by the overall environment.

6.2.2.3 Phosphorous

The presence of phosphorous in lake sediments is variable but the initial input through weathering of rocks is relatively small, and that seen in modern sedimentary materials for the most part is due to recent human input. The discovery of phosphates as a useful agricultural fertilizer in the mid-nineteenth century in Europe spreads the use of natural and treated phosphate deposits to many areas. The gradual increase in the use of this material increased the presence of P in sediments as one would suspect. Further, the use of phosphate compounds in cleansing agents in the mid-twentieth century increased the phosphorous content of freshwater sediments. Several chapters in Sly (1986, Chaps. 12, 13, 14, 15, 17) indicate the impact and complexity of analysis of the associated geochemical effects. One major problem in using chemical data for the assessment of the effect of phosphorous on trace metal movement is that the phases that are formed by phosphorous are usually ill defined and hence one does not know exactly if the elements concerned are associated with P or with another element such as Fe or Mn. Phosphorous is intimately related to organic matter in the initial stages of sedimentation and this might well be a vector for the P–minor element relationships, where the minor elements are in fact adsorbed onto organic molecules. The movement of waste materials at the surface is frequently related and can be traced by observing phosphorous abundance.

6.2.3 Diagenesis and Migration

The movement and redistribution of minor elements in freshwater sediments through their desorption and resorption on solid phases is an important aspect of sedimentation. This effect is a part of diagenesis. The motor for the migration is usually due to a change in oxidation state of the solids or at times the presence of dissolved material in the transport solutions or in the pore waters of the sediments. These changes in oxidation state are of course engendered by bio-action, chiefly bacterial in nature. The elements normally involved are iron and sulfur. Several authors give basic information to detail these effects (Alfaro-De la Torre and Tessier 2002; Audry et al. 2006; Carignan and Nriagu 1985; Vezina and Cornett 1990; Carignan and Tessier 1988; Huerta-Diaz et al. 1998; Gröger et al. 2011; Balistrieri et al. 1994 among many others). The observations of these authors do not always coincide concerning the migration of major element associations and different species of minor elements in the sediments. However some general guidelines can be established. The differences could very well be due to the relative concentrations of different carrier materials, such as iron oxyhydroxide, manganese oxyhydroxide, active and decaying organic matter, and the types of clays present. Another factor which can be important is the burial rate of the materials which changes the point of reference for the analysis of the sediments. If slow sedimentation occurs there is a certain probability of forming a more steady state of element migration whereas if high sedimentation rates occur or tidal changes are important, cycling will become a major effect. Most of the studies cited above observe pore water concentrations of metals and hence illustrate the mobility of elements in solution.

The major chemical changes which effect element movement are the reduction of iron at the oxy-reduction boundary in the sediment just below the water–sediment interface. Heavy and transition elements such as Pb, As, Cr, Co, Zn, and Ni are released into solutions where they can be adsorbed onto silicate minerals or organic matter. The reduction of sulfate ions produces in many instances sulfide minerals, for the most part iron based, which can trap or fix the heavy and transition metal ions in the sediment material. The precipitation mechanism will decrease the activity of the minor elements in the pore water solution with the result that some element migration will occur moving the elements into the layer of sediment where reduction occurs.

6.3 Sedimentation in Saltwater and Salt Marshes

Fine-grained sediments (mud in the early stages of sedimentation) are known to carry variable amounts of minor elements, for the most part as adsorbed ions. Comparing sediment concentrations for certain minor elements from 19 sites along the Atlantic Coast of France between the Loire and Gironde rivers (Velde 2006), it seems clear that the minor element content can vary significantly from one sample site to another for some elements but little for others. For example iron and manganese are relatively stable in abundance as are the transition metals such as Ni, Co, Cr, and V and heavy metals U, Th and Sb. Others such as Zn, Sn, Cu, As, and Cd are much more variable from one site to the other. Such a pattern suggests that an important part of some of the minor elements is due to localized input such as industrial activity or agricultural practice. For these minor elements, of stable or variable abundance, the concentrations in the fine-grained coastal sediments are several times higher than those found in soils (Kabata-Pendias and Pendias 1992). It seems clear that the tendency for fine particulate material in river suspension is to fix ions on the surfaces which are then deposited. However this material is not stable (the cation, oxoanion materials adsorbed on particulate surfaces) and the relations of abundance and fixation on or within phases is affected by the chemical environment of saltwater sedimentation.

River sediments eventually come into contact with saltwater, which changes the chemical and physical state of the materials transported (Sect. 6.2). Flocculation of some of the suspended particles occurs as a result of increase in ionic concentration of dissolved salts in the solution. Exchange of some adsorbed ions on the suspended matter occurs where the dissolved salts take the place of adsorbed heavy elements. In such a regime, some material is deposited and other material is transported further into the sea. Deposition in tidal estuaries is extremely dynamic. Audry et al. (2006) show that the movement of sediment creates essentially a two-layer structure where the deposited material (mud) is fluid at the surface and becomes soft mud below and eventually becomes consolidated and relatively immobile at greater depth. The soft mud–consolidated mud boundary shows a chemical content gradient for many heavy metal trace elements. Such elements as Co, Ni, U, Mo, Cu, Cd, V, and Mn show these effects. Oxidation reactions and transformations of organic matter in the soft mud layer are considered to be responsible for the release and migration of these elements downward in the fresh sediments. However some of the sediment material is again put into suspension and moved further off shore along the coastal areas through tidal action.

As one knows, industrial activity and urban input can determine the types of elements present in the sediments moved in rivers into the sea environment. Each river and series of sediments will contain a cortege of elements, which can be specific to the site. These materials, when moved along a coastline due to coastal currents and storm events, will become mixed and therefore become somewhat incoherent as to the relative concentrations of different elements. Therefore general patterns of element association are difficult to ascertain in coastal present day sediment materials.

There are several basic problems inherent in making general statements concerning minor element behavior in these saltwater sedimentary environments. The first is of course the initial content of the sediment material. The second is the plant regime present illustrated in Fig. 1.7. On coastal shore sedimentation sites one has what is called the slikke and the Shorre, where initial sedimentation occurs without the stabilization of sediments by plant roots (slikke) and the shorre where plants stabilize and interact with the elements in the sediments. The shorre is the salt marsh environment where grass-like plants are common, forming a covering mat of roots. Normally the upper portion of the root mat has a certain porosity and hence will have a contact with the ambient air masses and be as a result more or less oxidizing in its interaction with the pore water of the sediments in the root zone. Usually biomass production is important in this zone and the interaction of organic materials and different mineral elements is very intense. Among these is sulfur, an active component in salt marsh plants and sediment interface systems (Luther and Church 1992). Sulfur has an important role in freshwater systems as well but it is significantly less abundant. A second important factor in the disposition of minor elements in sedimented materials in salt marshes is iron. As in freshwater sediments these two elements, Fe and S, are strongly affected by oxidation potential, which in the salt marsh environment is also affected by bacterial action. As outlined above in various chapters, the movement of minor elements is largely determined by the availability of surface exchange sites on clays and hydroxy-oxides. The more permanent capture of minor elements can occur within the oxyhydroxide structures or sulfides such as pyrite, in substitution for iron, or at times in specific sulfide phases containing minor elements. Many published studies indicate these possibilities. However, even though they exist, the different phases and exchange sites need to be put in perspective in a natural setting, i.e., how much does each component affect the presence of each minor element and what are the more important variables in the attraction of different exchange sites for different minor elements?

Much information is lacking which would make it possible to make definitive statements concerning the more common minor element species present in the ocean shore sedimentation process. However, some information can be used to determine the relative behavior of certain elements.

6.3.1 Fe and S in Salt Marsh Sediments: Oxidation Effects

In salt marsh environments one finds a conjuncture of sedimentation, i.e., addition of new particulate material, and plant–biological (microbial) interaction. In the previously discussed freshwater sedimentary environments, sediments accumulate and show a redox zone, transition from surface waters where much material is in the oxidized state and a reducing zone somewhere below it. In salt marsh sites the influence of organic materials associated with the incoming sediment through algal association with particulate suspended matter and the influence of rooted organic material in the sediment strongly influence the oxidation state of the sediment. Burial of this material through sediment accumulation generally leads to an increase in reducing conditions with depth in the profile. However in certain cases of salt marsh sedimentation, the surface layer is one where plants are active, growing roots that need oxygen, and hence producing a porous network. This creates an oxidation zone that affects the chemical activity, redox potentials, and directs the bacterial action. As can be expected minor element abundance is affected by the state of solids that can form incorporating the elements into the solids or their fixation on clays or oxides. Thus two situations can occur at the surface in coastal salt Marsh sedimentation, one where there is a general change from slightly oxidizing conditions to reducing conditions with depth, and one where there is a strong increase in oxidation at the surface which is followed by more reducing conditions with depth.

The two major elements Fe and S are affected by changes in oxidation potential in the range commonly encountered in surface environments of salt marsh sedimentation. The changes in Eh are largely due to microbial activity, which extracts energy in changing the oxidation state in several elements. Carbon is primordial in that the overall tendency is for complex organic carbon-bearing molecules to be reduced to a carbon-rich product (humic material) and CO_2. Also the elements iron and sulfur are relatively easily oxidized or reduced under microbial activity. In many surface environments where plants are active one finds an oxidizing zone at the surface and at some depth a reducing zone. The interface is the redox layer which is largely controlled by plant root activity. Within the root zone, porosity due to various biological actions (worms, burrowing animals, and decaying roots leaving pore passages) maintains a zone of oxidation, which is necessary for plant activity (for most plants). The presence of iron and sulfur in dissolved or solid phases is governed by the oxidation state within the soil zone. These two elements can often form compounds or phases that attract minor elements either to the surface of the solid material or which incorporate them in small quantities into the crystalline structure. However under oxidizing conditions sulfur is released and Fe retained in the solid phases as a ferric oxide. Under these conditions S and Fe are separated into the solution and solid (particulate) phases.

6.3.1.1 Iron

Oxidation

Iron is known to precipitate from transport fresh waters and in some early stages of salt water mixing (see Sect. 6.2). The initial stages of sedimentation produce amorphous or very poorly crystalline materials. A study by Schäfer and Bauer (unpublished) on a French Atlantic coast salt marsh core indicates that the initial iron material sedimented at the top of a core in a salt marsh in the clay sized fraction is largely amorphous, about 40 % of total iron in the sediment material. This clay fraction material changes to a crystalline form at about 12 cm depth in the core studied, leaving 10 % of the iron in the amorphous state. Coarse iron oxides of detrital origin represent about 40 % of the total iron present in the sediment. Thus there is a strong change in the iron mineral carrier in the upper part of the sedimented material. The change in crystalline state of iron corresponds to a change in oxidation potential 50–350 mV, which increases in the upper 10 cm of the core and decreases back to an initial value below 30 cm depth. Essentially this upper 25 cm of sediment is where the root systems of the salt marsh grasses are active. There is a slight increase in iron content of the clay fraction in this upper zone of the core compared to the sediments at greater depth (Fig. 6.2). In the AT core Fe and S show contrasted concentrations in solids as oxidation occurs.

The classical associations of transition metals and iron indicate that the change in phase of iron, amorphous to crystalline, does not change the affinities of Ni, Cr, Co, and Mn. When correlated with iron content of the clay fraction of the core materials, these minor elements maintain the same distribution trends (Fig. 6.3). Hence whether the minor elements are associated by surface adsorption on amorphous material or in sites within the crystalline matrix, the relations of affinity are

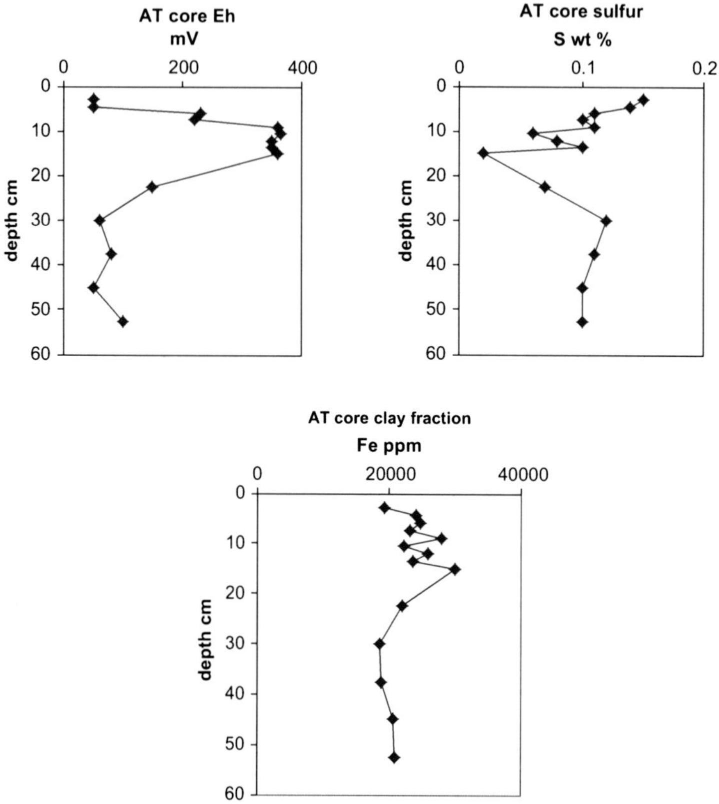

Fig. 6.2 Core AT, Eh, Fe, and S element trends. Data from Mizottes, baie de l'Aiguillon, Charente Maritime, France (unpublished data from Schafer and Bauer)

maintained. This occurs in a core where oxidation of the sediments occurs at the top of the sequence.

Reduction

Overall reducing environments are commonly produced in salt marsh sediments. One major result is the fixation of sulfur on organic materials with only a small amount forming a sulfide mineral (King et al. 1985). Luther and Church (1992) point out the complexity of the effect of the maturation of organic matter and the chemical affinities of sulfur in these environments. However the effect of oxidation or reduction is clearly seen in the presence of iron and its concentration in zones where there is a loss of iron to the solution through the formation of solid phases. This loss is compensated by ionic elemental migration to the zone of mineral precipitation and hence an overall concentration of iron in the zone of chemical

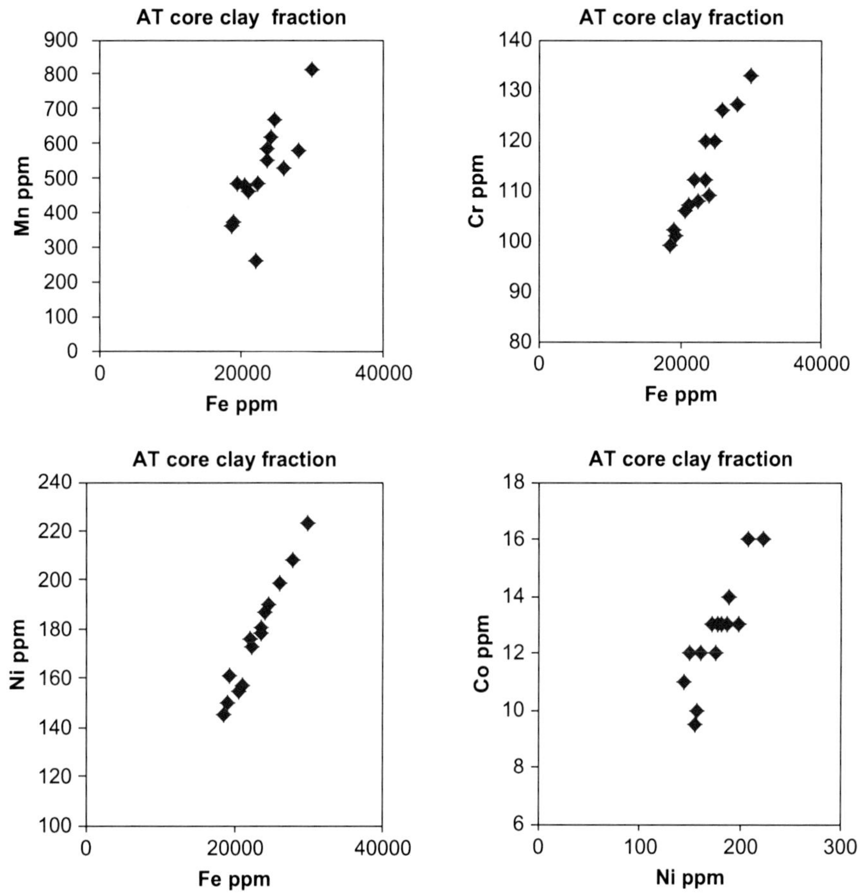

Fig. 6.3 AT core minor element trends compared to iron content under oxidizing conditions. Data from Mizottes, Baie de l'Aiguillon (AT), Charente Maritime, France (unpublished data from Schäfer and Bauer)

change. Change of chemical potentials from oxidizing conditions to those of reduction is a significant force in saltwater geochemical reactions at the surface in zones of deposition.

Minor element abundance compared to that of iron in the salt marsh cores is different depending upon the redox regime. Figure 6.4b indicates the relations of metal ions compared to Fe in the AT (surface oxidized zone) and ONR (reducing environment) cores. Reducing regimes show little correlation with iron whereas oxide regimes indicate a strong correlation of the minor element Ni and Cr metallic element ions with the iron oxide.

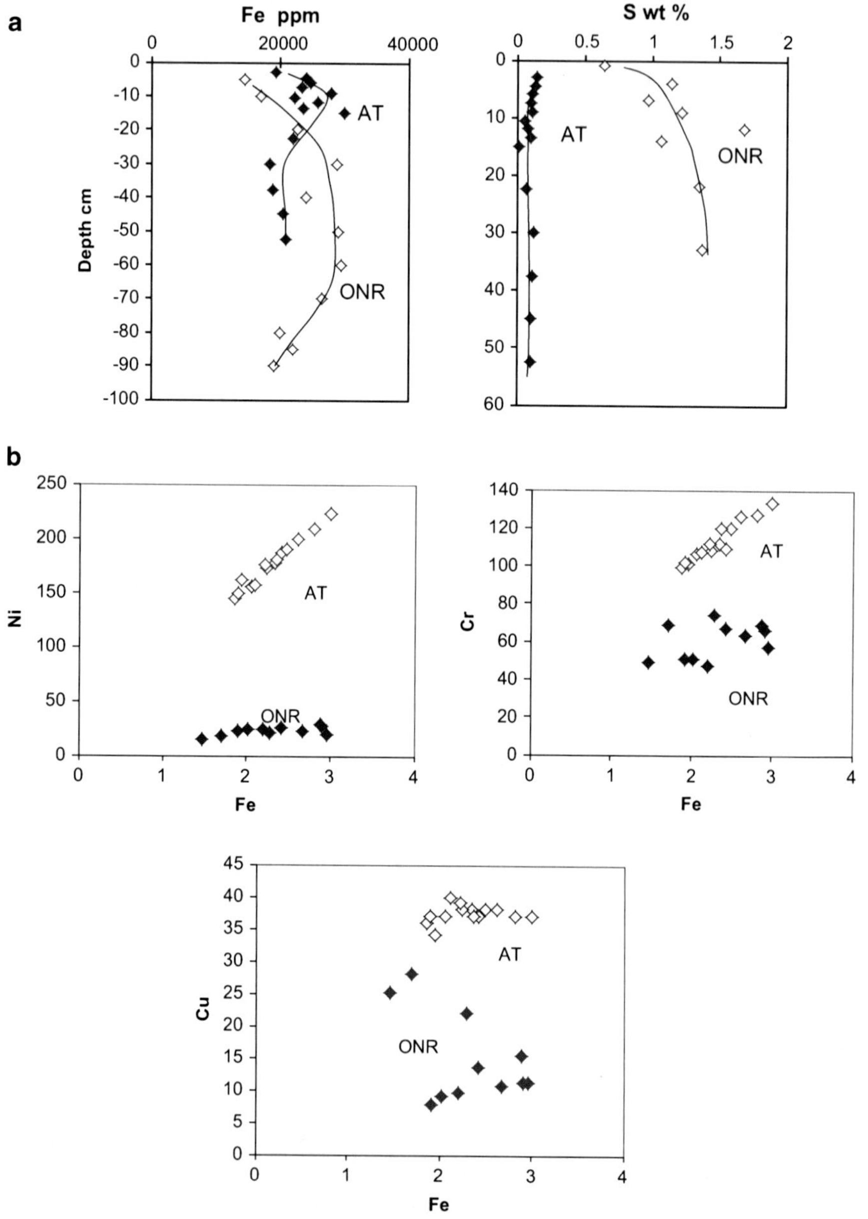

Fig. 6.4 (**a**) Comparison of iron and sulfur abundances in two salt marsh cores, one where oxidation occurs near the surface (AT) Baie d'Aiguillon, France, and another where reducing conditions prevail at depth (ONR) Delaware Bay, USA (data from Velde et al. 2003). (**b**) Comparison of Fe and minor elements in the two cores under reducing conditions (ONR) and an oxidizing regime (AT core)

6.3.1.2 Sulfur

Most soil sources of S are in the organic matter and are therefore concentrated in the topsoil. Sulfur is often considered to be a major actor in the chemistry of salt marsh sediments and sedimentation (Luther and Church 1992). The redox effects in the upper parts of the sediments are largely related to plant activity and bacterial action in salt marsh sediments, which determine the oxidation state of sulfur. Sulfur and iron are major sources of energy for bacteria as they change the oxidation state of these elements. The decayed plant material as humic substances is then further reacted by bacteria which changes the oxidation state of the sulfur (Ferdelman et al. 1991). These are quite complex reactions but important for the attraction and capture of many minor elements. This is especially so for the heavy metals, such as Cd, Sn, Zn, Pb, and Sb among others.

In general S is enriched in the geological alteration–sedimentation cycle, increasing tenfold from igneous to shale rock types (Mason 1966, p. 181). One sees a 17 % decrease in sulfur content of river water compared to seawater (Mason 1966, p. 198). Thus in general one can assume that the interaction of plants and bacteria in the surface sedimentation zone is a significant part of the concentration of sulfur in argillaceous sediments, which become shales.

The presence of sulfur appears to follow a reverse trend in the French coastal AT core, being decreased in the oxidizing zone and increasing to initial values in the core below the upper high Eh zone (Fig. 6.3). This is atypical for salt marsh sediments (see Luther and Church 1992). Since sulfide phases are assumed to contain minor elements, associated with iron or as individual sulfide phases (Audry et al. 2006), one can attempt to see what the minor element associations are in such a system where iron content increases and sulfur decreases.

Data for Cu and As do not show correlations with iron or manganese. However there is a reasonable positive correlation of As with total S in the salt marsh sediment core material studied and a negative relation between Cu and S in the materials of the core. Usually authors insist on a positive relation between As and Fe in soils and transport media but the relations of As and other elements are strongly affected by redox conditions upon sedimentation (Chaillou et al. 2003) where As is released from iron oxide-hydroxides upon reduction of the Fe oxidation state. Here when sulfur becomes an active part of the chemical system, it appears that it favors accumulation of arsenic in the iron sulfide phase instead of the oxide-hydroxyl material. Cu is not related to other mineral elements nor with organic carbon in the sediments. Frequently Cu is associated with organic carbon in surface materials but not in this instance.

It appears then that the redox changes induce changes in the solid components of sediments. The result is not important in the case of iron-and manganese-bearing phases concerning the affinities with transition metals commonly associated with the major element transition metals Fe and Mn as they remain similar under different conditions of oxidation state. However the heavy metals such as As, Pb, Sn, and Sb among others which do not seem to be associated with the Fe and Mn phases may well be controlled by their affinities for sulfur-bearing minerals. This is especially true in the case of crystallized sulfide minerals, iron-based, which can contain significant amounts of minor elements or form specific sulfides themselves. Sediments in a coastal lagoon lake in Thailand show similar relationships where Cu, Zn, Cd, and Pb are associated with sulfide concentrations but in this case As is bound to iron oxide-hydroxides and Co and Ni are associated with the clay (silicate) fraction (Pradit et al. 2010).

The study by Velde et al. (2003) where two types of salt marsh redox regimes occur in the sediments indicates that the minor element transition metals are associated with Fe under an oxidizing regime whereas there is less affinity under reducing conditions. Copper does not appear to be strongly fixed in iron-rich phases.

However the heavy metals such as Pb, Sn, and Zn do not follow the same inter relationship trends in the two types of cores, sulfur accumulation or sulfur loss. Here we see that transition metals follow iron when in an oxidized state and heavy metals follow sulfide minerals.

6.3.1.3 Clay Phases

Some minor elements are associated with clays, either being exchanged in internal (interlayer) sites as hydrated cations or fixed on variable charge edge sites or at times absorbed into the clay structure. This last type of association is typical of lithium, which diffuses into the internal octahedrally coordinated sites of aluminous clays (see Chap. 3). Correlations of aluminim with minor elements can give an insight to the relations of carrying agent and displacement of the minor element in question. In fact, the major consistent correlations with clays (Al) and minor elements are between alkali and alkaline earths and Al. For example Cs is well correlated in salt marsh sediments with Al. This is to be expected as Cs will be a hydrated cation attracted to the interlayer sites in clays. However Sr and Rb which would be expected to be present in interlayer sites in clays are not systematically present as a function of Al (clay mineral) content in the salt marsh core reported on here, the French Atlantic salt marsh core (AT). Usually one assumes a substitution of Sr for Ca in high temperature minerals and Rb usually accompanies K in high temperature minerals and in weathering phases (Figs. 3.9 and 3.10) and these elements are related in the dissolved fractions of rivers (Fig. 5.5). However these relations cannot be seen in the particulate heterogeneous materials brought in and deposited as sediments. Cs and Li are however related to Al suggesting substitution in the interlayer exchange sites of clay minerals and within the clay structures.

It seems that the silicate minerals are less strong carriers of metal elements of minor concentrations in sediments where iron and manganese oxide-hydroxides, phosphates, and sulfides become important carriers of these elements (Fig. 6.5).

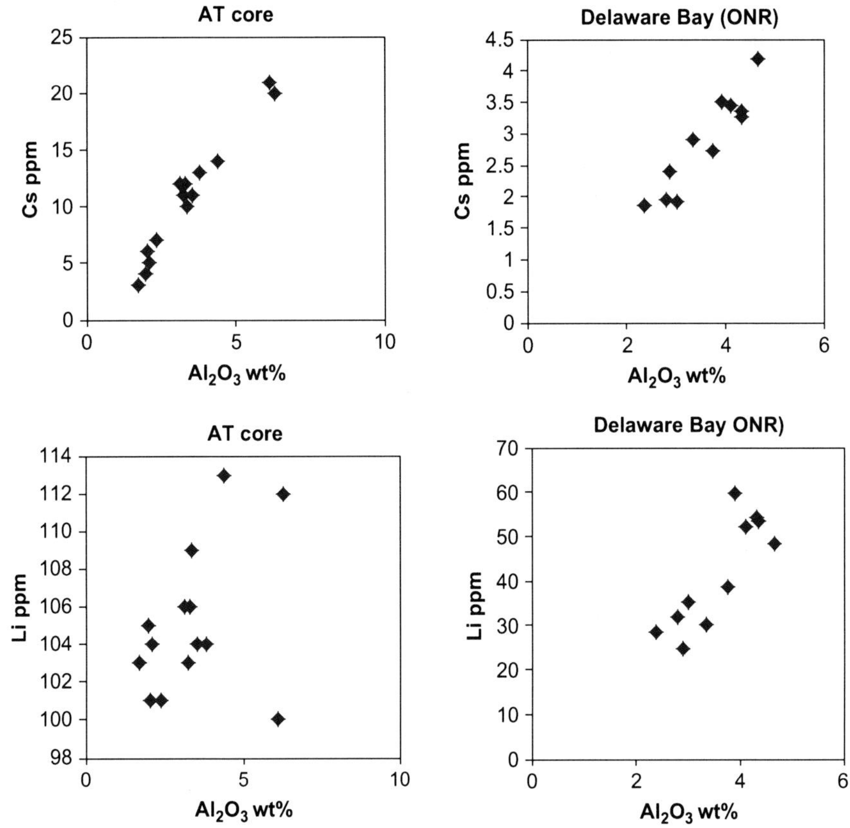

Fig. 6.5 AT French Atlantic Coast and Delaware Bay salt marsh core materials in the clay fraction showing the relations of Cs, an interlayer ion and Li an ion absorbed into the clay structure itself as correlated with Al, a major component of sedimentary alumino-silicate clays which occurs in materials of both cores (Velde et al. 2003)

6.3.1.4 Phosphorous

Phosphorous is an essential element for all forms of life. P in the earth's crust is mostly contained in soils and sediments. Soils contain both organic and inorganic P containing compounds. P is essential for plant growth and function.

P content decreases significantly in the transition from igneous rocks to the shale-weathering product. There is a loss of nearly 30 %. This phosphorous can be found again in secondary shallow deposits of organo-chemical precipitation from ocean water. However, the relatively recent (on a geologic time scale) abundance of phosphorous in surface environments has led to new minor element relationships (see Sect. 6.2.2.3). Church et al. (2006) indicate that the increased use of phosphorous as fertilizer in intensive agriculture and the use of phosphorous in domestic detergents can be traced in marsh sediments and can be correlated with heavy metal concentrations due to industrial and urban input. Velde (2006) found that the only minor element elemental correlations for heavy elements Cd, Sn, and As were related to P content in recent coastal marsh sediments.

It appears then that the relations of heavy metals in sediments can be associated with phosphorous in more recent sedimentation, which might well change the relations of these minor elements with other carriers such as sulfur phases.

The largest part of sedimentary phosphate has been formed on shallow continental shelf environments (Howarth et al. 1995). Estimates on abundance are difficult but it appears that most of the phosphorous deposited in sediments is on shallow continental shelf zones compared to deep sediments. Some of the phosphorous deposited by rivers is in the initial form of high temperature minerals for example apatite. However apatite is very easily dissolved in slightly acid solutions and plants use significant amounts in their metabolism such that most of the phosphorous transported has gone through a bio-cycle. Phosphorous is in part adsorbed on particulate matter in transport media, especially trivalent iron materials, some of which is desorbed in estuarine saline bodies of water. Here it enters into the bio-cycle as plankton and other life-forms.

On shallow continental shelf zones, upwelling of sediments rich in oxide-hydroxides fixes dissolved phosphorous and it is deposited on the sediment surface oxidized zone. Such situations can create significant deposits of phosphorous, in nodular form, which form the basis of phosphorous mineral resources.

Phosphate deposits tend to be enriched in As, Cd, Zn, Pb, U, and rare earth elements compared to average shales (Altschuler 1980). This is reflected in studies of modern salt marsh sediments where these heavy metals are frequently correlated with P content. Thus the eventual deposition of phosphorous is associated with heavy metals in different sedimentary environments. The trends found in modern polluted zones are just an amplification of a normal geochemical affinity.

6.3.1.5 Uranium in Salt Marsh Sediments

Uranium can be considered a special case, but it represents a spectrum of elements. The transuranides have similar behavior in surface environments due to their multiple oxidation states and large ionic size. These elements are present in small quantities in normal sediments and highly affected by oxidation conditions. Church et al. (1996) indicate that salt marsh sediments can be an important reservoir for uranium. Barnes and Cochran (1993) indicate that adsorbed uranium is closely related to the oxidation state of iron and manganese in sediment materials. In salt marsh sediments with a strong change in oxidation potential at a given depth in the core, a strong deposition of U at the redox interface (Cores AT and ONR, Fig. 6.6) occurs. Cores with a weaker change in oxidation potential show a gradual loss of U with depth. It does seem that the presence of U in the sediment materials such as the AT core can be strongly influenced by the redox reactions where reactions provoke precipitation of U in a reduced state such as is very well known in "roll front"

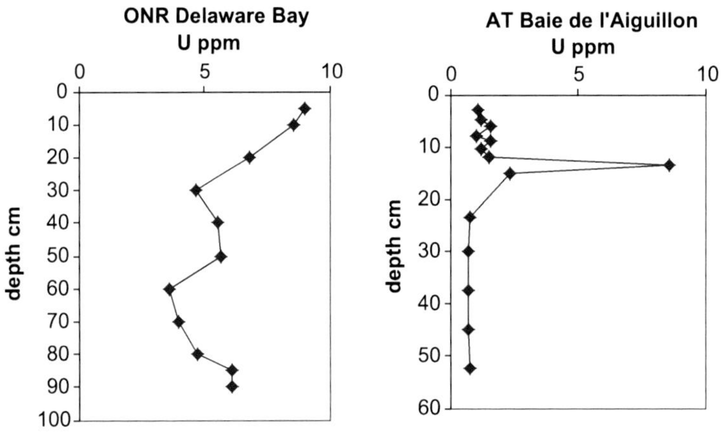

Fig. 6.6 U concentration versus depth in AT (French Atlantic coast salt marsh) and ONR (Delaware Bay salt marsh core, data from Velde et al. 2003)

deposits in sandstones such as those of the western United States sandstones (Gabelman 1977; Granger and Warren 1974 for example).

6.3.1.6 Mercury in Salt Marshes

Mercury is a highly toxic element that is released into the environment through a variety of human activities, including the burning of coal. Once emitted to the atmosphere it can be transformed in the environment by biotic and abiotic processes resulting in methylmercury. It is an element that is difficult to follow and hence rarely followed in geochemical cycles. Mercury occurs in several different geochemical forms, including elemental mercury [Hg(0)], ionic (or oxidized) mercury [Hg(II)], and a suite of organic forms, the most important of which is methylmercury (CH_3Hg^+). Methylmercury is the form most readily incorporated into biological tissues and most toxic to humans (Alpers and Hunerlach 2000). Hg is not easily incorporated in silicate minerals nor in or on oxides. It appears that in the surface geochemical cycle Hg is strongly associated with organic actions (Ehrlich and Newman 2009; Brinkman and Iverson 1975). However Hg is known to form a volatile phase in the seawater environment and can be moved from the surface of the sea. A striking relationship is between Hg and airborn particulate forms in aqueous suspensions. The possibility of transfer by wind-borne materials is of great importance (Weiss-Penzias et al. 2012).

Mercury can be accompanied by other minor elements such as Cd, Sn, Pb, or Th in some bacterial actions. Merritt and Amirbahman (2007) show that relatively strong mercury concentrations can be found in the near surface zone of estuarine sediments in a situation similar to other heavy metal elements in the redox zone where sulfur and iron are reduced to form sulfides. Although the path is strongly influenced by microbial activity, one can expect that Hg will follow much of the heavy metals as deposits in ocean edge sedimentation such as in salt marshes.

The cases of Hg and U in salt marsh environments indicate the strong role of the plant–sediment interface along the edges of continents where sediments are accumulated and eventually buried. However, if these repositories are disturbed, the redox balance will be broken and the heavy elements will be released into the environment. These problems are dealt with in more detail in Liu et al. (2012a) where the rather complex relations of elemental chemisorption, oxidation effects, and interactions with organic materials and organisms in soils and sediments are dealt with at length.

6.4 Element Concentration

6.4.1 Rare Earth Elements in the Alteration–Transportation–Deposition Cycle

Rare earth elements (REE) are frequently used as indicators of the origin of geological materials at the surface in that they are assumed to be weakly affected by alteration and transportation processes (Fleet 1984; Koppi et al. 1996; McLennan 1989; Hagedorn et al. 2011). This is attributed to the relative insolubility of the host minerals, in which REE are accumulated. However other authors find that REE can be displaced from solid phases and are then found either associated with solid oxide phases (Marker and de Oliviera 1990) or strongly dissolved and transported from soil alteration sites (Ma et al. 2011; Braun et al. 1998; Dupré et al. 1996; Bruque et al. 1980; Coppin et al. 2002). The dissolution effects and changing phase associations are considered to change the distribution patterns of heavy and light REE. One factor, which might affect the apparent change in light to heavy REE that leave a dissolving rock, is the possibility of selective dissolution of the constituent minerals. In high temperature rocks the REE distribution is not necessarily homogeneous in all of the phases that carry these elements. Phosphates and zirconates do not have the same selectivity for the different elements nor the same resistance to alteration. Since most rare earth studies compare the relative abundances between the elements in the series for a given rock, the subtlety of possible segregation is ignored. Chondrite, average shale, average basalt, etc. abundances and spectra do not consider inhomogeneity within the phases and the different phases present that can carry REE in the samples analyzed. When it does occur, different minerals will contain the elements in different abundances. If these minerals dissolve at different rates, the compositions of solution waters will be changed according to the minerals affected and their relative abundance compared to the phase remaining in the non-dissolved, particulate fraction of the alterite. Thus in fact one should expect to find some differences between dissolved and residual REE patterns compared to the initial rock materials. These effects have been demonstrated by Ma et al. (2011) to occur due to more or less minor differences in weathering intensity on hillsides of a limited area in the Susquehanna Hills, USA. Thus differential weathering of primary minerals can change the relative values of REE in alterites resulting in sediments, which reflect conditions of alteration and not necessarily chemical attractions or incorporation into alterite minerals.

These apparently contradictory observations reflect undoubtedly the complexity of the alteration processes, which can be affected by climate and biological interaction to produce different solid phases, which will attract the REE in different ways. This being the case one must be careful in using the REE elemental distributions as a signature of provenance of materials that have gone through the alteration and transportation process.

6.4.2 Sedimentary Iron Deposits

The most important metal from an economic and technical point of view is iron. Sedimentary iron deposits, from which almost all metallic iron is obtained, can therefore be viewed as one of the world's great mineral treasures. Two major types of deposits exist. The first, and by far the most important, is banded iron formations (BIFs), so called because they are finely layered alternations of cherty silica and an iron mineral, generally hematite, magnetite, or siderite. This material is precipitated from the aqueous transport medium and deposited in relatively shallow waters. Degens (1965) and Twenhofel (1961) cite the consensus of studies concerning sedimentary iron deposits. It has been deduced that the iron was brought into shallow shelf sedimentary areas of the ocean along with important amounts of silica in solution. Precipitation would have occurred in the near coastal zone. In most cases these iron deposits are very old, often of Pre-Cambrian age. The precipitation of Fe and Si is interesting but difficult to explain. The ability of iron to migrate increases in the absence of free oxygen. Appreciable amounts of iron can be dissolved in slightly acidic to neutral solutions. Under such conditions, changes in parameters such as pCO_2, pressure (= water depth), pH, and Eh (redox potential) determine whether iron is precipitated as a carbonate, oxide, silicate, or other salt. In contrast to ferruginous species, the solubility of silica is nearly independent of the acidity, ranging between 80 and 100 mg/l in a pH range of 2–9 to conditions at or near the silica saturation point (Siever 1992). Silica precipitation can then be achieved through evaporative supersaturation or coprecipitation with solid-phase iron minerals (Garrels 1960).

Obviously the pH and Eh conditions changed in order to form a precipitate of dissolved species. One would expect that the high silica and iron content represent soil alterite materials which have been highly altered. A likely explanation for such a chemical context at the surface is the lack of land plants in Pre-Cambrian times, which would otherwise fix silica in the soil zone. Lack of continental plants would result in more silica in solution, which would lead to an oversaturation in near shore oceanic waters producing the siliceous chert deposits associated with iron oxides in

Pre-Cambrian sediments. Lack of land plants will leave iron in the oxidized state to a large extent because there would be few reducing agents present at the surface, such as bacteria, which act on plant materials in post Pre-Cambrian surface environments. Thus the barren continental surfaces would present water–rock alterite materials that are directly moved into a sedimentary environment.

A change in pH from values below 6 upon the introduction of transport stream water containing divalent iron would form hematite precipitate (Garrels 1960). The present day occurrence of phases such as magnetite suggest that there was some organic material associated with the dissolved material which upon burial and microbial action affected the oxidation state of the precipitates as they crystallized. Since the materials are often of Pre-Cambrian age, the problem of oxidation state is discussed in that the old atmosphere is assumed to have contained less oxygen than today's atmosphere. Nevertheless there had to be some reducing agent present to transform the iron precipitate, largely oxidized, to much of the mineral content today which is in various states of reduction. It is deduced that in several instances that bacterial action is probably responsible for this chemical action (Degens 1965).

Perhaps the presence of massive sedimentary iron deposits is the manifestation of the last stages of alteration on the continents where the alterite is largely iron oxide and quartz, which might reflect on the probably somewhat low biotic continental masses in early Pre-Cambrian times. However the suggestion of plant life controlling the alteration rate is not necessarily upheld in that such iron deposits are also known in Cambrian sedimentary rocks.

Another, little studied, occurrence of iron oxide is in small surface deposits of recent age, often called bog iron. Such materials are oxides, very concentrated in iron, which are nodular in form. Generally it is assumed that they form in areas of iron concentration of special chemical character due to organic actions. These deposits were exploited locally from Roman times in Europe, and undoubtedly elsewhere. Given their small size, tens to hundreds of feet in surface area, and hence small economic dimension also, they have not been the subject of much geochemical interest. However such concentrations are undoubtedly important to an understanding of the concentration or iron under freshwater conditions.

6.5 Evaporites and Concentrated Saline Solutions

One surface geological environment whose geochemistry is frequently little discussed is that of evaporite basins. Ocean water is a major source of evaporite minerals. Roughly 3.5% of ocean water is composed of dissolved salts (Jensen and Bateman 1979; Bluch 1969). The dissolved cations are dominated by Na and Ca and the anions by Cl and CO_3. Concentration of these waters by evaporation engenders the formation of mineral precipitates in the order of their insolubility. In these environments one finds materials that have been important to the development of human activity, such as salt, and other materials used eventually in industrial processes.

Two major settings are known for the formation of evaporate deposits; those along continental margins, usually fed by subterranean or surface sources of sea water and those in closed intercontinental basins fed by rain water and its drainage (see Warren 2006 for a complete description of these deposits and their environments). On large, flat continental areas one often finds closed basins, i.e., those where rivers do not flow to the sea especially in areas of low rainfall. Closed continental basins can form two types of sediments, those formed below permanently present evaporation-concentrated water and those in "salt pans" where the incoming water is periodically totally evaporated within a yearly or pluri-annual cycle. In evaporite basins the input of dissolved alteration materials from rocks can be seen to form a series of new mineral deposits, which are a function of the relative solubility of the dissolved elements and their possible combination in the new phases. One thinks of salts or sulfates in evaporite deposits, but carbonates are most often present and silicates can be produced by the concentration of river waters or ground water that flows into the basins and is concentrated by evaporation.

Evaporite minerals are those associations of dissolved elements, which combine to produce phases that reach saturation during water evaporation in the solutions in which they are found, either fresh or saline. In general the solubility products for oxoanion phases (Lide 2000) such as those containing CO_3, SO_4, and PO_4 indicate that phosphates will be precipitated first, prior to carbonates and finally sulfates. Salinity values for the precipitation of Mg–Ca carbonates are in the range 0.2–0.5 g/kg while those for Ca sulfates are 5–20 g/kg and chlorides 100–4,000 g/kg total salts. Put another way, when the volume of seawater decreases by 50 % carbonate forms and when it has reached 10 % of its initial value salts form (Bluch 1969). This leads to segregation of the dissolved elements in new mineral precipitates according to their geologic disposition, i.e., layers of a specific type of mineral are found separated more or less from another type:carbonates from sulfates, and eventually salts. In a geographic context, carbonates will be found on the edges of an evaporate lake where the first precipitates form, followed by the sulfates more toward the center and finally salts in the central part of the lake. The salts of Cl and Br are the most soluble phases and hence the last to form in evaporite sequences. Along with the classical oxyanions of carbon, sulfur, and phosphorous, silica occurs in silicates that are often found in evaporite environments (see Millot 1970) between the carbonate and sulfate mineral facies. Also, and although present in minor quantities, boron and nitrates are systematically found in varying amounts in evaporite deposits. Figure 6.7 indicates the disposition of different sedimentary evaporite deposits as a function of geographic setting.

Usually evaporite deposits show concentrations by mineral type: carbonate, sulfate, salts which occur in layers of different thickness. In present day evaporate-depositing basins one finds surface areas succeeding one another over the basin where a high concentration of one type of minerals is present. This suggests that the evaporating water reaches a concentration level that deposits a given mineral and with further evaporation, and displacement towards the center of the basin at a lower geographic level, it precipitates another, more soluble species.

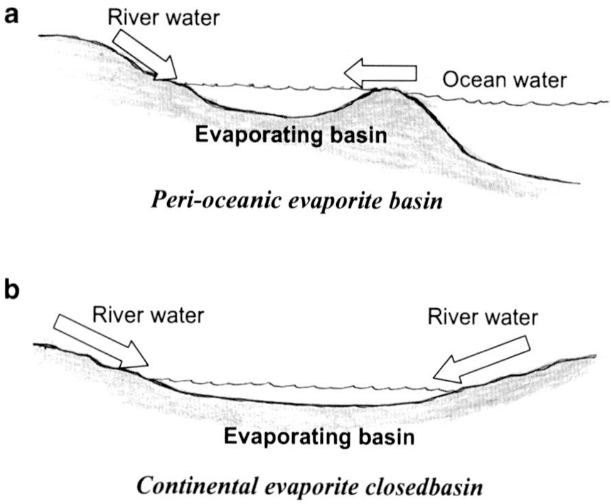

Fig. 6.7 (**a**) Schematic representation of peri-oceanic semi-closed basins forming evaporite deposits where water renewal occurs periodically from the sea. (**b**) Schematic representation of evaporite lake deposition in closed continental basins

Continued basin-filling and evaporation episodes then build up the series of mineral layers as a function of their distance from the center of the basin and solubility of the phases precipitated.

Evaporites produce concentrations of certain elements that have proved to be useful in industrial and agricultural applications. Phosphates and potassium salts were found to be important in producing increased agricultural yield in the nineteenth century, while borates were found to be useful in the twentieth century as a source of the chemical element boron.

Saline lakes formed in closed watershed basins are found on all continents (see Jones and Decampo 2004). The most important requirement is that the lake be closed to any seaward outlet and that the climate be alternately wet and dry. The composition in dissolved elements is determined by the rock compositions being altered in the drainage basin feeding the closed lake system. However one aspect that is important to note is that the content of CO_3 can be controlled by the atmosphere, whereas the content of phosphate, sulfate, and chlorides is fixed by the altering waters coming from the surrounding rock massifs. The cation populations guide the formation of the phases precipitated as the solution concentration increases. The ratios of these ions in the initial incoming waters coming from alteration or hydrothermal action can control the phases that form. Ca content is important and its ratio to Na is a key factor in determining the presence of sodium or calcium carbonates. Magnesium–calcium relations can be important in the formation of carbonate types and sulfates, but also in the formation of silicates, which are magnesium based. The phases that form in evaporate sequences depend upon the solubility of the minerals that could form and the relations of cation, anion and oxoanion concentrations in the incoming fluids which determine which minerals will be present. Precipitation of different phases as solute concentration increases follows the paths determined by the ratios of cations in solution as anions and oxoanions which combine to form solid phases. Evaporite basins related to

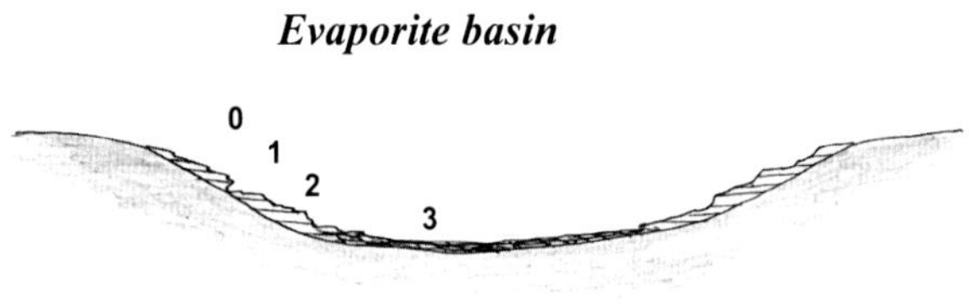

Evaporite Stage	*Major phases*	*Minor phases*
1	carbonate	phosphate, Mg silicates
2	sulfate	
3	salts	Na silicate, Na carbonate

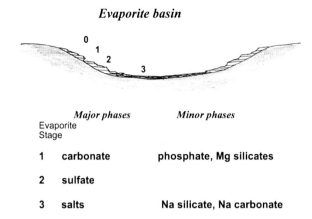

Fig. 6.8 Schematic representation of evaporite successions

seawater form more or less predictable sequences of evaporate minerals while those in more local basins due to the influx of rain water which has altered rocks or been in contact with volcanic gases will be more variable in initial composition and will give more varied sequences of evaporate minerals.

In following the compositions of incoming river waters and lakes of different evaporation intensities one can follow the effect of deposition of carbonates, sulfates, and salts (data presented by Jones 1966). The same deductions can be made for brines in evaporite lakes (Gueddari 1984). Generally speaking one finds the sequence of carbonate–sulfate–salt as evaporation intensity increases. As the sequence progresses the complexity of mineral types increases as more and more phases are formed from cation associations with anions and oxoanions.

A typical series of sediments at the periphery of an evaporite basin deposit is shown in Fig. 6.8. Very often the sequence is initiated by carbonate formation followed by sulfates and then salts of various types.

6.5.1 Carbonates

Carbonates are a major reservoir of carbon. They are often associated with materials that have precipitated from solution in ocean edge environments. In environments of high salt concentrations, the minerals formed and the constituent elements found in them are largely of inorganic origin except for carbonates which are most often influenced by different life-forms, either animal or vegetal, which used the carbonate material to form shells, or as is the case of algae which change the concentration of CO_3 in solution and can at times provoke carbonate precipitation (Krauskopf 1967, p. 54).

In considering the solubility constants of carbonates, one could expect the following sequence of carbonate types to form upon concentration of aqueous solution Pb–Co–Cd–Zn, Fe, Mn–Sr, Cu–Ba, Ca–Ni–Mg. Overall the minor element

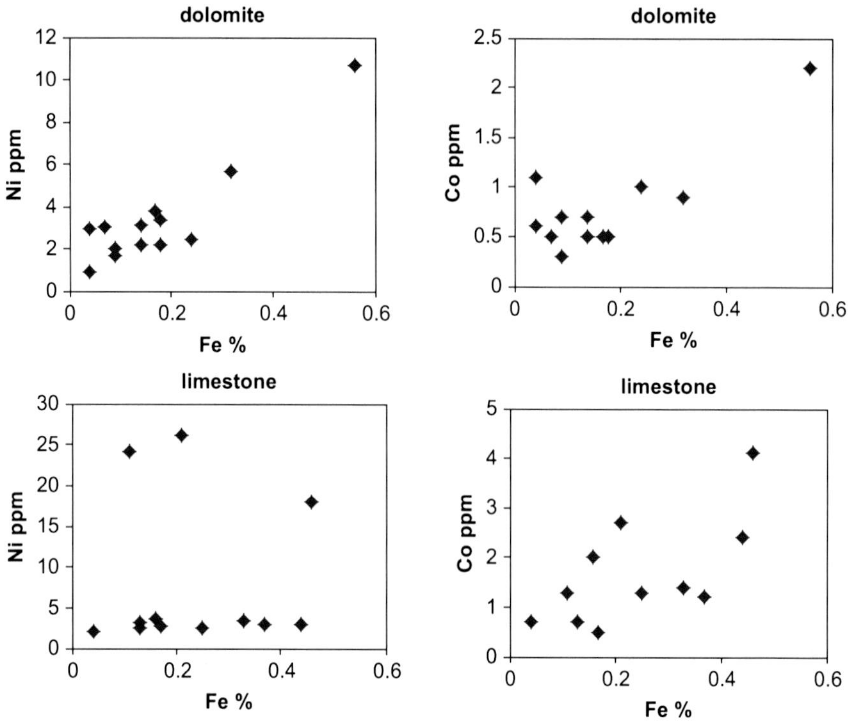

Fig. 6.9 Minor metal element concentrations in limestone and dolomite as a function of iron content (data from Karakaya et al. 2011)

carbonates are less soluble than the major element forms such as Ca and Mg carbonate. More complicated hydroxyl carbonates and compound cationic forms can be present also (Decampo 2010). Playa lake closed basin carbonates contain some minor elements in relatively high abundance such as Sr (100–2,500 ppm) whereas transition and heavy metals are of low abundance, on the order of 5–20 ppm (Karakaya et al. 2011; Trauth 1977). Thus despite their lower solubility and probable precipitation before calcium carbonates, the heavy element carbonates do not appear in evaporate carbonates to any great extent. In dolomites, where Mg is present accompanied by small amounts of Fe, transition elements Co and Ni can be reasonably well correlated with the iron content of the carbonate rocks. This is not the case for calcium carbonates (limestone, Karakaya et al. 2011) (Fig. 6.9). This would indicate that internal substitution in the minerals of Co and Ni for Fe is operative instead of the formation of separate carbonate phases. Rimstidt et al. (1998) indicate that many different trace elements should be able to be found in carbonates, based largely on thermodynamic calculations, such as Be, Cd, Co, Cu, Mn, Ni, and Zn, but it is suggested that they are rarely present due to dynamic constraints during crystallization. It would appear that transition elements follow iron in carbonates, and Sr is substituted for Ca but other possibilities are rare in natural materials although Cd and F have been noted to be incorporated in benthic foraminifera in warm water environments (Rosenthal et al. 1997).

Sodium carbonate is a somewhat special case for carbonate precipitation in that its solubility is high and it is one of the later forming precipitate minerals. Sodium carbonate, often called natron in former times, was the basis for a number of uses in chemical treatments. The Ancient Egyptians used it to embalm people to obtain mummies, usually of high standing in the community. Successive generations found another use for sodium carbonate, that of a fusing agent to form glass from quartz sands, discovered by the Greeks using their privileges during their occupation of the Nile delta where natron was found and used previously (Wadi Natron). Natron deposits are found in late stage evaporate deposits in most continents, and are often active in the present day geological contexts. In contrast to Ca or Mg carbonates, Na carbonate is a late stage precipitate.

6.5.2 Silicates

In evaporate basins one often finds that magnesium silicates have been formed by precipitation from solution where preexisting clays such as smectite have been incorporated into the newly formed clays (see Velde 1985 for a summary). The silicates are often associated, in separate layers, with carbonate deposits. They form in the early stages of evaporation deposition. Since the major elements present are Si, Al, and Mg one might expect to see some relations of the smaller amounts of Fe in the structures and minor elements. Such relations of Ni and Co with Fe are probable but not well defined in the data of Karakaya et al. (2011). There are no examples of current formation in alkaline lakes, and hence one must surmise the reactions and precursor minerals.

However in arid zones one finds magnesian silicates forming in subsurface layers of soils along with calcium carbonate minerals. The newly formed magnesium minerals are of two sorts: palygorskite, which contains a reasonable amount of aluminim, and sepiolite, which contains little aluminium. Trauth (1977) indicates that the aluminous magnesian mineral palygorskite contains much more Sr than the alumina poor sepiolite. The same is true for Ba. These magnesium silicates contain about two times more V, and Cu than Ni, Co, or Cr in their structures compared to carbonates. Karakaya et al. (2011) report data indicating a co-relationship of Pb, Ni, and Co with Fe content in sepiolite-bearing sediments. This indicates that the newly formed minerals were in chemical equilibrium with the dissolved elements in the brine solutions from which they were formed (Fig. 6.10).

A second type of silicate mineral is found in evaporate basins, based upon an association of sodium and silica in a hydrated phase. Such minerals (magadite, kenyaite, makatite, silhydrite, among others) are apparently rare but nevertheless they are found in deposits over the globe (Warren 2006). All are associated with high levels of silica in solution at high pH values, above 10. The conditions of their formation are at times difficult to establish but it appears that they are often related

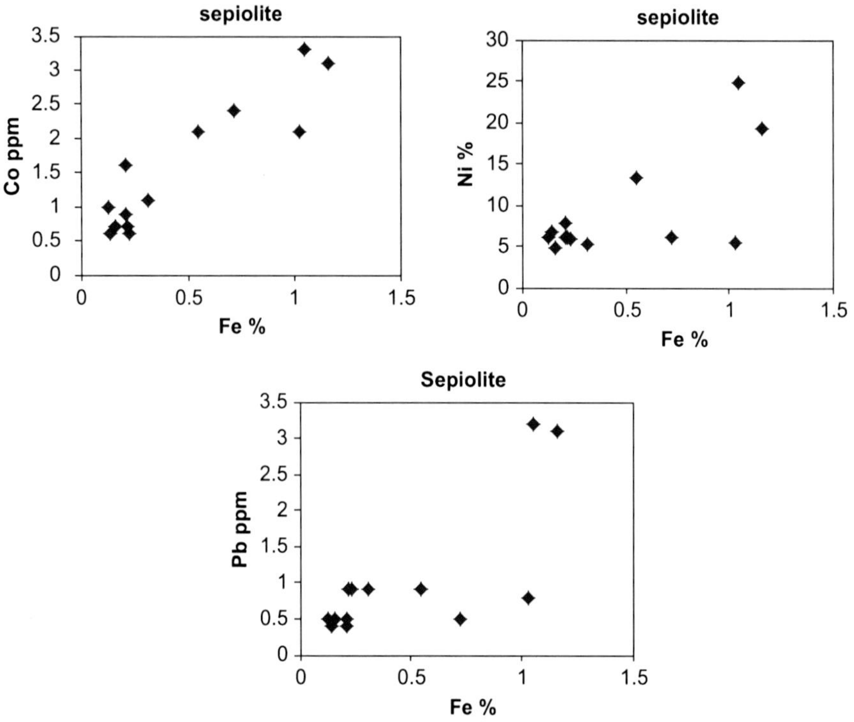

Fig. 6.10 Ni, Pb, Co vs. Fe in the evaporite magnesium silicate mineral sepiolite (data from Karakaya et al. 2011)

to hydrothermal action on volcano-clastic material in closed basins. The silicate minerals are often associated with the formation of sodium carbonates (trona and natron) and thus the incoming aqueous fluids are at the same time alkaline and carbonate rich. The sodium silicates are the late forming minerals in their precipitation sequences.

Thus one finds magnesium silicates to be among the first minerals formed in evaporite sequences and sodium silicates to be among the last formed minerals.

6.5.3 Phosphates

The exact conditions of phosphate formation, geographical and geochemical, are not well defined. It appears that phosphates can be found in oceanic environments or peri-oceanic semi-closed basins. Most often phosphate deposits are associated with carbonates. They are rarely associated with sulfates. This places their formation in the earlier stages of closed basin evaporitic formation. However, the source of phosphorous is generally assumed to be oceanic where the content of P is variable with time. It appears that phosphates are not formed under conditions of present day ocean water composition (Sheldon 1980). The source of phosphorous is in deep ocean waters which when brought to the surface by upwelling, become oversaturated or nearly so, and can form phosphatic precipitations. For the most part phosphate deposits are found on peri-continental zones. In some cases they are associated with the magnesian silicate evaporate mineral palygorskite (Boujo et al. 1980) and in many others they are associated with shallow water carbonate fossils (Giresse 1980). Also phosphates are found with glauconite (iron, potassic clay mineral) pellets forming on shallow continental oceanic platforms (Odin and Letolle 1980).

Phosphates appear to be formed in the early stages of seawater concentration along with the early forming carbonate phases on continental shelf platforms. They are also often associated with skeletal material of marine life. There is frequently a strong relationship between biological action and phosphate accumulation (Föllmi 1966). Altschuler (1980) gives compiled data for average minor element concentrations in phosphate marine deposits compared to average shale values. The "insoluble" elements Ag, Y, Yb, La, Mo are enriched in phosphates as are the more chemically active elements Cd, Pb, and U. Sr is found in increased abundance also, not surprising in the most phosphates are calcium based compounds. The association of REE with phosphates is important in that at times this can be used as a mineral resource. However the presence of Cd is not useful especially when phosphate deposits are sued as mineral fertilizer in agriculture.

Phosphates then in general collect various minor elements as they are deposited in marine or peri-marine environments which decreases the concentration of these elements in the aqueous seawater solutions. Altschuler (1980) indicates that phosphates in general are enriched in certain heavy elements relative to shales.

6.5.4 Sulfates

The most common sulfate is that of calcium, gypsum. It is characteristic of evaporate basin deposits, as the second or third mineral to form after calcite and when present sepiolite-palygorskite minerals. Another sulfate of sodium, thenardite, is also known to form later in the evaporite mineral sequence. Geochemically, calcium sulfates tend to concentrate Sr in their structure relative to carbonates or silicates (Trauth 1977). Ba and Pb content is higher than in the other major evaporites formed in the early stages of mineral precipitation (before salts). These elements would be substituting in the structural position of calcium. Dean and Tung (1974) find positive correlations between Mg, Fe, Mn, and K in sulfate minerals in Permian sediments in Arizona. Gueddari (1984) indicates that fluorine content in brines is affected by the precipitation of gypsum, indicating that some F is fixed in this mineral.

More complex sulfate minerals containing both Ca and Na (glauberite) or hydrated Na forms occur (Mirabalite) or hydrated Mg sulfates as well as Na–Mg sulfates can be found in various deposits of evaporite minerals in continental basin deposits associated with salts, chlorides, and bromides. Thus the most common and early forming sulfate (gypsum) is not the only phase to form due to solution concentration of dissolved elements. The stability of the various hydrated sulfates is often determined by temperature and exchange of cations in the structures such as Na for Mg occurring between chlorides and sulfates (see Braitsch 1971). As a result it is very difficult to establish major and minor element relations according to the mineral species present especially when looking at sediments that have been buried and subjected to diagenetic conditions, i.e., temperatures in excess of 50 °C.

Several elemental relations have been found: For example Bromine (from Greek brómos, meaning "stench") is not found in sulfates nor in borates. Sr sulfate is less soluble that Ca sulfate and would be expected to form an independent and early phase as sulfates form from evaporating brines.

The more complex sulfate phases Na, Mg, or K tend to occur in the later stages of evaporate deposition, being associated with salt precipitation. This is reflected in the solubility of the different sulfates, where Ca forms are far less soluble than alkali or Mg forms (Spencer 2000).

6.5.5 Salts

The last phases to form, being the most soluble, are the salts where the Cl or Br anion combines with various cations. The most common is in fact salt, i.e., NaCl, Bromine substitutes for Cl in salts and is strongly concentrated (a factor of 10) in salts compared to the solutions from which it is precipitated (Holser 1966). These phases can be hydrated at times (see Braitsch 1971). Since the salts are the last minerals to form from evaporating brines, one would expect that they would contain residual cationic elements such as K, Rb, and Li and perhaps Boron as reported by Gueddari (1984). Dean and Tung (1974) indicate that the cations Mn, Mg, Sr, Ca, and K are interrelated in abundance in halite deposits suggesting incorporation in salt minerals. Potassium salts tend to incorporate Rb (Garrett 1996). These can be potassium (sylvite, KCl) or mixed salts of Mg and K such as kainite ($MgSO_4{\cdot}KCl{\cdot}3H_2O$) or carnallite $KMgCl_3{\cdot}6(H_2O)$. The major chemical elemental associations in salts are cation–anion in nature. Thus there is a wide variety of dissolved elements in simple cationic form associated with chlorine and bromine in salt deposition.

6.5.6 Oxyonions in the Last Stages of Evaporite Mineral Formation

The last stages of evaporite formation includes salts which form a complex group of minerals of varying compositions and states of hydration. In inland closed basins, where sources of silica can be found either from volcanic hydrothermal sources or amorphous silica deposits (diatomaceous earth), one can find the extremes of geochemical separations and concentrations due to the formation of the oxyanion minerals nitrates, borates, potassium chloride, and sodium carbonate. Oxoanions such as Perchlorate (ClO_4^-), Perbromate (BrO_4^-), Periodate (IO_4^-), Permanganate (MnO_4^-), Chlorate (ClO_3^-), Bromate (BrO_3^-), and Iodate (IO_3^-) tend to form solid phases with different cations in the concentrating solutions.

6.5.6.1 Boron

Borax-($Na_2B_4O_7 \cdot 10H_2O$) and boron-rich minerals (there are some 230 natural mineral forms) are special cases of late stage concentration of rare elements in evaporating waters. Boron minerals in sedimentary deposits are most often hydrated or hydroxylated forms of cation–oxoanion structures. Different cations are found such as Na and Ca, which are the major boron forming agents in concentrated surface waters. As it turns out, borate minerals occur with clays, carbonates, and chlorides thus at various stages of the evaporite mineral sequences (Garrett 1998). In fact the presence of borate minerals appears to be due to the relative abundance of boron in waters feeding evaporating basin waters which is a function of the rocks or hydrothermal sources that feed the rivers that find their way into the evaporate basins. This abundance is largely due to volcanic activity and hence boron tends to be concentrated in special geological environments (Garrett 1998). Evaporite solutions and brines, tending to be of high pH, will lose their boron content rather easily as evaporation proceeds under alkaline conditions common to evaporate solutions.

Nitrate deposits are rather mysterious in that the vector of concentration and transport is poorly known to say the least. Nitrates are formed from a primordial atmospheric component, nitrogen, which is captured by plants through biological activity in the root zone of soils. The use of nitrogen to form complex organic molecules indicates that it will remain strongly bound in the organic chemical sphere. Certainly nitrogen compounds are lost to rainwater and moved in river systems eventually to the sea. The incorporation of nitrogen into or on suspended matter is not documented. However nitrates, dominantly sodic, are known to form in arid continental basins, some of which are rather exotic, such as the Altiplano deposits in Chile. Here it is assumed that the source of nitrate material comes from mists coming from the sea, some thousands of meters below the zone of deposition. The transformation of the dissolved nitrates of unknown chemistry into sodium nitrate is assumed to occur under catalytic conditions engendered by bio-mineral reactions (Jensen and Bateman 1979). Whatever their origin, these nitrate deposits have been exploited for over a hundred years to the great satisfaction of preindustrial farmers over the world.

Other nitrate deposits are found in more continental closed basins where it is assumed that volcanic activity concentrates nitrate ions in hydrothermal solutions that feed ground waters flowing into the evaporate basins or in other closed basins (Sonnefield 1984).

Other minerals form in late stage evaporite deposits. They can be extremely varied in composition. This is the result of being the last materials to be concentrated in solution. As concentrations of the remaining elements increase they approach the saturation point, where exotic cations and anions can combine. The lists of these salts and other phases is very long, hundreds of species of chemical combinations occur for elements such as B, N. The variety reflects the fact that most minerals precipitated before this stage do not accommodate these elements into their structures leaving them to form specific phases in the very late stage concentration cycles.

6.5.7 Mineral Associations in Evaporite Deposits

Considering the three major classes of evaporite depositional environments, one can attempt to observe the types of minerals formed and their sequence of formation. In general, evaporite minerals form as the concentration of constituent elements reach levels where the aqueous solution becomes saturated with these components compared to a possible solid phase. However some phases are present in one environment and not in another, which would lead one to think that the dissolved element composition of the evaporating solutions is not the same in all cases. Risacher (1992) indicates the possible sequences of mineral precipitation from evaporating brines as a function of the relations of dissolved elements. Initially the first mineral to form is calcite in all instances. This occurs at 0.3 g/l salinities. After Ca–Mg carbonates have been formed, the first criterion is the relation between alkalies and Ca in the brines. Where alkali content and Mg are important there is a tendency to form magnesium silicate minerals, sepiolite and/or attapulgite. If alkali content is high enough, the next minerals to form are salts and natron (sodium carbonate). This occurs at high salinities, above 100 g/l. These are the alkaline facies.

At lower alkali contents, where Ca is dominant after carbonate crystallization, gypsum is the second phase to form at salinities of near 10 g/l. If sulfate remains after the formation of gypsum, one will find other sulfate minerals in the stage of salt formation.

6.6 Summary

We have chosen to consider continental and peri-continental sediments and sedimentation as the last stage of surface geochemistry. These stages of movement at the surface are for the most part waterborne. We fully recognize the importance of wind-borne materials, which have been treated in Chap. 4 as a stage of transport.

Materials deposited from aqueous media are of two types: particulate and dissolved forming sediments and eventually evaporites. Evaporites can be found on land and at the edges of landmasses in contact with seawater. Here the deposited materials formed and reacted at the surface are concentrated and usually eventually become coved by other sediments to become sedimentary rocks at greater depths of burial. These rocks buried to various depths are usually cycled by tectonic events to become the materials of further rock–surface interaction cycles. A large portion of rock materials on present-day continents has already gone through one or more cycles of surface geochemical interaction. Thus the extent of chemical transformation as water and air interact with rock is a function of the origin of the rocks in contact with surface geochemical forces. The largest changes in phases and overall chemistry occur when the rocks have been subjected to high temperatures and pressures, which create mineral phases that are far out of equilibrium with surface conditions. However all rocks will interact and be modified to a certain extent by surface chemical forces.

6.6.1 Particulate Material Sediments

The deposition of alterite materials, dissolved and suspended, occurs at various stages of transport of alterite from the site of its formation to the sea, the ultimate depository. When the energy of the transportation medium is low enough, suspended material falls out of suspension and forms sediments. The largest particles are released first and the smallest last, forming sand beaches on the shoreline of a water body and clay sediments further from the shoreline. This separation forms strong chemical segregation of the materials. For the most part the sand materials carry little adsorbed elemental material while the fine materials carry very much, either on the silicates or the oxides present. However large amounts of dissolved ionic material are still present in the transport waters after sedimentation of particulate material.

The formation of these sediments can occur from freshwater or salt (sea) water transport solutions. This gives what are called continental sediments and marine sediments. The major differences are the relative concentrations of the dissolved elements in fresh- and seawater fluids. However, the major cycling of dissolved and suspended material can be found on the continents or along their edges.

A critical area in the salt marsh environment is where sediment material is captured by marsh grasses and held in reserve. In this environment the change in oxidation state of iron and sulfur is critical to the presence of different heavy metals held in adsorbed conditions. Other cations are fixed within the solids in absorbed states or within the crystalline structure such as in sulfides. Bacterial action is a very important factor in the capture or release of heavy metals and other elements.

Redox reactions driven by organic agents are of fundamental importance to the presence of heavy metal elements in these environments.

6.6.2 Evaporites and Organically Precipitated Materials

Evaporite deposits, forming either in closed continental basins from freshwater sources or peri-continental basins from seawater, represent the ultimate depositions of transport materials in that the dissolved constituents will be the last to be deposited, long after the suspended matter has been dropped from the aqueous solutions. These phases can select minor elements to incorporate them into their structures or they can reject such material to form nearly pure mineral components. This is especially true for carbonates, and to a large extent true for sulfates. Salts tend to form phases, which include minor elements compatible with the highly ionic character of the major component ions, both cationic and anionic. Phosphates seem to concentrate heavy elements from the evaporating solutions but they are for the most part of minor element abundance, of less than a percent weight. One must remember that even though evaporate minerals form at the last stages of solution evaporation, still occasional rainfall on the sedimented materials can wash away "left over" ions which might eventually escape deposition. Peri-oceanic basins, not totally closed to fluid movements, will have waters that can carry away the elements not found in the precipitates.

Chemically, carbonate and chlorides dominate in seawater producing limestone and salt deposits. However other oxoanions such as silicon, boron, sulfur, and phosphorus are present and play a strong role in the geochemical distribution and concentration of minor elements under some geological circumstances. The role of plant and animal life is frequently of great importance in the formation of deposits concentrated from solutions such as carbonates and phosphates.

6.7 Useful References

Braitsch O (1971) Salt deposits: their origin and composition. Springer, Heidelberg, pp 297

Degens E (1965) Geochemistry of sediments: a brief survey. Prentice Hall, Upper Saddle River, NJ, pp 342

Garrett D (1996) Potash. Chapman Hall, Londan, pp 734

Garrett D (1998) Borates. Academic Press, London, pp 483

Horowitz A (1985) Sediment-trace element chemistry. Lewis, Chelsea, MI, pp 136

Nissenbaum A ed (1980) Hypersaline brines and evaporitic environments. In: Developments in sedimentology 28. Elsevier, Amsterdam, pp 270

Pye K (1987) Aeolian dust deposits. Academic Press, London, pp 334

Sly P (ed) (1986) Sediments and water interactions. In: proceed of 3rd international symposium interaction between sediments and water. Springer, Heidelberg, pp 517

Sonnefield P (1984) Brines and evaporates. Academic Press, London, pp 612

Twenhofel W (1961) Treatise on sedimentation. National Academic Science Publications, Washington, DC, pp 458

Summary

Geochemical Cycle at the Surface: Origins of Chemical Change and the Resulting Movement of Elements

Major Actors in Chemical Change: Water, Air, and Biological Action

Water is the major vehicle of change in the surface transformations of solid phases contained in rocks. Essentially geological forces bring rock masses to the surface and move them to heights above the surface of the ocean. Water in the form of rain affects erosion, which brings the solids from mountains into the sea and to a position of physical equilibrium. What goes up must come down. However, the true agent that allows this movement is the transformation of rock matter into new minerals that are also in equilibrium with the contacting solutions. Re-equilibration of silicate and other minerals with aqueous solutions under surface conditions results in the formation of new minerals, usually of small grain size, and removal of some material as dissolved ions. Surface alteration reduces the mechanical properties of rocks and they tend to move under the forces of gravity aided by abundant rainfall, producing mass transport to lower altitudes. If the slope is not great, the altered material remains in place and produces an ever deepening alteration profile.

The motor of change in surface reactions is the exchange of hydrogen ions for elements of cationic character, which are present in the initial rock minerals. Hydrogen ions diffuse into the minerals, especially silicates, and are exchanged for elements held by non-covalent bonding. New minerals are formed which contain the hydrogen. These are most often called clay minerals.

During the interaction process between water and rock minerals, the oxygen dissolved in surface water has an effect on many minerals where oxidation of metallic elements, especially Fe and Mn, occurs. When the oxidation state of an element changes the electronic equilibrium of its surrounding elemental

A. Bauer and B.D. Velde, *Geochemistry at the Earth's Surface*,
DOI 10.1007/978-3-642-31359-2,

environment changes and the initial mineral is often destabilized. This contributes to the formation of new mineral phases. Often, rock minerals which contain a high amount of Fe or Mn are transformed to oxide phases during the alteration process. A second source of change in oxidation state of elements in rocks is via bio-cycle, bacteria, and plant action. Elements such as Fe and S are used as a source of energy in the bio-cycle as they change oxidation state, by either oxidation or reduction of the electronic state. Solubilities of solid phases change greatly with the relevant oxidation state, and some tend to be adsorbed on surfaces from the aqueous medium upon change to a lower oxidation state.

The reactions due to mutual contact between aqueous solution and rock minerals are largely controlled by the residence time and the quantity of water in contact with rock materials. The more water present, the lower the amount of dissolved solids in the solution and the greater the tendency to dissolve more solids from rocks in order to reach chemical equilibrium. If water is of low abundance and remains in contact with the solids, the tendency to dissolve the rock material lowers. Thus not only is the presence of water important but also its relative abundance determines the rates of reaction and the amount of material affected by surface reactions. A second aspect of high importance is the contact time between water and rock minerals. The longer the time water is present in contact, the more surface reactions will occur. Therefore, in climates of regular and repeated rainfall more dissolution and eventual formation of alterite materials will occur. In areas of infrequent but abundant rainfall dissolution alteration will occur to a lesser extent. The more water the more reaction and the more erosion. The longer the contact time the more reaction and the more erosion.

The altered material is then of two sorts, elements dissolved in aqueous solution and the residue of alteration which is of small grain size and which is eventually transported in rivers as particulate material, solids, in the aqueous solutions. The alteration process occurs in two situations: (1) where the rainwater reaches the rock material forming new minerals and dissolves matter into the aqueous medium. A second and very important process occurs, (2) where plants interact at the surface with the alteration zone in soils. In this second surface environment the overall balance is one of retention of certain elements and particulate matter. In the surface bio-zone the retention of many elements is a major effect. There is of course loss of matter in this zone, but the general trend is one of retention and conservation. The stabilization of altered material in this area is of primordial importance. The pattern of retention is a general trend, but some elements are retained more strongly than others and the effect in the bio-zone on retention or release of elements of minor abundance is crucial. Minor elements are extremely important to the life cycles of plants, and especially those affecting humans. Some elements are beneficial while others are difficult to handle such as Cu, As, Pb, Zn, U, and so forth.

Geochemical Principles

Surface geochemistry deals essentially with natural processes of chemical interaction and the balance of the different elements relative to the source materials, rocks and minerals of high temperature origin. However one must consider the processes, which govern elemental abundance as an overriding trend determined as a natural chemical balance but in cases of human activity the normal system can be overloaded with one or several elements, which are usually of minor abundance in alteration sequences. In some cases of heavy overloading in surface environments with minor elements, the overall chemical reactions and forces can be modified significantly. Here it is necessary to understand the normal trends and equilibria which will tend to correct local, in time and space, elemental overloads given enough time under normal conditions of chemical interaction.

One major change for many elements of minor abundance is that of being initially incorporated within a mineral or crystalline phase, formed at high pressure and temperature, to being essentially excluded from the internal mineral structures of new phases and hence being subject to changing chemical conditions such as pH and aqueous solution dilution at the surface. Many elements of minor abundance are found in specific mineral phases at high temperature, such as Pb, Cd, and other heavy elements where they occur in sulfides and oxides of low abundance in rocks, whereas they are not present within individual mineral phases at the surface, but they are found on surfaces of minerals stable at earth surface conditions. This instead results at the same time in greater mobility and greater instability in their mineral or phase attachment. Minor elements migrate more easily under surface conditions and are at times included in new mineral phases formed under changing surface conditions where new phases can occur.

Two actions are essential: (1) the incorporation of elements into new mineral phases derived from other different preexisting phases as alteration redistributes the elements under conditions of surface alteration and (2) the adsorption of elements from solution on the surfaces of the new phases that are transported with aqueous solutions. The altered materials are twofold: dissolved ions in solution and particulate matter found in alterites and soils or suspended in moving aqueous media. The reorganization of rock elements involves incorporation or binding of the different elements into or onto the solid phases where they remain relatively stable on their transport path. The new phases are usually oxides and silicates, which contain almost exclusively major elements. These phases also have the capacity to adsorb elements onto their surfaces where they are temporarily part of the solid transport movement and the dispersion in the aqueous phase. There are three types of situations in which elements of minor abundance can be found subsequently: within minerals, chemisorbed on minerals or as free/complexed ions in solution. The chemisorption state is governed by the types of solid phases present and the chemical constraints in the aqueous solution.

Some general principles can be mentioned. First two types of ions are present in solutions: those negatively charged (anions) and those positively charged (cations).

As a first approximation such ions will be attracted to sites on solids of opposite charge on crystals, or large molecular compounds where net charge is present (i.e., uncompensated charges on the surface or unfulfilled charges on their surface or surfaces where inter-crystalline substitutions leave residual charges on the crystal surface). The best known and studied phenomenon is that of cation attraction to negatively charged surfaces on solids (permanent charge). These surface charges are diffuse on the mineral surface. They originate from ionic substitution of elements of different valence within the mineral leaving a residual and dispersed charge on the surface oxygens of the structure. Cations are attracted to these surfaces in relation to their charge density (charge divided by ion radius). The higher its charge density the more strongly an ion will be attracted by electrostatic forces. Another possibility is diffusion of solutes into a solid, where they replace matrix ions within the mineral structure. A third possibility is the chemisorption of ions in solution on sites at the surface of the mineral at variable charge surface sites where the major contributions to accumulation of ions are chemical and electrostatic interactions between ions in solution and ions with unsatisfied charges due to incomplete structural formation. These are edge sites where either cation or anion elements in the mineral structure are not completely chemically bonded. The extents of ion accumulation can be described by mass-law equations corrected for electrostatics and are therefore governed by the relative abundance of the involved solutes in solution. The presence of an anion on an exchangeable site depends on its relative attraction, its inherent charge/diameter ratio, and its relative abundance in the surrounding aqueous solution compared to other anions in solution as is the case for cations.

The laws of mass action control the attraction and fixation of elements from solution at the various ion attracting sites of clay minerals. Hydrogen ions are not normally attracted to the permanent charge sites (diffuse surface charge) except at low pH values where H_3O^+ ions are present in sufficiently high concentrations to compete with cations for adsorption sites. However on variable charge sites (incomplete structural bonding) pH is a major factor in determining the presence of cations (replaced by hydrogen ions) or anions (replaced by hydroxyl ions).

New minerals formed at the earth's surface by alteration processes normally do not contain elements of minor abundance. Silicates formed at the surface tend to contain within their structure few minor elements. They are composed essentially of Si, Al, and O atoms. Minor amounts of Fe and Mg with some K and Ca ions are the major constituents of these surface mineral phases. However oxides of Fe and Mn can contain other elements, especially transition metal ions. Phosphate minerals formed at the surface frequently contain heavy metal ions such as Pb, As, Bi, and others. The carbonate and sulfate minerals formed at the surface under conditions of alteration contain few elements other than the major constituents. All of these minerals have active surfaces where elements are chemisorbed.

Chemical Trends of Alteration

The basic geochemical principal in alteration processes at the surface of the earth is one of the loss of minor elements from the high temperature minerals as they form new phases. The dominant minerals formed are silicates and oxides in the stages of transformation from high temperature minerals into those stable at the surface. Sulfide and phosphate minerals in rocks, for example, are destabilized and the oxoanion portion becomes a dissolved ion in aqueous solution. The alteration process is one of chemical simplification: complex associations of minerals and phases are reduced to more simple minerals and phases and the solid solution or substitution of different elements within minerals is greatly reduced. In surface minerals, the amount of variability of elements in different minerals is also greatly reduced. Many mineral phases are lost from the mineral types initially present and major as well as minor elements are moved into different phases or in the majority of cases they are dissolved into aqueous solution from where they may or may not become attached to the minerals stable at the surface. The two types of minerals stable under surface alteration conditions are aluminous silicates (clay minerals) and oxyhydroxides of iron and manganese. The overall effect of alteration chemistry is that of a transfer of alkali and alkaline earth ions to the solution (Na, Mg, Ca, and eventually K). However the dominant clay minerals in temperate climate alteration tend to retain potassium, while sodium is almost systematically released. This is why the sea is salty (NaCl) and why plants, which need alkali ions for their metabolism, are capable of using potassium in their metabolic cycles on land areas and sodium in oceanic or peri-ocean environments. Animals never understood the change in chemistry between seawater and land-based alterites and maintain a need for sodium instead of potassium, which is abundant in terrestrial plant materials. The remaining major elements (Al, Si, Fe, and Mn) form either silicates or oxides. Minor elements are rarely found within silicates stable at the surface and only some elements are found in the complex oxyhydroxide forms of Fe and Mn (Co, Ni, Cu, Zn, Cr, Cd, and Pb can be found on internal sites of the crystal structures of goethite, a hydrous ferric mineral). Gradually, as alteration intensity increases, the stable minerals change to oxyhydroxides of Al and Fe, silica being gradually lost to the aqueous solution. The correlative increase of Al and Fe under weathering conditions of various intensities illustrates the overall tendency to form surface deposits of these elements.

The various stages of weathering are characterized by the transformation of most minerals and the resistance of some to transformation. Notably quartz, a form of pure silica, is little affected by surface chemistry and present in alterites, soils, and sediments of various zones of alteration. Other minerals, such as zirconates and Fe–Ti oxides, resist alteration in temperate climate alteration, and are found in the transported materials of streams and rivers in an essentially un-altered form. The chemical abundance relations of Fe and V, Cr, Ti, Zn in sedimentary material of rivers and streams are typically related to the abundance in initial high temperature minerals that have not reacted to alteration at the surface and are transported in the more coarse size fraction of particulate matter.

Under conditions of stronger alteration one finds that heavy metals are associated with silicate clays where these elements are attached chemically to the clay surfaces instead of being within the clay structure. The presence of different minor elements adsorbed onto silicate clays or iron oxides is subject to changes in concentration of these elements in altering and transport solutions as well as the pH of the solutions. Acid conditions tend to release the surface-bound cationic elements to the solution. The acidity of altering solutions is controlled by the plant regime pertaining at the surface of an alteration profile and the overall chemistry of the minerals present. Tropical forest plants for example tend to produce acid substrates in the leaf litter of the soil profiles. In tropical alteration facies minor elements are associated with iron oxide minerals (transition metals essentially) whereas the silicates and aluminum oxides have much less binding capacity to hold ionic elements.

Overall the chemistry of alteration is one of the reduction of chemical variability where minor elements are lost to solution and certain major elements are also lost, such as alkalis and alkaline earths. The elements lost are transported in mobile fluids, in soils, streams, and rivers, until they are lost from the transporting medium or eventually retained in seawater. Depending upon the minerals formed in the alteration process, which to a large extent depend upon the intensity of alteration (i.e., the amount of rainwater arriving on the rocks that are being altered) the new minerals formed can retain more or less of the elements released in the recrystallization process.

An important factor in the alteration process is the amount of rainfall, which affects the rock-alteration process. High input ensures low concentrations of elements in solution and high alteration effects. The overall tendency to transform rocks with eight major elements (Na, Mg, Al, Si, K, Ca, Fe, Mn) into material composed almost exclusively of Al and Fe oxyhydroxides is the overall chemical path of surface alteration. The majority of materials transformed and transported to a site of sedimentation do not go this far in the course of chemical simplification. However the major trends of concentration of Al and Fe are visible even in the early stages of alteration. The distribution of highly altered and little altered sedimentary material is largely a function of climate. Climate controls the undersaturation of altering solutions but also the type of plants present at the surface and these agents affect the chemistry of the surface solutions significantly. For example they control the pH of entering solutions, which is a determining factor in the retention of minor elements on or in alterite materials. The pH of soils, river waters, or sedimentary pore solutions is a determining factor concerning the presence of minor elements.

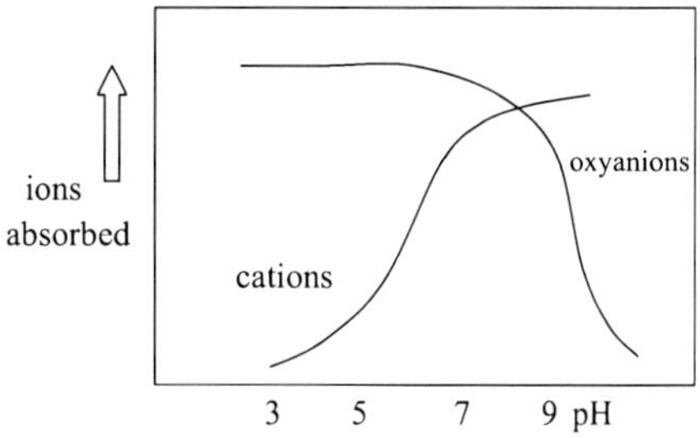

Fig. A.1 Illustration of the relative amounts of ions fixed on variable charge absorbers as a function of pH for cation and oxyanions species

Soils

The chemistry of alteration is to a large extent initiated and to a certain extent controlled by soil interactions, i.e., those with plants and rock alteration materials. Overall the chemical trends of soil–plant interaction are those of uplift of certain elements into the surface root zone affected by plants. The major elements are potassium and silica. These elements are useful for plants (K) and stabilize siliceous clay minerals, which retain the potassium, and stabilize the soils in aggregates that hold capillary water. Also minor elements are brought to the surface by plant root action and incorporation into the organic matter of the plants; among these elements are P, Cu, Zn, and B. An element not found in significant abundance in rocks is nitrogen, which occurs in soils by being introduced from the atmosphere and converted by bacterial action, and subsequently moved by dissolution and oxyanion fixation onto clays as they move with transport media of streams and rivers to be eventually deposited in sediments. Carbon of course comes from the atmosphere and is cycled in the soil zone with some of it moved out of the soil in groundwater as organic matter. Nitrogen is also an element from the atmosphere cycled in soils. Soils then can accumulate elements from the atmosphere and fix them in the geochemical cycle of alteration and transport.

In soils, oxidation–reduction is engendered by bacterial action involving organic matter from the plant cover. This action can strongly affect the oxidation state of certain elements, such as Fe, As, S, Sb. Depending on the system this creates oxyanions or cations from the same elements of certain redox states. The behavior of these ions as they are attached to clays and other variable charge substrates is extremely important. Oxyanions are strongly attracted to substrate surfaces at low pH values (below 8 or so) whereas at higher solution pH values they are released to the solution. The inverse is the case for cations, which are fixed to substrates at pH values of about 4–5 and above with the majority being adsorbed at pH values above 7 (Fig. A.1).

If an element changes oxidation state significantly, it can become associated with oxygen ions forming an anion whereas in a more reduced state it is present as a cation. The effect is very important concerning the tendency of the element to be fixed to particulate material in solutions or to persist as a dissolved species in solution depending upon Eh and pH conditions. The same element can be a cation when in a low oxidation state and an anion when in a higher state of oxidation.

The tendency of ions to be retained in soils by their surface charge, positive or negative, is to a large extent a function of soil solution pH. This is especially important for minor elements that are not incorporated into mineral structures, silicates or oxides.

Movement of elements within soil-alteration profiles is twofold. In percolating solutions within soils alterite material is present either as dissolved material or to a lesser extent as fine-grained particulate material. Very much of the alterite material remains in place as solids, which are further modified at the surface in the soil zone. Essentially Ca, Mg, and Na, major elements, are lost to solutions in a dissolved form. Fe, Mn, Si, Al, and K are lost to a lesser extent and remain in the newly formed minerals. Phosphorous and sulfur are also largely lost to solutions in the alteration stages and in soils. Iron, manganese, and Al–Si complexes remain for the most part in the alterite material. At the surface of soils one can have cases of erosion when plants do not retain the soil material sufficiently or when the soil zone is destabilized physically due to a landslide of massive transport. These events can be significant in areas of high relief, or of much lower intensity under normal conditions of vegetation where material is displaced by creep mechanisms. However in arid climates transport and loss of fine material can occur regularly due to a lack of sufficient vegetation forming deposits of fine-grained material in stream beds or in lake bottoms (playas).

The pH of soils is largely a function of the climate or bedrock type and hence the type of plants present. In confider forests, and to a lesser extent in broad-leaved temperate climate forests, the pH of the soils is acidic to slightly acidic. Under prairie regimes the pH is basic. In tropical climate forests, broad-leaved evergreen forests, the pH is acidic. These regimes determine to a large extent the retention or release of minor elements such as heavy metals and transition metal elements. One would expect a higher loss of minor and heavy elements from acidic forest soils than from prairie soils. A general relationship exists between rainfall and soil pH for soils in the central United States (Jenny 1994), which will be reflected in the amount of minor elements retained as a function of climate of greater amplitude.

An extremely important factor in the chemistry of percolating solutions and those due to erosion is the content of organic matter in the solutions. Dissolved organic matter in soils plays an important role in the biogeochemistry of carbon, nitrogen, and phosphorus, in pedogenesis, and in the transport of pollutants in soils.

Transport of Alterite Materials

If water–rock interaction is the basic phase of surface geochemistry followed by plant–alterite interaction in soils, the transport of this altered material is a second extremely important step in the geochemical process affecting surface materials. Here various terrains (bedrock of the alterite actions) and materials from different stages of alteration due to geographic differences (high mountain with low alteration zones and low lands with highly altered materials) are mixed and transported together to a zone of deposition and sedimentation. Some continental areas are formed of rather homogeneous rock materials, such as the central parts of the United States with its large areas of sedimentary rock substrates, but a certain portion of the materials in the draining rivers comes nonetheless from peripheral mountain ranges. The mixture of different types of alterite determined by the chemistry of the underlying rock substrate can create a strongly nonhomogeneous mixture of suspended and dissolved material in the transport medium. As the waters mix the one underlying factor common to the chemistry of the dissolved, adsorbed, and suspended materials is the pH of the river waters. As streams enter the transport system from different climate areas the pH can change significantly and with it the amount and types of adsorbed elements. One observation that tends to substantiate this point of view is the relation of Fe and Al relative to Si in the dissolved fractions of transporting river fluids. Low pH (<5) shows an increase in dissolved Fe and Al while higher pH (>7) shows a significant increase in dissolved Si. This reflects the presence of dissolved minerals in the transport media and will also affect the relative abundance of elements in solution or fixed on particulate minerals. If for example a river drains a mountain chain with conifer forests the initial materials will have much of the metallic minor elements in the dissolved phase while when this river reaches the plains at the base of the mountains the prairie soils, with higher pH, will tend to fix these elements on the particulate matter in suspension.

In general one finds correlations of transition metals among themselves for suspended particulate materials in rivers that are related to the iron oxyhydroxide carrier minerals, either on the surface or within the mineral structures. Reasonable correlations are found between major element alkaline earths and the accompanying minor elements such as the pair Ca–Sr in the dissolved fraction of river waters. However the relations between major element alkalis, K and Na, indicate that part of the material in the dissolved and suspended fractions is not in chemical equilibrium: some of the material in the particulate suspended material has not reacted during the transport period to attain chemical equilibrium.

The vector of wind transport is not one of chemical equilibrium. Materials brought from alterites and deposited at the surface are redeposited in various sites, either of similar chemical contexts or in sites of very different chemistry after wind transport for dust transported from peri-glacial zones. Here the alteration of the low temperature rock minerals is largely intact. It can also come from volcanic emissions that are deposited in various zones such as prairies in temperate climates, in semiarid zones, or sedimented in aqueous bodies such as lakes or the oceans. These new materials react to the environment that they find at the surface, usually one where biologic activity is important. The initial water–rock interaction is largely bypassed and the plants create the chemistry of the alteration zone. Some differences between water–rock interaction and plant–rock interaction can be seen. However, the overall pattern of chemical release upon phase change of the solids is more or less respected. The major elements Fe and Al increase in relative abundance with chemical transformations. Many other elements tend to be retained in the process of alteration by plant dominated chemical solutions, more so than in the water–rock interactions without plant interactions.

Materials transported by wind are very important, especially those of the recent past, for human agricultural activity in that large parts of the northern hemisphere agricultural areas in Europe, North America, and Asia are covered with layers of loessic deposits, usually fine-grained material of sedimentary rock origin and transported by wind events. These areas are reputed for their fertility, to a large extent where grasses have grown for long periods but one might also consider that the initial rock material present which is the substrate for the plant root systems is less depleted in minor elements which might well be an aid in forming strong plants and enhancing the productivity of those which were species selected for human consumption.

The transport phase of surface geochemical displacement, by wind or water, can be considered as one of inhomogeneity. Different mineral assemblages are mixed in waters that can change composition as the rivers of transport flow to the sea. It will be difficult to find consistent geochemical relations between water and solids. However the more homogeneous the river system the more the elements in suspension and dissolved in the aqueous solution will represent an equilibrium of chemical reactions. This is true for example in tropical climates where the intensity of alteration tends to flatten out the chemical differences that are normally present in temperate climate systems. Wind-borne materials are hardly reacted and will show a contrast with the material transported by rivers.

As transported material encounters saltwater, the effect of solution concentration of sodium in the seawater tends to strip the adsorbed elements and the adsorbed elements from the particulate matter. Salinity transfers much of the minor element content adsorbed on or in solids to solution by exchange with sodium ions.

Sedimentation

In this environment one finds a tendency to homogenization of the transported materials. The contrasts of transport materials which are put in the same system encounter chemical forces which tend to create reactions between the various minerals and adsorbed materials on and within the particulate matter and the dissolved materials in the sedimenting systems. The major agents of homogenization are the reactions due to reduction and oxidation produced by bacteria and other organisms, either in freshwater sediments or in saltwater sediments. The vehicles of change are the elements that can change oxidation state and which are relatively abundant, Fe and S. These two elements control the oxidation state of much of the sediment material and they determine the fate of many minor elements. Iron is a major component of the non-silicate materials of alterites where it is in the oxidized state, combined to a large state with hydroxyl groups. It occurs as an independent phase in sediments. This material binds many of the transition and heavy metals that have been released from solid phases in rocks in different high temperature phases both on its surface and within the structure of the hydroxides. Redox reactions can also affect the oxyanions of such elements as As and Bi. Oxidized iron is relatively insoluble while reduced iron is more soluble. Thus the substrate for many adsorbed ions would be lost as the iron is reduced but where it is found in an oxidized state, it can be concentrated through diffusion in the sediments themselves. However in the presence of sulfur, reduced iron forms a sulfide phase which accommodates many transition metal and heavy metal elements fixing them in solid and stable phases. Thus the element couple of iron and sulfur in fact concentrate minor elements in the sulfide phases at the surface under sedimentation conditions due to the reduction of iron in the presence of reduced sulfur. In sediments, freshwater and saline, one finds a layer of iron–sulfur enrichment in the upper levels of the sedimentary sequence. This material is of course subject to subsequent destruction if oxidizing conditions prevail. Burial rate and relative amounts of biologic activity determine the persistence or destruction of this chemical equilibrium. Since this chemical system is near the surface of sedimentation in freshwater deposits and along ocean–continent boundaries (salt marsh wetlands), the capture or release of minor elements is extremely important for the living species present, animal and others. Salt marsh and nearshore sediments are potential reservoirs of minor elements when the system is in an overall reducing environment but one of release of minor elements if it becomes oxidizing. Extensive reworking of salt marsh areas by human activity will tend to release these elements into the nearshore biosphere producing undesired effects on marine life, especially shell fish and fish life.

Evaporites

One normally thinks of sediments as particulate matter deposited when transport media have achieved low energy status, deltaic and nearshore deposits. This leaves much of the products of chemical transformation, alterites, in solution as dissolved ions, notably Ca, Mg, Na, P, and S among the major elements. However, a significant part of these dissolved elements find their way into sediments as solid phases, through evaporation or partial evaporation of transport water. Closed basin waters frequently produce evaporites or mineral accumulations due to the precipitation of dissolved materials from solution. Various types of evaporites form depending upon the specific geographic structure of the basin. They can form either in a continental setting or a peri-oceanic setting where initial "fresh" water is evaporated or seawater is evaporated. Basically the sequence of deposition is similar in most settings although local input from continental materials can change the sequence of material deposited or even create new types.

Initially the first formed mineral type is a carbonate, coming from dissolved CO_2 from the air and concentrations of Ca and eventually Mg in solution. Phosphates can be found at this stage of mineral formation. Also at times these early mineral precipitates are accompanied by magnesium silicate materials. The next stage is calcium sulphate precipitation. Under conditions of high evaporation the highly soluble salts can form such as sodium chloride among many others. Borates and nitrates are among the more rare materials found in evaporites that are formed in continental basins. Finally under conditions of strong evaporation the sodium can be combined with CO_2 in the solutions and Na silicates can form.

Late stage evaporites are sources of chemical elements that are easily extracted and have been used as sources of chemical agents for some time, especially NaCl. Many minor elements can be extracted from evaporite materials, hence their geochemical interest.

The general structure of surface geochemistry is one of chemical reorganization of the elements present in rocks, the transport of these elements and their deposition as sediments. The effect is to separate the major elements from high temperature rocks where they occur together in minerals into several types of new rock deposits where elements are more concentrated. First the low solubility quartz grains form sand deposits and then sandstone often of almost pure SiO_2. Dissolved calcium is extracted from transport and sedimentary solutions largely by plant and animal interaction but at times by pure precipitation to make carbonate rock. The alterite residue of new minerals, clays and oxides, is deposited to make what are called shales as sedimentary rock types. This material represents to a large extent the results of surface chemical interaction (hydration and oxidation) to form new minerals during the initial alteration process. Elements lost from the solid phases to solution are recovered as solids to a certain extent in evaporite deposits. Thus the starting and end products can be shown schematically as:

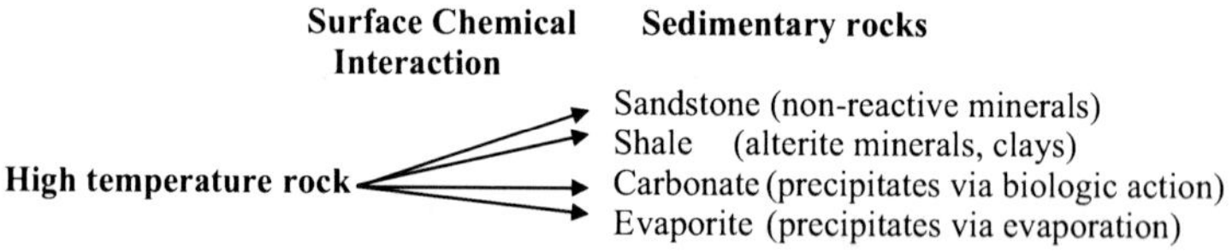

The process of alteration and transportation to the sediment phase is strongly influenced by surface chemical effects of living materials, plants and bacteria. These effects often control the pH and oxidation state of surface materials, in soils as well as in the initial stages of burial of sedimentary materials, either freshwater or saltwater.

Obviously the geochemistry of the surface depends on the chemical forces acting at a given point at the surface. Since geology is varied from one site to another (rock types of different chemical trends) and topography is also variable, as well as climate which is dependent upon the site on the globe being considered and the topography of the site, the chemistry of alteration will be quite varied from one point to another at the surface. This is a very important factor to consider when making a geochemical assessment of materials at the surface. Maximum variability occurs at the site of initial water–rock contact, as a function of rock type and rainfall intensity. Further the effect of plant regimes controls much of the upper level chemical conditions in an alteration profile. As materials move due to displacement by the vector of water movement, the chemical individuality of each site is gradually effaced to become more of an average value. The last stage of transport, sedimentation, mixes the varied materials into a more homogeneous (on a macro-scale) material. Thus a study of sediments compared to initial rock types loses much of the variability of surface geochemistry. Averages are averages. Understanding the complexity of each geochemical site is the basis of understanding the nature of the surface environment. The factor of aqueous solution undersaturation with respect to solids is a major factor in alteration. The more undersaturated a solution, the greater its altering capacity. Areas of low rainfall tend to have percolating fluids near saturation with the altered material and they show a lower tendency to interact to form new mineral phases. Within the same rock one can find different stages of alteration depending upon the flow of water through the rock itself, i.e., contrasting alterite mineral phases (see Meunier and Velde 2008). The geochemical (chemical) forces operative on a local scale can produce specific effects of retention or loss of elements found in the minerals of high temperature rocks. Surface geochemistry is more complex than that of rock geochemistry due to the greater number of variables operative which control the chemistry of the system. The basic difference from rock systems is the existence of fluid flow which produces an open system in a thermodynamic sense, one which changes it chemical forces as a function of time. Thus the basic premises of surface geochemistry are different from those of classical high temperature geochemistry, which treat materials occurring in more or less closed systems.

Bibliography

Adamo P, Violante P (2000) Weathering of rocks and neogenesis of minerals associated with lichen activity. Appl Clay Sci 16:229–256

Aiken W (2002) Global patterns: climate, vegetation and soils. University of Oklahoma Press, Norman, OK, 435 pp

Alfaro-De la Torre C, Tessier A (2002) Cadmium deposition and mobility in sediments of an acidic oligotrophic lake. Geochim Cosmochim Acta 66:3549–3562

Allard B (1995) Groundwater. In: Salbu B, Steinnes E (eds) Trace elements in natural waters. CRC, Boca Raton, FL, pp 151–172, 302 pp

Aller R, Michalopoulos P (1999) Tropical mobile belts as global diagenetic reactors. In: Armansson H (ed) Geochemistry of the Earth's Surface. Balkema, Rotterdam, pp 289–292, 574 pp

Alloway B (ed) (1995) Heavy metals in soils. Blackie Academic, London, 368 pp

Alpers CN, Hunerlach MP (2000) Mercury contamination from historic gold mining in California: U.S. Geological Survey Fact Sheet FS-061-00

Altschuler Z (1980) The geochemistry of trace elements in marine phosphorites. Part I: Characteristic abundances and enrichment. In: Bentor Y (ed) Marine phosphorites – geochemistry, occurrence and genesis. Society of Economic Paleontologists and Mineralogists special publication 29, Tulsa, OK, pp 19–30, 350 pp

Amato M, Migliozi A, Mazzoleni S (2004) Il sistema suolo vegetazione. Liguori Editore, Napoli, 350 pp

ANPA (2000) SEMINAT. Long-term dynamics of radionuclides in semi-natural environ-ments: derivation of parameters and modelling. Final report 1996-1999, European Commission-Nuclear Fission Safety Programme

Arai Y (2010) Ch 16 Arsenic and antimony. In: Hooda P (ed) Trace elements in soils. Wiley, Chichester, UK, pp 396–435, 596 pp

Arnfalk P, Wasay S, Tokunaga S (1996) A comparative study of Cd, Cr(III), Cr(IV) Hg and Pb uptake by minerals and soil materials. Water Air Soil Pollut 87:131–148

Aubert H, Pinta M (1977) Trace elements in soils. Elsevier, Amsterdam, 395 pp

Aubert D, Stille P, Probst A (2001) REE fractionation during granite weathering and removal by waters and suspended loads: Sr and Nd isotopic evidence. Geochim Cosmochim Acta 65:387–406

Audry S, Blanc G, Schäfer J, Chaillou G, Robert S (2006) Early diagenesis of trace metals (Cd, Cu, Co, Ni, U, Ma and V) in the freshwater reaches of a macrotidal estuary. Geochim Cosmochim Acta 70:2264–2282

Avena MJ, De Pauli CP (1998) Proton adsorption and electrokinetics of an Argentinean montmorillonite. J Colloid Interface Sci 202:195–204

DOI 10.1007/978-3-642-31359-2,

Baeyens B, Bradbury MH (1997) A mechanistic description of Ni and Zn sorption on Na-montmorillonite: Part I. Titration and sorption measurements. J Contam Hydrol 27:199–222

Bain D, Tait J (1977) Mineralogy and origin of dust fall on Skye. Clay Miner 12:353–355

Balistrieri L, Mirray J, Paul B (1994) The geochemical cycling of trace elements in a biogenic meromictic lake. Geochem Cosmochim Acta 58:3993–4008

Barkoudah Y, Henderson J (2006) Plant ashes from Syria and the manufacture of ancient glass: ethnographic and scientific aspects. J Glass Stud 48:297–321

Barnes C, Cochran J (1993) Uranium geochemistry in estuarine sediments: controls on removal and release processes. Geochim Cosmochim Acta 57:555–569

Bartelson G (1971) The chemical investigation of recent lake sediments from Wisconsin lakes and their interpretation. US Environmental Protection agency, 279 pp

Beauvais A, Mazaltarim D (1988) Etude des cuirasses latéritiques dans la région de Dembia-Zémio en Centrafrique, Pétrograpie, minéralogy et géochimie. Sci Géol Bull 41:47–69

Belzile N, Tessier A (1990) Interactions between arsenic and iron oxyhydroxides in lacustrine sediments. Geochim Cosmochim Acta 54:103–109

Bergseth H (1980) Selektivitat von Illit, Vermiculit und Smectit gegenüber Cu^{2+}, Pb^{2+}, Zn^{2+}, Cd^{2+}, und Mn^{2+}. Acta Agricult Scand 30(4):460–468

Besnus Y (1977) Etude géochimique comparative de quelques gisements supergènes de fer. Sci Géol Mem 47, 214 pp

Bickmore BR, Rosso KM, Cygan RT, Nagy KL, Tadanier CJ (2003) Ab initio determination of edge surface structures for dioctahedral 2:1 phyllosilicates: implications for acid-base reactivity. Clay Clay Miner 51:359–371

Birkland P (1999) Soils and geomorphology. Oxford University Press, New York, 430 pp

Black C (1957) Soil-plant relationships. Wiley, New York, 792 pp

Bluch B (1969) Introduction to sedimentology. In: Eglinton G, Murphy M (eds) Organic geochemistry. Springer, New York, pp 245–259, 829 pp

Boeglin J-L, Mazaltarim D (1989) Géochimie, degrés d'évolution et lithodépendance des cuirasses ferrugineuses de la région de Gaoua au Birkina Faso. Sci Géol 42:27–44

Bonifacio E, Zanini E, Boerro V, Franchini-Angela M (1997) Pedogenesis in a soil catena on serpentinite in north-western Italy. Geoderma 75:33–51

Bonneau M, Souchier B (1979) Constituants et propriétés du sol. Masson, Paris, 455 pp

Borrman B, Wang D, Bormann F, Benoit G, April R, Snyder M (1998) Rapid plant-induced weathering in an aggrading experimental ecosystem. Biogeochemistry 43:129–155

Bouchez J, Gaillardet J, France-Lanord C, Maurice L, Dutra-Maia D (2011) Grain size control of river suspended sediment geochemistry: clues from Amazon River depth profiles. Geochem Geophys Geosyst 12:24

Boujo A, Faye B, Gigot C, Lucas J, Manvitt H, Monciardini C, Prevot L (1980) The early Eocene of Lake Guires (Western Senegal) – reflections on some characteristics of phosphate sedimentation in Senegal. In: Bentor Y (ed) Marine phosphorites – geochemistry, occurrence and genesis. Society of Economic Paleontologists and Mineralogists special publication 29, Tulsa, OK, pp 207–213, 350 pp

Bourg A (1983) Role of fresh water/sea water mixing on trace metal adsorption phenomena. In: Wong C, Boyle E, Bruland K, Burton J, Goldberg E (eds) Trace elements in sea water. Plenum, New York, pp 195–208, 918 pp

Bourg A (1995) Speciation of heavy metals in soils and groundwater and implications of their natural and provoked mobility. In: Salmons W, Förstner U, Mader P (eds) Heavy metals – problems and solutions. Springer, Berlin, pp 19–32, 412 pp

Bradbury M, Baeyens B (2005) Modelling the sorption of Mn(II), Co(II), Ni(II), Zn(II), Cd(II), Eu (II), Am(III), Sn(IV), Th(IV), Np(V) and U(VI) on montmorillonite: linear free energy relationships and estimates of surface binding constants for some selected heavy metals and actinides. Geochim Cosmochim Acta 69:875–892

Brady PV, Cygan RT, Nagy KL (1996) Molecular controls on kaolinite surface charge. J Colloid Interface Sci 183:356–364
Brady N, Weil R (2002) The nature and properties of soils, 13th edn. Prentice Hall, Upper Saddle River, NJ, 960 p
Braitsch O (1971) Salt deposits: their origin and composition. Springer, Berlin, 297 pp
Brantley S, White A (2009) Approaches to modelling weathered regoliths. In: Oelkers E, Schott J (eds) Thermodynamics and kinetics of water–rock interaction, vol 70, Reviews in mineralogy and geochemistry. Mineralogical Society of America, Washington, DC, pp 435–464, 569 pp
Brantley S, Kulbicki J, White A (eds) (2008) Kinetics of water-rock interaction. Springer, New York, 833 pp
Braun J-J, Pagel M, Herbillon M, Rosin C (1993) Mobilization and redistribution of REE and thorium in a syenite laterite profile. Geochim Cosmochim Acta 57:4419–4434
Braun J-J, Viers J, Dupré B, Polve M, Ndam J, Muller J-P (1998) Solid/liquid REE fractionation in the lateritic system of Goyoum, East Cameroon; The implication for the present dynamics of the soil covers in the humid tropical regions. Geochim Cosmochim Acta 62:273–299
Brinkman F, Iverson W (1975) Chemical and bacterial cycling of heavy metals in estuarine systems. In: Church TC (ed) Marine chemistry in the coastal environment. ACS Symposium 18. American Chemical Society, Washington, DC, pp 319–342, 710 pp
Broadley M, Bowen H, Cotterill H, Hammond J, Meacham M, Mead A, White P (2004) Phylogenic variation in the shoot mineral concentration of angiosperms. J Exp Bot 56:321–336
Brody N, Weil R (2008) Nature and properties of soils. Prentice Hall, Upper Saddle River, NJ, 965 pp
Brookins DG (1988) Eh-pH diagrams for geochemistry. Springer, New York
Brookins D (1989) Aqueous geochemistry of rare earth elements, Ch 8. In: Lipin B, McKay G (eds) Geochemistry and mineralogy of rare earth elements, Reviews in mineralogy. Mineralogical Society of America, Washington, DC, pp 201–225, 348 pp
Brüchert V (1998) Early diagenesis of sulphur in estuarine sediments: the role of sedimentary humic and fulvic acids. Geochim Cosmochim Acta 62:1567–1586
Bruggeman C, Maes N, Christiansen BC, Stipp SLS, Breynaert E, Maes A, Regenspurg S, Malstrom ME, Liu X, Grambow B, Schaefer T (2012) Redox-active phases and radionuclide equilibrium valence state in subsurface environments-new insights from 6th EC FP IP FUNMIG. Appl Geochem 27:404
Bruque S, Mozas T, Rodrigues A (1980) Factors influencing retention of lanthanide ions by montmorillonite. Clay Miner 15:413–420
Bryant J, Dixon J (1964) Clay mineralogy and weathering of a red-yellow podzolic soil from quartz mica schist in the Alabama piedmont. In: Ingerson E (ed) Clays and clay minerals, Monograph 19. Pergamon, New York, pp 509–521, 691 pp
Buckman H, Brady N (1969) The nature and properties of soils. Macmillan, New York, 651 pp
Byrne R, Lee J, Bingler L (1991) Rare earth complexation by $PO^{3-}{}_4$ ions in aqueous solution. Geochim Cosmochim Acta 55:2729–2735
Caillaud J, Proust D, Philippe S, Fontaine C, Fialin M (2009) Trace metals distribution from a serpentinite weathering at the scales of the weathering profile and its related weathering microsystems and clay minerals. Geoderma 149:199–208
Caquineau S, Gaudicher A, Gomes L, Magonthier M-C, Chatenet B (1998) Saharan dust: clay ratio as a relevant tracer to assess the origin of soil-derived aerosols. Geophys Res Lett 25:983–986
Carignan R, Nriagu J (1985) Trace metal deposition and mobility in the sediments of two lakes near Sudbury, Ontario. Geochim Cosmochim Acta 49:1753–1764
Carignan R, Tessier A (1988) The co-diagenesis of sulphur and iron in acid lake sediments in southwestern Quebec. Geochim Cosmochim Acta 52:1179–1188
Catalano JG, Brown GE (2005) Uranyl adsorption on montmorillonite: evaluation of binding sites and carbonate complexation. Geochim Cosmochim Acta 69(12):2995–3005
Catt J (1988) Loess: its formation, transport and economic significance. In: Lerman A, Meybeck M (eds) Physical and chemical weathering in geochemical cycles. Kluwer Academic, Dordrecht, pp 113–142, 375 pp

Chaillou G, Schäfer J, Anschutz P, Lavaux G, Blanc G (2003) The behaviour of arsenic in muddy sediments of the Bay of Biscay (France). Geochim Cosmochim Acta 67:2993–3003

Chester R, Nimmo M, Keyse S (1996) The influence of Saharan and middle eastern desert-derived dust on the trace metal composition of Mediterranean aerosols and rainwaters: and overview. In: Guerzoni S, Chester R (eds) The impact of desert dust across the Mediterranean. Kluwer, Dordrecht, pp 253–273, 278 pp

Chesworth W, Dejou J, Larroque P (1981) The weathering of basalt and relative mobility's of major elements at Belbex, France. Geochim Cosmochim Acta 45:1235–1241

Chisholm-Brause C, Conradson SD, Buscher CT, Eller PG, Morris DE (1994) Speciation of uranyl sorbed at multiple binding sites on montmorillonite. Geochim Cosmochim Acta 58(17):3625–3631

Church T, Sarin M, Fleisher M, Ferdelman G (1996) Salt marshes: an important coastal sink for dissolved uranium. Geochim Cosmochim Acta 60:3879–3887

Church T, Sommerfield C, Velinsky D, Point D, Benoit C, Amouroux D, Plass D, Donard O (2006) Marsh sediments as records of sedimentation, eutrophication and metal pollution in urban Delaware Estuary. Mar Chem 102:72–95

Church T, Fontaine C, Sedwick P, Velde B (2010) Diagnostic source minerals of eolian transport in Bermuda dust. In: Proceedings of Ocean Sciences Meeting 2010, Portland, OR

Collignon C (2011) Facteurs controllant l'altération biologique des minéraux dans l'rhizosphère des écosysètems forestieres. Thesis, Univ Nancy, France, 450 pp

Coppin F, Berger G, Bauer A, Castet S, Loubet M (2002) Sorption of lanthanides on smectite and kaolinite. Chem Geol 182:57–68

Cornelis J-Y, Ranger J, Eserentant I, Delvaux B (2010) Tree species impact the terrestrial cycle of silicon through various uptakes. Biogeochemistry 97:231–245

Correns C (1969) The discovery of the chemical elements, the history of geochemistry. In: Wedephol H (ed) Definitions of geochemistry, vol 1, Handbook of geochemistry. Springer, Berlin, pp 1–11

Dähn R, Scheidegger AM, Manceau A, Schlegel ML, Baeyens B, Bradbury MH, Morales M (2002a) Neoformation of Ni phyllosilicate upon Ni uptake by montmorillonite. A kinetics study by powder and polarized extended X-ray absorption fine structure spectroscopy. Geochim Cosmochim Acta 66:2335–2347

Dähn R, Scheidegger AM, Manceau A, Curti E, Baeyens B, Bradbury MH, Chateigner D (2002b) Th uptake on montmorillonite: a powder and polarized extended X-ray absorption fine structure (EXAFS) study. J Colloid Interface Sci 249:8–21

Dähn R, Scheidegger AM, Manceau A, Schlegel ML, Baeyens B, Bradbury MH, Morales M (2003) Structural evidence for the sorption of metal ions on the edges of montmorillonite layers. A polarized EXAFS study. Geochim Cosmochim Acta 67:1–15

Daux V, Vtobidirt J, Hemond C, Petit J-C (1994) Geochemical evolution of basaltic rocks subjected to weathering: fate of the major elements, rare earth elements and thorium. Geochim Cosmochim Acta 58:4941–4954

Dean W, Tung A (1974) Trace and minor elements I anhydrite and halite, Supai formation (Permian) East-Central Arizona. In: Coogan A (ed) Fourth symposium on salt, Northern Ohio Geological Society, pp 287–301, 530 pp

Decampo D (2010) The geochemistry of continental carbonates. In: Alonso-Zarza A, Tanner L (eds) Carbonates in continental settings: geochemistry; diagenesis and applications, vol 62, Developments in sedimentology. Elsevier, Amsterdam, pp 1–45, 319 pp

Decarreau A (1981) Mesure expérimentale des coefficients de partage solide/solution pour les éléments de transition A^{2+} dans les smectites magnésiennes (A = Ni, Co, Zn, Fe, Cu, Mn). C R Acad Sci Paris 292:459–462

Decarreau A (1985) Partitioning of divalent transition elements between octahedral sheets of trioctahedral smectites and water. Geochim Cosmochim Acta 49:1537–1544

Degens E (1965) Geochemistry of sediments: a brief survey. Prentice Hall, Englewood Cliffs, NJ, 342 pp

Dethier D (1986) Weathering rates and chemical flux from catchments in the Pacific Northwest. In: Coleman S, Dethier D (eds) Rates of chemical weathering of rocks and minerals. Academic, Dordrecht, pp 503–530, 630 pp

Dixon J, Weed S (eds) (1996) Minerals in soil environments. Soil Science Society of America, Madison, WI, 997 pp

Dosseto A, Bourdon B, Gaillardet J, Allègre C, Filizola N (2006) Time scale and conditions of weathering under tropical climate: study of the Amazon with U-series. Geochim Cosmochim Acta 70:71–89

Drever J (1982) The geochemistry of natural waters. Prentice Hall, Englewood Cliffs, NJ, 436 pp

Dupré B, Gaillardet J, Rousseau D, Allègre C (1996) Major and trace elements of river-borne material: the Congo Basin. Geochim Cosmochim Acta 60:1301–1321

Egli M, Fitze P, Mirabella A (2001) Weathering and evolution of soils formed on granitic, glacial deposits: results from chronosequences of Swill alpine environments. Catena 45:19–47

Ehrlich H, Newman D (2009) Geomicrobiology. CRC, New York, 606 pp

Evans L, Barabash S, Lumsdon D, Gu X (2010) Application of chemical speciation modelling to studies on toxic element behaviour in soils. In: Hooda P (ed) Trace elements in soils. Wiley, Chichester, UK, pp 210–214, 596 pp

Fehrenbacher J, White J, Ulrich H, Odell R (1965) Loess distribution in South-eastern Illinois and South-western Indiana. Soil Sci Soc Proc 29:566–579

Feng J-L (2010) Behaviour of rare earth elements and yttrium in ferromanganese concretions, gibbsite spots and the surrounding terra rosa over dolomite during chemical weathering. Chem Geol 271:112–132

Ferdelman T, Church T, Luther G (1991) Sulphur enrichment of humic substances in a Delaware salt marsh sediment core. Geochim Cosmochim Acta 55:979–988

Ferrage E, Tournassat C, Rinnert E, Lanson B (2005) Influence of pH on the interlayer cationic composition and hydration state of Ca-montmorillonite: analytical chemistry, chemical modelling and XRD profile modelling study. Geochem Cosmochim Acta 69:2797–2815

Fitzgerald W, Gill G, Hewitt A (1981) Air–sea exchange of mercury. In: Wong C, Boyle E, Bruland K, Burton J, Goldberg E (eds) Trace metals in sea water. Plenum, New York, pp 297–316, 920 pp

Fleet A (1984) Aqueous and sedimentary geochemistry of the rare earth elements. In: Henderson P (ed) Rare earth element geochemistry. Elsevier, Amsterdam, pp 343–371, 575 pp

Föllmi K (1966) The phosphorous cycle, phosphorogenesis and marine phosphate-rich deposits. Earth Sci Rev 40:55–124

Fontenaud A (1982) Les faciès d'altération supergène des roches ultrabasiques. Etude de deux massifs de lherzolites (Pyrénes France). Thesis, Univ Poitiers, 103 pp

Förstner U (1986) Metal speciation in solid wastes – factors affecting mobility. In: Lardner L (ed) Speciation of metals in water, sediment and soil systems, Lecture notes in earth sciences. Springer, Berlin, pp 13–27, 189 pp

Foth D (1990) Fundamentals of clay science. Wiley, New York, 360 pp

Frenet M (1981) The distribution of mercury, cadmium and lead between water and suspended matter in the Loire estuary as a function of the hydrological regime. Water Res 15:1343–1350

Freyssinet P (1991) Géochimie et minéralogie des latérites du Sud-Mali. Doc BRGM 203, Editions BRGM, 290 pp

Froelich PN, Klinkhammer GP, Bender ML, Luedtke NA, Heath GR, Cullen D, Dauphin P (1979) Early oxidation of organic matter in pelagic sediments of the eastern equatorial Atlantic: suboxic diagenesis. Geochim Cosmochim Acta 43:1075–1090

Funare L, Vailonis A, Strawn D (2005) Polarized XANES and EXAFS spectroscopic investigation into copper (II) complexes on vermiculite. Geochim Cosmochim Acta 69:5219–5231

Fuss C, Driscoll C, Johnson C, Petras R, Fahey T (2011) Dynamics of oxidized and reduced iron in a northern hardwood forest. Biogeochemistry 104:103–119

Gabelman J (1977) Migration of uranium and thorium – exploration significance, Studies in geology, no 3. American Association of Petroleum Geologists, Tulsa, OK

Gac J-Y (1980) Géochimie du basin du Lac Tchad: Bilan de l'altération, de l'érosion et de la sedimentation. Mem ORSTOM, 250 pp

Gaiero D, Prost J-L, Depertris P, Didart S, Leleyter L (2003) Iron and other transition metals in Patagonian riverborne and windborne materials: geochemical control and transport to the South Atlantic Ocean. Geochim Cosmochim Acta 67:3603–3623

Gaillardet J, Dupre B, Louvat P, Allegre CJ (1999) Global silicate weathering and CO2 consumption rates deduced from the chemistry of large rivers. Chem Geol 159:3–30

Gaillardet J, Viers D, Dupré C (2004) Trace elements in river waters. In: Holland H, Turkian K (eds) Treatise on geochemistry, vol 5. Elsevier, Oxford, pp 225–260, Ch 509

Gao Y, Mucci A (2001) Acid base relations, phosphate and arsenate complexation and the competitive adsorption at the surface of goethite in 0.7 M NaCl solution. Geochim Cosmochim Acta 65:2361–2378

Garcia-Garcia S, Wold S, Jonsson M (2009) Effects of temperature on the stability of colloidal montmorillonite particles at different pH and ionic strength. Appl Clay Sci 43:21–26

Garrels R (1960) Mineral equilibria at low temperature and pressure. Harper, New York, 535 pp

Garrett D (1996) Potash. Chapman Hall, New York, 734 pp

Garrett D (1998) Borates. Academic, New York, 483 pp

Geckeis H, Rabung T (2002) Solid–water interface reactions of polyvalent metal ions at iron oxide–hydroxide surfaces. Marcel Dekker, New York

Geckeis H, Lützenkirchen J, Polly R, Rabung T, Schmidt M (2013) Mineral–water interface reactions of actinides. Chem Rev 113(2):1016–1062

Gibbs RJ (1967) The geochemistry of the Amazon River basin. Part 1. The factors which control the salinity and the composition and concentration of the suspended solids. Geo Soc Am Bull 78:1203–1232

Gill R (1996) Chemical fundamentals of geology. Chapman Hall, London, 290 pp

Giresse P (1980) The Maastrichtian phosphate sequence on the Congo. In: Bentor Y (ed) Marine phosphorites – geochemistry, occurrence and genesis. Society of Economic Paleontologists and Mineralogists special publication 29, Tulsa, OK, pp 193–205, 350 pp

Glauccum R, Prospero J (1980) Saharan aerosols over tropical north Atlantic – mineralogy. Mar Geol 37:295–321

Goldschmidt V (1954) Geochemistry. Oxford University Press, Oxford, 730 pp

Gorsline DS (1984) A review of fine-grained sediment origins, characteristics, transport and deposition. In: Stow D, Piper D (eds) Fine grained sediments: deepwater processes and facies. Geological Society special publication no 15, London

Goudie A, Middleton N (2001) Saharan dust storms: nature and consequences. Earth Sci Rev 56:179–204

Graham R, Southard A (1983) Genesis of Vertisol and associated Mollisol in Northern Utah. Soil Sci Soc Am J 54:1682–1690

Granger H, Warren C (1974) Zoning in the altered tongue associated with roll-type uranium deposits. In: Proceedings of International Atomic Energy Agency: formation of uranium ore deposits, pp 185–216

Grauby O (1993) Nature et etendue des solutions octaedriques argileuses. Approche par synthese minerale. Ph.D., Universite de Poitiers, Poitiers

Grauby O, Petit S, Decarreau A, Barronnet A (1993) The beidellite-saponite series: an experimental approach. Eur J Miner 5:623–635

Greathouse J, Stellalevinsohn H, Denecke M, Bauer A, Pabalan R (2005) Uranyl surface complexes in a mixed-charge montmorillonite Monte Carlo computer simulation and polarized XAFS results. Clay Clay Miner 53:278–286

Grim R (1953) Clay mineralogy. McGraw Hill, New York, 375 pp

Gröger J, Proske U, Hanebuth T, Hamer K (2011) Cycling of trace metals and rare earth elements (REE) in acid sulfate soils in the Plain of Reeds, Vietnam. Chem Geol 288:162–177

Gueddari M (1984) Géochimie et thermodynamique des évaporites continentales. Sci Géol Mem 76, 143 pp

Guegueniat P (1986) Comportement géochimique des éléments à l'état de traces dans l'estuaire de la Seine. In: Colloque La Baie de Seine IFREMER Actes de Colloques 4, pp 247–290, 531 pp
Guieu C, Thomas A (1996) Saharan aerosols from the soil to the ocean. In: Gluerzoni S, Chester R (eds) The impact of desert dust across the Mediterranean. Kluwer, Dordrecht, pp 207–216, 287 pp
Gupta V, Miller JD (2010) Surface force measurements at the basal planes of ordered kaolinite particles. J Colloid Interface Sci 344:362–371
Hagedorn B, Cartwright I, Raveggi M, Maas R (2011) Rare earth element and strontium geochemistry of the Australian Victorian Alps drainage system: evaluating the dominance of carbonate vs. aluminosilicate weathering under varying runoff. Chem Geol 284:105–126
Hartmann E, Baeyens B, Bradbury M, Geckeis H, Stuympf T (2008) A spectroscopic characterisation and quantification of M(III) clay mineral outer-sphere complexes. Environ Sci Technol 42:7601–7606
Hayes KF, Katz LE (1996) Application of x-ray absorption spectroscopy for surface complexation modeling of metal ion sorption. In: Brady PV (ed) Physics and chemistry of mineral surfaces. CRC Press: Boca Raton, FL, p 147
Hayes KF, Traina SJ (1998) Metal speciation and its significance in ecosystem health. In: Huang PM (ed) Soil chemistry and ecosystem health. Soil Science Society of America, Madison, WI, pp 45–84 (SSSA special publication, 52)
Hayes MHB, MacCarthy P, Malcolm RL, Swift RS (1989) (eds) Humic substances II: in search of structure. Wiley-Interscience, New York
He Y, Li D, Velde B, Yang Y, Huang C, Gong Z, Zhang G (2008) Clay minerals in a soil chronosequence derived from basalt on Hainan Island China. Geoderma 148:206–212
Hennig C, Reich T, Dähn R, Scheidegger AM (2002) Structure of uranium sorption complexes at montmorillonite edge sites. Radiochim Acta 90(9–11):653–657
Herwitz S, Muhs D, Prospero J, Mahan S, Vaughan B (1996) Origin of Bermudan clay-rich Quarternary paleosols and their paleoclimatic significance. J Geophys Res 101:23389–23400
Hillier S (1995) Erosion, sedimentation and sedimentary origin of clays. In: Velde B (ed) Origin and mineralogy of clays. Springer, New York, pp 162–213, 335 pp
Hitchcock DR (1975) Biogenic contributions to atmospheric sulphate levels. In: Proceedings of the second national conference on complete water re-use, May. American Institute of Chemical Engineers and U.S. Environmental Protection Agency, Chicago, IL, 291 pp
Hodgeman C, Weast R, Selby S (1959) Handbook of chemistry and physics. Chemical Rubber, Cleveland, OH, pp 9–76
Hodson M, White P, Mead A, Broadley M (2005) Phylogenic variation in silicon compositions of plants. Ann Bot 96:1027–1046
Holloway J, Dahlgren R (2010) Geologic nitrogen in terrestrial biochemical cycling. Geology 27:567–570
Holmes C, Miller R (2004) Atmospherically transported elements and deposition in the Southeastern United States local or transoceanic? Appl Geochem 19:1189–1200
Holmgren G, Meyer M, Chaney R, Daniels R (1993) Cadmium, lead, zinc, copper and nickel in agricultural soils of the United States of America. J Environ Qual 22:335–348
Holser W (1966) Bromide geochemistry of salt rocks. In: Rau J (ed) Second symposium on salt, vol 1, Northern Ohio Geological Society, pp 248–263, 443 pp
Hong J, Calman W, Förstner U (1995) Interstitial waters. In: Salbu B, Steinnes E (eds) Trace elements in natural waters. CRC, Boca Raton, FL, pp 117–150, 302 pp
Hooda P (ed) (2010) Trace elements in soils. Wiley, Chichester, UK, 596 pp
Horowitz A (1985) Sediment-trace element chemistry. Lewis, Chelsea, MI, 136 pp
Howarth R, Jensen H, Marino R, Postma H (1995) Transport to and processing of P in near-shore and oceanic waters. In: Tiessen H (ed) Phosphorous in the global environment; SCOPE 54. Wiley, Chichester, UK, pp 323–346, 735 pp
Huang P, Gobran G (eds) (2005) Biogeochemistry of trace elements in the rizosphere. Elsevier, Amsterdam, 465 pp

Huang C, Gong Z, He Y (2004) Elemental geochemistry of a soil chronosequence on basalt on northern Hainan Island, China. Chin J Geochem 23:245–254

Huang J-H, Iilgen G, Matner E (2011) Fluxes and budgets of Cd, Zn, Cu, Cr and Ni in a remote forested catchment in Germany. Biogeochemistry 103:59–70

Huerta-Diaz M, Tessier A, Carignan R (1998) Geochemistry of trace metals associated with reduced sulphur in freshwater sediments. Appl Geochem 13:213–233

Huertas FJ, Chou L, Wollast R (1998) Mechanism of kaolinite dissolution at room temperature and pressure: Part 1. Surface speciation. Geochim Cosmochim Acta 62(3):417–431

Huminicki D, Hawthorne F (2002) The crystal chemistry of phosphate minerals. In: Kohn M, Rakovan J, Hughes J (eds) Phosphates: geochemical, geo-biological and materials importance, vol 48, Reviews in mineralogy and geochemistry. Mineralogical Society of America, Washington, DC, pp 63–390, 740 pp

Ildefonse P (1978) Mécanismes de l'altération d'une roche gabbroique du massif du Pallet (Loire Atlantique). Thesis, Univ Poitiers, France, 142 pp

Isaure M-P, Manceau A, Geoffroy N, Laourdigue A, Tamura N, Marcus M (2005) Zinc mobility and speciation in soil covered in contaminated dredged sediment using micrometer-scale and bulk average X-ray fluorescence absorption and diffraction techniques. Geochem Cosmochim Acta 69:1173–1198

Isaure M-P, Sarret G, Harada E, Choi Y-E, Marcus M, Faraka S, Geoffry N, Pairis S, Susini J, Clements S, Manceau A (2010) Calcium promotes elimination as vaterite grains by tobacco trichomes. Geochim Cosmochim Acta 74:5817–5834

Jahan N, Guan H, Bestland E (2011) Arsenic remediation by Australian laterites. Environ Earth Sci 64:247–253

Jenny H (1994) Factors of soil formation. Dover, New York, 281 pp

Jensen M, Bateman A (1979) Economic mineral deposits. Wiley, New York, 593 pp

Jobbagy EG, Jackson RB (2004) The uplift of soil nutrients by plants: biogeochemical consequences across scales. Ecology 85:2380–2389

Johannesson K, Stetzenbach K, Hodge V (1997) Rare earth elements as geochemical tracers of regional groundwater mixing. Geochim Cosmochim Acta 61:3605–3618

Johnson C (1986) The regulation of trace element concentrations in river and estuarine waters contaminated with Acid mine drainage. The absorption of Cu and Zn on amorphous Fe oxyhydroxides. Geochim Cosmochim Acta 50:2433–2438

Jones B (1966) Geochemical evolution of closed basin water in the Western Great Basin. In: Rau J (ed) Second symposium of salt, Northern Ohio Geological Society, pp 181–193, 443 pp

Jones B, Decampo D (2004) Geochemistry of saline lakes. In: Holland H, Turkian K (eds) Treatise on geochemistry, vol 5. Elsevier, Oxford, pp 393–422

Jouanneau JM, Latouche C (1981) The Gironde estuary. In: Fürchtbauer H, Lisitzyn AP, Millerman JD, Seibold E (eds) Contribution to sedimentology. E. Schweizerbart'sche Verlagsbuchhandlung, Stuttgart, pp 1–115

Kabata-Pendias A, Pendias H (1992) Trace elements in soils and plants. CRC, Boca Raton, FL, 364 pp

Karakaya M, Karakaya N, Temel A (2011) Mineralogical and geochemical characteristics and genesis of the sepiolite deposits at Polatli Basin (Ankara, Turkey). Clay Clay Miner 59:286–314

Kauer N, Grafe M, Singh B, Kennedy B (2009) Simultaneous incorporation of Cr, Zn, Cd, and Pb in the goethite structure. Clay Clay Miner 57:244–250

Kawano M, Tomita K (1991) Dehydration and rehydration of saponite and vermiculite. Clay Clay Miner 39(2):174–183

Keene W, Montag J, Maben J, Southwell M, Leonard J, Church T, Moody J, Galloway J (2002) Organic nitrogen in precipitation over Eastern North America. Atmos Environ 36:4529–4540

Keilhack K (1975) The riddle of loess formation. In: Smalley I (ed) Loess: lithology and genesis, vol 26, Benchmark papers in geology. Dowden Hutchinson and Ross, Stroudsburg, PA, pp 47–50, 429 pp

Kerr S, Shafer M, Overdier J, Armstrong D (2008) Hydrologic and biochemical controls on trace element export from northern Wisconsin wetlands. Biogeochemistry 89:273–294

Khan M, Zaide A, Goel R, Musarrat J (eds) (2011) Biomanagement of metal-contaminated soils, vol 20, Environmental pollution. Springer, Dordrecht, 512 pp

King GM, Howes BL, Dacey JWH (1985) Short term end products of sulfate reduction: formation of acid volatile sulfides, elemental sulfur, and pyrite. Geochim Cosmochim Acta 49:1561–1

Kirpichikova T, Manceau A, Lanson B, Marcus M, Jacquet T (2003) Speciation and mobility of Zn, Cu and Pb in truck farming soil contaminated by sewage irrigation. J Phys Chem 107:695–698

Knecht M, Goransson A (2004) Terrestrial plants require nutrients in similar proportions. Tree Physiol 24:447–469

Koppi A, Edis R, Field D, Geering H, Klessa D, Cockayne D (1996) Rare earth element trends and cerium–uranium–manganese associations in weathered rock from Koongarra, Northern Territory, Australia. Geochim Cosmochim Acta 60:1695–1707

Koren R, Mezuman V (1981) Boron absorption by clay minerals using a phenomenological equation. Clay Clay Miner 29:198–204

Korshinski D (1959) Physicochemical basis of the analysis of the paragenesis of minerals. Consultants Bureau, New York, 143 pp

Kosmulski M (2009) pH-dependent surface charging and points of zero charge. IV. Update and new approach. J Colloid Interface Sci 337(2009):439–448

Kraepiel A, Chiffoleau J-F, Martin J-M, Morel F (1997) Geochemistry of trace metals in the Gironde estuary. Geochim Cosmochim Acta 61:1421–1436

Krauskopf K (1967) Introduction to geochemistry. McGraw Hill, New York, 721 pp

Kühn W (1982) Ausbreitung radioaktiver Stoffe im Boden. In: Arbeitsgemeinschaft für Umweltfragen (ed) Das Umweltgespräch. Tagungsprotokoll: Radioökologie-symposium vom 15/16, Okt. 1981, Univ. Stuttgart, pp 76–99

Kulik DA, Aja SU, Sinitzsyn VA, Wood SA (2000) Acid-base surface chemistry and sorption of some lanthanides on K+ saturated Marblehead illite: II. A multi-site-surface complexation modeling. Geochim Cosmochim Acta 64(2):195–213

Lanson B, Drits V, Gaillot A-C, Silvester E, Plançon A, Manceau A (2002) Structure of heavy-metal sorbed birnessite: Part I results from X-ray diffraction. Am Miner 87:1631–1645, 69:1173–1198

Lanson B, Marcus M, Farka S, Pafili F, Geoffroy N, Manceau A (2008) Formation of Zn-Ca phyllomanganate nanoparticles in grass roots. Geochim Cosmochim Acta 72:2478–2490

Lee S, Anderson PR, Bunker GB, Karanfil C (2004) EXAFS study of Zn sorption mechanisms on montmorillonite. Environ Sci Technol 38:5426–5432

Lemarchand E, Schott J, Gaillardet J (2005) Boron isotopic fractionation related to boron absorption on humic acid and the structure of surface complexes formed. Geochim Cosmochim Acta 69:3519–3533

Lemarchand E, Schott J, Gaillardet J (2007) How surface complexes impact boron isotope fractionation: evidence from Fe and Mn oxide experiments. Earth Planet Sci Lett 260:277–296

Leumbe Leumbe O, Bitom D, Tematio P, Temgoua E, Luca Y (2005) Etude des sols ferrallitiques à charactères andiques sur trachytes en zone de montagne humide tropical. Etude et Gestion des Sols 12:313–326

Lide D (1999) Handbook of chemistry and physics. CRC Press, New York

Lide D (2000) Handbook of chemistry and physics. CRC, Boca Raton, FL, pp 9–75

Lienemann C-P, Taillefert M, Pierret D, Gaillard J-F (1997) Association of cobalt and manganese in aquatic systems: chemical and microscopic evidence. Geochim Cosmochim Acta 61:1437–1446

Lindqvist-Reis P, Klenze R, Schubert G, Fanghänel T (2005) Hydration of Cm^{3+} in aqueous solution from 20 to 200°C. A time-resolved laser fluorescence spectroscopy study. J Phys Chem B109:3077–3083

Lipin B, McKay G (eds) (1989) Geochemistry of rare earth elements, vol 21, Reviews in mineralogy. Mineralogical Society of America, Washington, DC, 348 pp

Liu Y, Laird D, Barak P (1997) Release and fixation of ammonium and potassium under long-term fertility management. Soil Soc Am J 61:310–314

Liu G, Cai Y, O-Driscoll N (2012a) Environmental chemistry and toxicology of mercury. Wiley, Hoboken, NJ, 579 pp

Liu Z, Wang H, Hantoro W, Sathaimurthy E, Colin C, Zhao Y, Li J (2012b) Climatic and tectonic controls on chemical weathering in tropical Southeast Asia (Malay Peninsula, Borneo and Sumatra). Chem Geol 291:1–12

Lord C, Church T (1983) The geochemistry of salt marshes: sedimentary ion diffusion, sulphate reduction and pyritization. Geochim Cosmochim Acta 47:1381–1391

Loughnan F (1969) Chemical weathering of the silicate minerals. Elsevier, New York, 154 pp

Lowe D (1986) Controls on the rates of weathering and clay mineral genesis in airfall tephras: a review and New Zealand case study. In: Coleman SM, Dethier DP (eds) Rates of chemical weathering of rocks and minerals. Academic, New York, pp 265–319

Luther G, Church T (1992) An overview of the environmental chemistry of sulphur in wetland systems. In: Howarth R, Stewart J, Ivanov M (eds) Sulphur cycling on the continents, SCOPE. Wiley, Chichester, UK, pp 125–139, 386 pp

Lützenkirchen J, Preocanin T, Bauer A, Metz V, Sjöberg S (2012) Net surface proton excess of smectites obtained from a combination of potentiometric acid–base, mass and electrolyte titrations. Colloids Surf A Physicochem Eng Asp 412:11–19

Ma L, Jin L, Brantley S (2011) How mineralogy and slope affect REE release and fractionation during shale weathering in the Susquehanna/Shale Hills Critical Zone Observatory. Chem Geol 290:31–39

Mackie D, Boyd G, Tindale N, Westberry T, Hunter K (2008) Biogeochemistry if iron in Australian dust: from eolian uplift to marine uptake. Geochem Geophys Geosystems 9(3):1–24

Maher K, Bargar JR, Brown GE Jr (2012) Environmental speciation of actinides. Inorg Chem 52 (7):3510–3532

Manceau A, Schlegel M, Musso M, Sole V, Gauthier C, Petot S, Troelard F (2000) Crystal chemistry of trace elements in natural and synthetic goethite. Geochim Cosmochim Acta 21:3643–3661

Manceau A, Marcus M, Tamura N, Proux O, Geoffroy N, Lanson B (2004) Natural speciation of Zn at the micrometer scale in clayey soil using X-ray fluorescence adsorption and diffraction. Geochim Cosmochim Acta 68:2467–2483

Manceau A, Tommasso C, Rims S, Geoffroy N, Chataigner D, Schlegel M, Tisserand D, Marcus M, Tamura N, Chen Z (2005) Natural speciation of Mn, Ni, Zn at eh micrometer scale in a clayey paddy using X-ray fluorescence, absorption and diffraction. Geochim Cosmochim Acta 69:4007–4014

Manceau A, Lanson M, Geoffroy N (2007) Natural speciation of Ni, Zn, Ba and As in ferromanganese coatings on quartz using X-ray fluorescence, adsorption and diffraction. Geochim Cosmochim Acta 71:95–128

Manceau A, Nagy K, Marcus M, Lanson M, Geoffroy N, Jacquet T, Kirpichtchikova T (2008) Formation of metallic copper nanoparticles at the soil–root interface. Environ Sci Technol 42:1766–1772

Mareschal L, Bonnaud P, Turpault M-P, Ranger J (2010) Impact of common European tree species on the chemical and physiochemical properties of fine earth: an unusual pattern. Eur J Soil Sci 61:14–23

Mariano A (1989) Economic geology of rare earth elements. In: Lipin B, McKay G (eds) Geochemistry and mineralogy of rare earth elements, Reviews in mineralogy. Mineralogical Society of America, Washington, DC, pp 309–337, 347 pp

Mariotti A (1982) Apports de la Géochimie isotopique a la connaissance du cycle de l'Azote. Thesis, Univ Paris VI, 472 pp

Marker A, de Oliviera J (1990) The formation of rare earth element scavenger minerals in the weathering products derived from alkaline rocks of S-E Bahia, Brazil. In: Geochemistry of the Earth's surface and mineral formation. 2nd International Symposium Aix en Provence, pp 373–374

Markert B (1998) Instrumental multi-element analysis in plant material: a modern method in environmental chemistry and tropical system research. In: Wasserman J, Silvia-Filho E, VillasBoos R (eds) Environmental geochemistry in the tropics. Springer, Berlin, pp 75–95, 305 pp

Marques FM, Baeyens B, Daehn R, Scheinost AC, Bradbury MH (2012) U(VI) sorption on montmorillonite in the absence and presence of carbonate: a macroscopic and microscopic study. Geochim Cosmochim Acta 93:262–277

Martin C, McCulloch M (1999) Nd/Sr isotopic and trace geochemistry of rive sediments and soils in fertilized catchment, New South Wales, Australia. Geochim Cosmochim Acta 63:287–305

Martínez Cortizas A, Garcia-Rodeja Gayoso E, Novao Munuoz J, Ponteverdra Pombal X, Buurman P, Terrible F (2003) Distribution of some selected major and trace elements in four Italian soils developed from the deposits of the Gauro and Vico volcanoes. Geoderma 117(3–4):215–226

Mason B (1958) Principles of geochemistry. Wiley, New York, 329 pp

Mason B (1966) Principles of geochemistry, 3rd edn. Wiley, New York, 329 pp

Matocha C, Grove J, Karathanasis A (2010) Nitrogen fertilizer effects on soil mineralogy in an agrosystem. In: Abstracts annual ASA, CSSA, SSSA meeting, number 509

McBride MB (1981) Forms and distribution of copper in solid and solution phases in soil. In: Lonergan JF, Robson AD, Graham RD (eds) U Copper in soils and plants. Academic, Australia, Str. 25–45

McFarlane M (1976) Laterite and landscape. Academic, London, 151 pp

McKay G (1989) Partitioning of rare earth elements between major silicate minerals and basaltic melts. In: Lipin B, McKay G (eds) Geochemistry and mineralogy of rare earth elements, Reviews in mineralogy. Mineralogical Society of America, Washington, DC, pp 45–77, 348 pp

McLennan S (1989) Rare earth elements in sedimentary rocks: influences of provenance and sedimentary processes. In: Lipin B, McKay G (eds) Geochemistry and mineralogy of rare earth elements, vol 21, Reviews in mineralogy. Mineralogical Society of America, Washington, DC, pp 169–200, Ch 7, 348 pp

McQueen K (2008) Regolith geochemistry. In: Scott K, Pain C (eds) Regolith science. Springer, Dordrecht, pp 73–104, 461 pp

Meade RH, Parker RS (1985) Sediment in the rivers of the United States. US Geol Surv Water Supply Pap 2275:49–60

Meijer E, Burrman P (2003) Chemical trends in a perhumid soil catena on the Turrialba volcano (Costa Rico). Geoderma 117:185–201

Melegy A, Slaninka I, Paces T (2011) Weathering fluxes of arsenic from small catchment in Slovak Republic. Environ Earth Sci 6:549–555

Merritt K, Amirbahman A (2007) Mercury dynamics in sulphide-rich sediments: geochemical influence on contaminant mobilization within the Penobscot River estuary, Maine, USA. Geochim Cosmochim Acta 71:929–941

Meunier A (1980) Les mécanismes de l'altération du granites et le rôle des microsystèmes, Etude des arènes du massif granitique de Parthenay (Deux Sevres). Mém Soc Géol Fr 146:1–80

Meunier A, Velde B (2008) Origin of clay minerals in soils and weathered rocks. Springer, Berlin, 406 pp

Mileti FA, Langella G, Prins MA, Vingiani S, Terribile F (2013) The hidden nature of parent material in soils of Italian mountain ecosystems. Geoderma 207–208:291–309

Milliman J, Meade R (1983) World-wide delivery of river sediments to the oceans. J Geol 91:1–21

Millot G (1970) Geology of clays. Springer, New York, 429 pp

Millot R, Gaillardet J, Dupré B, Allègre C (2003) Northern latitude chemical weathering rates: clues from the Mackenzie River basin Canada. Geochim Cosmochim Acta 67:1305–1329

Millward G, Turner A (1995) Trace elements in estuaries. In: Salbu B, Steinnes E (eds) Trace elements in natural waters. CRC, Boca Raton, FL, pp 223–246, 302 pp

Molinaroli E (1996) Mineralogical characterisation of Saharan dust with a view to its final destination in Mediterranean sediments. In: Guerzoni S, Chester R (eds) The impact of desert dust across the Mediterranean. Kluwer Academic, Dordrecht, pp 153–162, 452 pp

Moon S, Huh Y, Qin J, van Pho N (2007) Chemical weathering on the Hong (Red) River basin: rates of silicate weathering and their controlling factors. Geochim Cosmochim Acta 71:1411–1430

Moore DM, Hower J (1986) Ordered interstratification of dehydrated and hydrated Na-smectite. Clay Clay Miner 34:379–384

Morel J-L, Mench M, Guckert A (1987) Dynamique des métaux lourds dans la rhizosphere: rôle des exudats racinaires. Rev Ecol Biol Sol 24:485–492

Mosser C (1980) Etude géochimique de quelques éléments traces dans les argiles des altérations et des sédiments. Sci Géol Mem 83, 386 pp

Motta MM, Miranda CF (1989) Molybdate adsorption on kaolinite, montmorillonite and illite: constant capacitance modelling. Soil Sci Soc Am J 53:380–385

Muhs D, Budahn J, Prospero J, Carey S (2007) Geochemical evidence for African dust inputs to soils of western Atlantic islands: Barbados, the Bahamas and Florida. J Geophys Res 112:148–227

Navarrete I, Tsutsuki K, Kondo R, Asio V (2008) Genesis of soils across a late Quaternary volcanic landscape in the humid tropical island of Leyte, Philippines. Aust J Soil Res 46:403–414

Nesbit H, Markovics G (1997) Weathering of granodioritic crust, long-term storage of elements in weathering profiles and petrogenesis of siliclastic sediments. Geochim Cosmochim Acta 61:1653–1670

Nissenbaum A (ed) (1980) Hypersaline brines and evaporitic environments, vol 28, Developments in sedimentology. Elsevier, Amsterdam, 270 pp

Nozaki Y, Lerche D, Alibo D, Snidvongs A (2000) The estuarine geochemistry of rare earth elements and indium in the Chao Phraya river, Thailand. Geochim Cosmochim Acta 64:3983–3994

Odin G, Letolle R (1980) Glauconitization and phosphatization environments: a tentative comparison. In: Bentor Y (ed) Marine phosphorites – geochemistry, occurrence and genesis. Society of Economic Paleontologists and Mineralogists special publication 29, Tulsa, OK, pp 227–237, 350 pp

Oelkers E, Schott J (2009) Thermodynamics and kinetics of water–rock interaction, vol 70, Reviews in mineralogy and geochemistry. Mineralogical Society of America, Washington, DC, 569 pp

Oh N, Richter D (2005) Elemental translocation and loss from three highly weathered bed-rock profiles in the southeastern United States. Geoderma 126:5–25

Ohta A, Kawabe I (2001) REE(III) adsorption onto Mn dioxide (δ-MnO_2) and Fe oxyhydroxide: Ce(III) oxidation by δ-MnO_2. Geochim Cosmochim Acta 65(5):695–703

Onishi H (1969) Arsenic, Ch 33. In: Wedephol H (ed) Handbook of geochemistry. Springer, Berlin

Palissy B de (1563) Discours admirables de la nature, 158 pp, editor unknown, re-published under Oeuvres Completes de Bernard Palissy P-A Cap Lib. Aci. Blanchard Paris 1961

Panfili F, Manceau A, Sarret G, Spadini L, Kirpichtchikova T, Bert V, Laboudigue A, Marcus MA, Ahamdach N, Libert MF (2005) The effect of phytostabilization on Zn speciation in a dredged contaminated sediment using scanning electron microscopy, X-ray fluorescence, EXAFS spectroscopy and principal components analysis. Geochim Cosmochim Acta 69:2265–2284

Pauling L (1947) Principles of chemistry. Freeman, San Francisco, CA, 596 pp

Peackock C, Sherman D (2004) Vanadium (V) adsorption onto goethite (FeOOH) at pH 1.5 to 12: a surface complexation model based on *ab initio* molecular geometries and EXAFS spectroscopy. Geochim Cosmochim Acta 68:1723–1733

Pedro G (1966) Essai sur la caractérisation géochimique des différents processus zonaux résultant de l'altération des roches superficielle. C R Acad Sci D 262:1828–1831
Peretyazhko T, Sposito G (2005) Iron (III) reduction and phosphorous solubilisation in humid tropical forest soils. Geochim Cosmochim Acta 69:3643–3652
Perkins S, Filippelli G, Souch C (2000) Airborn trace metal contamination of wetland sediments at Indiana Dunes National Lakeshore. Water Air Soil Pollut 122:231–260
Piccolo A (ed) (1996) Humic substances in terrestrial ecosystems. Elsevier, New York
Pierce F, Dowdy R, Grigal D (1982) Concentrations of six trace metals in some major Minnesota soil series. J Environ Qual 11:416–422
Pion J-C (1979) Altération des massifs cristallins basiques en zone tropical sèche. Sci Géol Mem 57, 230 pp
Pösfai M, Dunin-Borkowsky R (2006) Sulfides in biosystems. In: Vaughan D (ed) Sulfide mineralogy and geochemistry, vol 61, Reviews in mineralogy and geochemistry. Mineralogical Society of America, Washington, DC, pp 679–711, 714 pp
Pradit S, Wattayakorn G, Anguspanich S, Baeyens W, Leermakers M (2010) Distribution of trace elements in sediments and biota of Songkhla Lake, Southern Thailand. Water Air Soil Pollut 206:155–174
Probst J-L (1990) Géochimie et hydrogéologie de l'érosion continentale. Sci Géol, 161 pp
Prospero J, Ginoux P, Tores O, Nicholson SE, Gill TE (2002) Environmental characterisation of global sources of atmospheric soil dust identified with NIMBUS7. Rev Geophys 40:1002
Proust D (1976) Etude de l'altération des amphibolites de la Roche-l'Abeille: évolutions chimiques et minéralogiques des plagioclases et hornblendes. Thesis, Univ Poitiers, 197 pp
Proust D (1985) Amphibole weathering in a glaucophane schist (Ile de Groix, Morbhian). Clay Miner 20:161–170
Pye K (1987) Aeolian dust and dust deposits. Academic, San Diego, CA, 334 pp
Pye K (1995) The nature, origin and accumulation of loess. Quaternary Sci Rev 14:653–667
Rabenhorst M, Foss J, Fanning D (1982) Genesis of Maryland soils formed from serpentine. Soil Sci Soc Am J 46:607–616
Rai D (1986) Chemical attenuation rates, coefficients, and constants in leachate migration. Vol. 1: A critical review. Report to Electric Power Research Institute, Palo Alto, CA, by Battelle Pacific Northwest Laboratories, Richland, WA (Research Project 2198-1)
Rasmussen C, Dahlgren R, Southard R (2010) Basalt weathering and pedogenesis across an environmental gradient in southern Cascade Range California. Geoderma 154:473–485
Righi D, Cauvel A (1987) Podzols and podzolisation. AFES and INRA Publication, Paris, 227 pp
Righi D, Reisänän M, Gillot F (1997) Clay mineral transformations in podzolised tills in central Finland. Clay Miner 32:531–544
Righi D, Huber F, Keller C (1999) Clay formation and podzol development from postglacial moraines in Switzerland. Clay Miner 34:319–332
Rimstidt J, Balog A, Webb J (1998) Distribution of trace elements between carbonate minerals and aqueous solutions. Geochim Cosmochim Acta 62:1851–1863
Risacher F (1992) Géochimie des lacs salés et croûts des sel de l'altiplano Bolivien. Sci Géol 45:140–219
Ritchie JC, Rudolph WK (1970) Distribution of fallout and natural gamma radionuclides in litter, humus and surface mineral soil layers under natural vegetation in the Great Smoky Mountains, North Carolina-Tennessee. Health Phys 18:479–489
Rizkalla EN, Choppin GR (1994) Lanthanides and actinides hydration and hydrolysis. In: Gschneidner KAJ, Eyring L (eds) Handbook on the physics and chemistry of rare earths. Elsevier, Amsterdam
Rosenthal Y, Boyle E, Slowey N (1997) Temperature control on the incorporation of magnesium, fluorine And cadmium in benthic foraminiferal shells from the Little Bahama Bank: prospects for thermocline paleoceanography. Geochim Cosmochim Acta 61:3633–3643
Ross CS (1946) Sauconite – a clay mineral of the montmorillonite group. Am Miner 31:411–424
Ross S (1994) Retention, transformation and mobility of toxic metal in soils. In: Ross S (ed) Toxic metals in soil–plant systems. Wiley, Chichester, UK, pp 94–210, 466 pp

Roy S, Gaillardet J, Allègre C (1999) Geochemistry of dissolved and suspended loads of the Seine river, France: anthropogenic impact, carbonate and silicate weathering. Geochim Cosmochim Acta 63:1277–1292

Rozalén M, Brady PV, Huertas FJ (2009) Surface chemistry of K-montmorillonite: ionic strength, temperature dependence and dissolution kinetics. J Colloid Interface Sci 333:474–484

Ruhe R (1984) Soil-climate systems across the prairies in Midwestern USA. Geoderma 54:201–219

Sawhney BL (1972) Selective sorption and fixation of cations by clay minerals: a review. Clay Clay Miner 20:93–100

Salbu B, Steinnes E (1995) Trace elements in natural waters. CRC, Boca Raton, FL, 302 pp

Salomons W, Förstner U (1984) Metals in the hydrocycle. Springer, Berlin, 349 pp

Salomons W, Förstner U, Mader P (eds) (1995) Heavy metals: problems and solutions. Springer, Berlin, 412 pp

Sato T, Watanabe T, Otsuka R (1992) Effects of charge, charge location, and energy change on expansion properties of dioctahedral smectites. Clay Clay Miner 40:103–113

Scheinost AC, Kretzschmar R, Pfister S (2002) Combining selective sequential extractions, x-ray absorption spectroscopy, and principal component analysis for quantitative zinc speciation in soil. Environ Sci Technol 36:5021–5028

Schindler PW, Fürst B, Dick R, Wolf PU (1976) Ligand properties of surface silanol groups I. Surface complex formation with Fe^{3+}, Cu^{2+}, Cd^{2+} and Pb^{2+}. J Colloid Interface Sci 55:469–475

Schlegel ML, Descostes M (2009) Uranium uptake by hectorite and montmorillonite: a solution chemistry and polarized EXAFS study. Environ Sci Technol 43:8593–8598

Schlegel ML, Manceau A (2006) Evidence for the nucleation and epitaxial growth of Zn phyllosilicate on montmorillonite. Geochim Cosmochim Acta 70:901–917

Schlegel ML, Manceau A, Hazemann J-L, Charlet L (2001a) Adsorption mechanisms of Zn on hectorite as a function of time, pH, and ionic strength. Am J Sci 301:798–830

Schlegel ML, Manceau A, Charlet L, Chateigner D, Hazemann J-L (2001b) Sorption of metal ions on clay minerals. III. Nucleation and epitaxial growth of Zn phyllosilicate on the edges of hectorite. Geochim Cosmochim Acta 65:4155–4170

Schultz M, Vivit D, Schultz C, Fitzpatrick J, White A (2010) Biologic origin of iron nodules in a marine terrace chronosequence, Santa Cruz, California. Soil Sci Soc Am J 74:550–564

Scott K, Pain C (2008) Regolith science. Springer, Dordrecht, 461 pp

Selby MJ (1984) Hillslope sediment transport and deposition. In: Pye K (ed) Sediment transport and depositional processes. Blackwell, London, pp 61–87

Serret G, Isaure M-P, Marcus M, Harada E, Choi Y-E, Pairis S, Fakhara S, Manceau A (2007) Chemical forms of calcium in Ca, Zn and Ca, Cd containing grains excreted by tobacco trichomes. Can J Chem 88:738–746

Shaheen S (2009) Sorption and lability of cadmium and lead in different soils from Egypt and Greece. Geoderma 153:61–68

Sheldon R (1980) Episodicity of phosphate deposition and deep ocean circulation – a hypothesis. In: Bentor Y (ed) Marine phosphorites – geochemistry, occurrence and genesis. Society of Economic Paleontologists and Mineralogists special publication 29, Tulsa, OK, pp 239–247, 350 pp

Shiller A, Boyle E (1991) Trace elements in the Mississippi river delta outflow region: behavior at high discharge. Geochim Cosmochim Acta 55:3241–3251

Sholkowitz E (1993) The geochemistry of rare earth elements in the Amazon River estuary. Geochim Cosmochim Acta 57:2181–2199

Shotyk W, Goodsite M, Roos-Barraaclough F, Givelet N, Le Roux G, Weiss D, Cheburkin A, Knudsen K, Heinemeir J, van Der Knaap W, Norton S, Lohse C (2005) Accumulation rtes and predominant atmospheric sources of natural and anthropogenic Hg and Pb on the Faroe Islands. Geochim Cosmochim Acta 69:1–17

Siever R (1992) The silica cycle in the Precambrian. Geochim Cosmochim Acta 56(8):3265–3272

Siffermann G (1973) Les sols de quelques régions volcaniques du Cameroun. Mémoires ORSTOM, no 66, 182 pp
Simonetti A, Gariépy C, Carignan J (2000) Pb and Sr isotopic evidence for sources of atmospheric heavy metals and their deposition budgets in Northeastern North America. Geochim Cosmochim Acta 64:3439–3452
Singh P, Rajamani V (2001) REE geochemistry of recent clastic sediments from Kaveri floodplains, southern India: implications to source area weathering and sedimentary processes. Geochim Cosmochim Acta 65:3093–3108
Sly P (ed) (1986) Sediments and water interactions. In: Proceedings of the 3rd international symposium on the interaction between sediments and water. Springer, 517 pp
Smalley I (1975) Loess: lithology and genesis, Benchmark papers in geology, vol 26. Dowden Hutchinson and Ross, Stroudsburg, PA, 429 pp. Keilback K The riddle of loess formation, 47–50
Soil Science Society of America (1984) Glossary of soil science terms. Soil Science Society of America, Madison, WI
Sonnefield P (1984) Brines and evaporates. Academic, New York, 612 pp
Sparks DL (1998) Kinetics of sorption/release reactions on natural particles. In: Huang PM, Senesi N, Buffle J (eds) Structure and surface reactions of soil particles. Wiley, New York, pp 413–448
Spencer R (2000) Sulfate minerals in evaporate deposits. In: Alpers C, Jambor J (eds) Sulfate minerals: crystallography, geochemistry and environmental significance, vol 40, Reviews in mineralogy and geochemistry. Mineralogical Society of America, Washington, DC, pp 173–192, 608 pp
Sposito G (1989) The chemistry of soils. Oxford University Press, New York, 277 pp
Sposito G (1994) Chemical equilibrium and kinetics in soils. Oxford University Press, Oxford, 269 pp
Sposito G, Skipper N, Sutton R, Park S-H, Soper A (1999) Surface geochemistry of the clay minerals. Proc Natl Acad Sci U S A 96:3358–3364
Squire HM, Middleton LJ (1966) Behavior of Cs-137 in soils and pastures. A long term experiment. Radiat Bot 6:413–423
Steinberg M, Treuil M, Touray JC (1979) Géochimie—principes et méthodes. Doin, Paris, 353 pp
Strawn DG, Sparks DL (1999) The use of XAFS to distinguish between inner- and outer-sphere lead adsorption complexes on montmorillonite. J Colloid Interface Sci 216:257–269
Stumm W (1992) Chemistry of the solid–water interface. Wiley, New York, 427 pp
Stumpf T, Bauer A, Coppin F, Kim J (2001) Time resolved laser fluorescence spectroscopic study of the sorption of Cm(III) onto smectite and kaolinite. Environ Sci Technol 35:3691–3694
Stumpf T, Hening C, Bauer A, Denecke M, Fanghänel T (2004) An EXAFS and TRLFS study of the sorption of trivalent actinides onto smectite and kaolinite. Radiochim Acta 92:133–138
Stumpf S, Stumpf T, Dardenne K, Henning C, Foerstendorf H, Klenze R, Fanghänel T (2006) Sorption of Am(III) onto 6-line-ferrihydrite and its alteration products: investigation by EXFAS. Environ Sci Technol 40:3522–3528
Sverjensky DA (1994) Zero-point-of-charge prediction from crystal chemistry and solvation theory. Geochim Cosmochim Acta 58:3223–3329
Sylwester ER, Hudson EA, Allen PG (2000) The structure of uranium (VI) sorption complexes on silica, alumina, and montmorillonite. Geochim Cosmochim Acta 64:2431–2438
Taillefert M, Lienemann C-P, Gaillard J-F, Perret D (2000) Speciation, reactivity and cycling of Fe and Pb in a meromictic lake. Geochim Cosmochim Acta 64:169–183
Takahashi Y, Higashi M, Furukawa T, Mitsunobu S (2001) Change in iron species and iron solubility in Asian dust during the long range transport from western china to Japan. Atmos Chem Phys 11:11237–11252
Tan M (1998) Principles of soil chemistry. Marcel Dekker, New York, 513 pp
Tan XL, Fang M, Wang XK (2010) Sorption speciation of lanthanides/actinides on minerals by TRLFS, EXAFS and DFT studies: a review. Molecules 15:8431

Tang J, Johanesson K (2005) Absorption of rare earth elements onto Carrizo sand: experimental investigations and modelling with surface complexation. Geochim Cosmochim Acta 69:5247–5261

Tardy Y (1993) Pétrologie des latérites et des sols tropicaux. Masson, Paris, 457 pp

Taylor G, Eggleton R (2001) Regolith geology and geomorphology. Wiley, Chichester, UK, 376 pp

Temgoua E (2002) Cuirassement ferrugineux de bas versants en zone forestière du Sud-Cameroun. Mem Geol (Lausanne) 38, 134 pp

Tertre E, Beaucaire C, Coreau N, Juery A (2009) Modelling Zn(II) absorption onto clayey sediments using a multi-site ion-exchange model. Appl Geochem 24:1852–1861

Tertre E, Prêt D, Ferrage E (2011) Influence of the ionic strength and solid/solution ratio on Ca(II)-for-Na^+ exchange on montmorillonite. Part 1: Chemical measurements, thermodynamic modelling and potential implications for trace element geochemistry. J Colloid Interface Sci 353:248–256

Teutsch N, Erel Y, Halicz L, Chadwick OA (1999) The influence of rainfall on metal concentration and behavior in the soil—evidence from 210Pb and stable Pb isotopes. Geochim Cosmochim Acta 63(21):3499–3511(13)

Tinnacher RM, Zavarin M, Powell BA, Kersting AB (2011) Kinetics neptunium(V) sorption and desorption on goethite: an experimental and modeling study. Geochim Cosmochim Acta 75:6584

Tournassat C, Gailhanou H, Crouzet C, Braibant G, Gautier A, Lassin A, Blanc P, Gaucher EC (2007) Two cation exchange models for direct and inverse modelling of solution major cation composition in equilibrium with illite surfaces. Geochim Cosmochim Acta 71(5):1098–1114

Tournassat C, Gailhanou H, Crouzet C, Braibant G, Gautier A, Gaucher EC (2009) Cation exchange selectivity coefficient values on smectite and mixed-layer illite/smectite minerals. Soil Sci Soc Am J 73:928–942

Trauth N (1977) Argiles évaporitiques dans la sedimentation carbonate continentale et épicontinentale Tertaire. Sci Géol 49, 195 pp

Troup R, Bricker O (1975) Processes affecting the transport of materials from continents to oceans. In: Church T (ed) Marine chemistry in the coastal environment. American Chemical Society symposium series 18, Washington, DC, pp 133–151, 710 pp

Tungsheng L (1985) Loess in China. Springer, Berlin, 224 pp

Twenhofel W (1961) Treatise on sedimentation. National Academy of Sciences, Washington, DC, 458 pp

Usher C, Michel A, Grassian V (2003) Reactions in mineral dust. Chem Rev 703:4883–4939

Valeton I (1972) Bauxites, vol 1, Developments soil science. Elsevier, Amsterdam, 226 pp

Van Laer L, Degryse F, Leynen D, Smolders E (2010) Mobilisation of Zn upon waterlogging riparian Spodosols is related to reductive dissolution of Fe minerals. Eur J Soil Sci 61:1014–1024

Velde B (1985) Clay minerals: a physico-chemical explanation of their occurrence. Elsevier, Amsterdam, 427 pp

Velde B (2006) Preliminary study of heavy metal chemistry of shore and slikke clay deposits in the Brouage region: concentration of Cd, Sn and As as related to P. Cahiers Biol Mar 47:93–102

Velde B, Barré P (2010) Soils, plants and clay minerals. Springer, Berlin, 344 pp

Velde B, Church T (1999) Rapid clay transformations in Delaware Salt marshes. Appl Geochem 14:559–568

Velde B, Meunier A (2008) The origin of clay minerals in soils and weathered rocks. Springer, Berlin, 405 pp

Velde B, El Moutaouakkil N, Iijima A (1991) Compositional homogeneity in low-temperature chlorites. Contrib Mineral Petrol 107:21–26

Velde B, Church T, Bauer A (2003) Contrasting trace element geochemistry in two American and French salt marshes. Mar Chem 83:131–144

Vezina A, Cornett R (1990) Iron transport and distribution between freshwater and sediments over different time scales. Geochim Cosmochim Acta 54:2635–2644

Walker L, Del Moral R (2003) Primary succession and ecosystem rehabilitation. Cambridge University Press, Cambridge, 456 pp

Wang YH, Huang CB, Hu YH, Hu YM, Lan Y (2008) Beneficiation of diasporic-bauxite ore by selective flocculation with a polyacrylate flocculant. Miner Eng 21(9):664–672

Wang S, Lin C, Cao X (2011) Heavy metals content and distribution in the surface sediments of the Guangzhou section of the Pearl river, Southern China. Environ Earth Sci 64:1596–1615

Wanner H, Albinsson Y, Karnl O, Wieland E, Wersin P, Charlet L (1994) The acid/base chemistry of montmorillonite. Radiochim Acta 66/67:157–162

Warren J (2006) Evaporites: sediments, resources and hydrocarbons. Springer, Berlin, 1035 pp

Wedephol H (1969) Handbook of geochemistry, vol I. Springer, New York

Weiss-Penzias PS, Ortiz C Jr, Acosta RP, Heim W, Ryan JP, Fernandez D, Collett JL Jr, Flegal AR (2012) Total and monomethyl mercury in fog water from the central California coast. Geophys Res Lett 39:3

Westrich B, Förstner U (eds) (2007) Sediment dynamics and pollutant mobility in rivers. Springer, Berlin, 430 pp

White AF, Brantley SL (eds) (1995) Chemical weathering rates of silicate minerals, vol 31, Reviews in mineralogy. Mineralogical Society of America, Washington, DC, 584 pp

White A, Blum A, Schultz M, Vivit D, Stonestrum D, Larsen M, Murphy S, Eberl D (1998) Chemical weathering in a tropical watershed, Luquillo Mountains Puerto Rico: I. Long-term versus short-term weathering fluxes. Geochim Cosmochim Acta 62:209–236

White A, Schultz M, Vivit D, Blum A, Stonestrom A, Anderson S (2008) Geochemical weathering of a marine chronosequence, Santa Cruz California: interpreting rates and controls based upon soil concentration depth profiles. Geochim Cosmochim Acta 72:36–68

Williams LB, Hervig RL (2002) Intracrystalline boron isotope variations in clay minerals: a potential low-temperature single mineral geothermometer. Am Mineral 87:1564–1570

Yaron B, Dror I, Berkowitz B (2012) Soil-subsurface change. Springer, Berlin, 366 pp

Zahrn D, Johnson A (1995) Nutrient accumulation during primary succession in a montane tropical forest, Puerto Rico. Soil Sci Soc Am J 59:1444–1452

Zheng J, Aono T, Uchida S, Zhang J, Honda M (2012) Distribution of Pu isotopes in marine sediments in the Pacific 30 km off Fukushima after the Fukushima Daiichi nuclear power plant accident. Geochem J 46:361–369

Zibold G, Drissner J, Klemt E, Konopleva AV, Konoplev AV, Miller R (1997) Biologische Verfügbarkeit von Cäsium-Radionukliden in Waldgebieten des nördlichen und südlichen Voralpenlandes. In: Honikel KO, Hecht H (eds) Radiocäsium in Wald und Wild, vol 2. Veranstaltung, Kulmbach, 10/11 June 1997